Advances in Numerical Mathematics

Andreas Prohl

Projection and Quasi-Compressibility Methods
for Solving the Incompressible Navier-Stokes
Equations

Projection and Quasi-Compressibility Methods for Solving the Incompressible Navier-Stokes Equations

Von Dr. rer. nat. Andreas Prohl
University of Minnesota

 Springer Fachmedien Wiesbaden GmbH 1997

Dr. rer. nat. Andreas Prohl

Geboren 1968 in Lübeck. Von 1988 bis 1993 Studium der Mathematik und Physik an der Ruprecht-Karls-Universität Heidelberg,1993 Diplom in Mathematik. Von 1993 bis 1996 Stipendiat des Graduiertenkollegs „Modellierung und Wissenschaftliches Rechnen in Mathematik und Naturwissenschaften" am Interdisziplinären Zentrum für Wissenschaftliches Rechnen (IWR) in Heidelberg, 1996 Promotion. Seit 1996 Forschungsstipendiat der DFG am Institute for Mathematics and its Applications (IMA) der University of Minnesota, Minneapolis (USA).

Die Deutsche Bibliothek – CIP-Einheitsaufnahme

Prohl, Andreas:
Projection and quasi-compressibility methods for solving the incompressible Navier-Stokes equations / von Andreas Prohl. –
(Advances in numerical mathematics)
ISBN 978-3-519-02723-2   ISBN 978-3-663-11171-9 (eBook)
DOI 10.1007/978-3-663-11171-9

© Springer Fachmedien Wiesbaden 1997
Ursprünglich erschienen bei B.G.Teubner Stuttgart 1997

Einband: Peter Pfitz, Stuttgart
Umschlagbild: Mit freundlicher Unterstützung von Dr.Turek

Dedicated to my parents and my sister

Differentibus quaestionibus modi solvendi differentes

Preface

The numerical treatment of the evolutionary incompressible Navier-Stokes equations, which determine many practically relevant fluid flows, is an area of considerable interest for industrial as well as scientific applications. Important for drawing further conclusions for the behavior of certain flows in diverse disciplines such as (astro-)physics, engineering, meteorology, oceanography, or biology is a reliable, robust and efficient numerical model. The goal of computing highly complex flows requires the development of sophisticated algorithms. In general, numerical schemes which do not cause high computational cost, often suffer from stability or reliability problems and vice versa. So, it demands a numerical and physical a-priori knowledge from the user in order to select the "best fitting algorithm" for a particular problem under consideration. The use of knowledge about physical phenomena appearing in a specific problem allows the relaxation of some robustness-conditions that otherwise need to be imposed on the numerical scheme in order to ensure reliability with respect to the convergence behavior. To this end, this leads to permittance of numerical models simulating continuous flows which do not satisfy severe stability restrictions that lead to robust schemes, with the advantage of lower computational costs necessary to obtain the same accuracy. A major part of this book is devoted to such schemes that are of great importance: *classical projection methods of higher order* and *nonstationary quasi-compressibility methods*.

Projection methods have been initially introduced by Chorin as semi-discretization schemes to efficiently compute the velocity and pressure functions at each time-step independently. Due to their superiority from the computational point of view compared with other time discretizations they are wide-spread in engineering sciences. Moreover, modifications like the Van Kan scheme, which combines the projection idea with a time-discretization ansatz of second order, are often used for practical computations. Especially, over the last ten years there appeared a considerable quantity of publications that deal with the study of projection methods. However, related error analyses suffer from nonoptimal convergence statements for the iterates of the scheme and/or a restricted application of the theoretical results to model situations with a limited practical relevance. One objective of this book is to figure out the acting mechanisms of the diverse projection schemes and to present *optimal* convergence results for *general* flow constellations.

The error analysis of the classical *Schemes of Chorin and Van Kan* in the subsequent chapters classifies them as pointwise first and second order semi-discretizations in time, respectively. Nevertheless, the objective of the rather large analysis is not only the verification of this important result for its own sake. Moreover, the proofs provide an insight into the underlying error-mechanisms of these schemes that determine drawbacks of the methods visible in computational test calculations, like boundary layers arising in the pressure approximation or a reduction of convergence order in the case of Van Kan's scheme for general fluid flows with non-compatible initial data. The mathematical understanding of these features of the splitting schemes serves as the basis for the construction of new numerical models which do not suffer from these drawbacks, like the so-called *Chorin-Uzawa scheme* or the *multi-component projection schemes*, and further the proposal of *projection schemes on certain time-grid structures*. The construction and analysis of these new projection schemes rank among the main contributions of the book.

The mathematical approach for a rigorous analysis of the projection schemes is based on their reformulation as semi-explicit quasi-compressibility methods. These methods relax the incompressibility constraint, thus providing a more favorable environment for standard numerical finite element discretizations. We will distinguish between *stationary ansatzes* (penalty, pressure stabilization) and *nonstationary* ones (pressure correction, artificial compressibility). This distinction depends on the quasi-compressibility constraint being algebraic or evolutionary, respectively. These more or less classical quasi-compressibility methods are widely used for practical computations. Despite a vast quantity of theoretical investigations carried out for them there are no optimal error estimates available that classify the different algorithms. The rigorous analysis of them is the second objective to be accomplished here, as well as the proposal of certain modified and combined new quasi-compressibility methods with improved stability properties.

The organization of the book is as follows: It starts with the analysis of the various stationary and nonstationary quasi-compressibility methods, which forms the first part of the book. These results form the basis for the investigation of different projection schemes constituting the second part. Due to the former statements, they are semi-explicit quasi-compressibility methods. Therefore, in the error analysis of the different schemes we can employ many arguments that work in the proofs of the corresponding quasi-compressibility methods. From this observation, we can focus on the differences of both classes of methods that stem from the explicit treatment of the pressure function in the momentum equation and their time discretization with identification of the quasi-compressibility parameter and the time discretization parameter.

Last but not least, the author is very grateful to Prof. Dr. R. Rannacher for his continuing support, his valuable suggestions and unfailing encouragement, without which this work would not have been possible. Special acknowledgments also have to go to all the members of the Heidelberg Numerical Mathematics Group, providing me with substantial help in both computational as well as technical areas. Finally, it is a pleasure to thank Dr. M. Gobbert for his careful reading of the work and suggestions to improve the readability of the book.

The results presented here are extended results of the diploma and the Ph.D. theses [26], [27] of the author. The latter work has been supported by a DFG scholarship of the author as a member of the Graduiertenkolleg "Modellierung und Wissenschaftliches Rechnen" at the IWR Heidelberg. — The hope is that this work will stimulate further research in this interesting field of numerical analysis for computational fluid dynamics.

Contents

List of Figures

List of Tables

Chapter 1

Introduction

The mathematical description of viscous fluid flows is given by the Navier-Stokes equations, a system of partial differential equations that result from the conservation laws for energy, momentum, and mass. The following mathematical model for physically relevant incompressible flows in a bounded domain $\Omega \subset \mathbb{R}^d$, $d = 2, 3$ is the starting point for the subsequent numerical algorithms,

$$
\begin{aligned}
&u_t - \nu \Delta u + (u \cdot \nabla)u + \nabla p = f, \qquad \forall \, (x,t) \in Q_T := \Omega \times (0,T], \\
&\operatorname{div} u = 0, \qquad \forall \, (x,t) \in Q_T, \\
&u(0) \equiv u_0, \qquad u|_{\partial \Omega} = 0.
\end{aligned}
\tag{1.1}
$$

Using this notation, $u \equiv u(x,t)$ denotes the velocity field, and $p \equiv p(x,t)$ the scalar pressure function. $f \equiv f(x,t)$ is the given volume force, $\nu > 0$ the viscosity factor and u_0 the given initial velocity of the fluid flow that is governed by (1.1) for positive times.

From the numerical point of view, the incompressibility constraint in (1.1) couples both functions u and p, thus causing a high computational effort. As an example, the application of the implicit Euler scheme as a semi-discretization scheme in time leads to the following equations to be solved,

$$
\begin{aligned}
&\frac{1}{k}\{u^{m+1} - u^m\} - \nu \Delta u^{m+1} + (u^{m+1} \cdot \nabla)u^{m+1} + \nabla p^{m+1} = f^{m+1}, \\
&\operatorname{div} u^{m+1} = 0, \\
&u^0 \equiv u_0,
\end{aligned}
\tag{1.2}
$$

using a partition $0 = t_0 < \ldots \le t_{m+1} \le \ldots \le t_{M+1} \equiv T$, with $k = t_{m+1} - t_m$ as the time-step parameter. If we apply a stable spatial discretization, using a finite element method at every time t_{m+1}, there arise systems of linear equations of large complexity. Therefore, a significant reduction of the computational effort is necessary to profitably operate the algorithm for the computation of complex flows.

For that purpose, splitting schemes have been proposed already in the late 1960's that aim at computing the tuple $\{u^{m+1}, p^{m+1}\}$ in *separate steps*, thus yielding a drastic reduction of computational work. One major objective of the present monograph is the presentation and investigation of well-known as well as the proposal of new, improved projection schemes.

The most famous projection scheme has been formulated by Chorin in 1968 in [3] and is of the following form,

1. Start with an appropriate initial guess $u^0 \approx u_0$.

2. For $m \ge 0$, find $\tilde{u}^{m+1}$ as the solution of

$$\frac{1}{k}\{\tilde{u}^{m+1} - u^m\} - \nu\Delta\tilde{u}^{m+1} + (u^m \cdot \nabla)\tilde{u}^{m+1} = f^{m+1},$$

$$\tilde{u}^{m+1}|_{\partial\Omega} = 0. \qquad (1.3)$$

3. Provided with $\tilde{u}^{m+1}$, determine the tuple $\{u^{m+1}, p^{m+1}\}$ as the solution of

$$\frac{1}{k}\{u^{m+1} - \tilde{u}^{m+1}\} + \nabla p^{m+1} = 0,$$

$$\mathrm{div}\, u^{m+1} = 0, \qquad u^{m+1}|_{\partial\Omega} \cdot n = 0. \qquad (1.4)$$

Notice that the second step is necessary to ensure the incompressibility of the computed actual velocity field. It can be reformulated as a problem solely for the pressure function. In doing so, we apply the divergence operator to the first equation in (1.4) and get

$$-\Delta p^{m+1} = -\frac{1}{k}\mathrm{div}\,\tilde{u}^{m+1}, \qquad \partial_n p^{m+1}|_{\partial\Omega} = 0. \qquad (1.5)$$

The actual solenoidal velocity field is calculated by means of a simple update,

$$u^{m+1} = \tilde{u}^{m+1} - k\nabla p^{m+1}. \qquad (1.6)$$

Therefore, each iteration step splits into two elementary problems: a linear convection-diffusion equation with homogeneous Dirichlet data for the field $\tilde{u}^{m+1}$ and a Laplace-Neumann problem for the pressure iterate p^{m+1}. — The significant gain of efficiency of the Chorin scheme implies the prescription of unphysical boundary conditions with respect to the pressure function p^{m+1} in (1.5). This requirement leads to the question of whether p^{m+1} is a sufficiently accurate approximation of the actual pressure $p(t_{m+1})$.

First investigations of this scheme are due to Chorin [4] and Temam [40] in 1969. Nevertheless, the error estimates derived in [4] are sub-optimal and only applicable to restricted model configurations, i.e., problems with periodic boundary conditions. The first striking error bounds for practically relevant fluid flows have been proved by J. Shen in a series of articles, see [33], [35]. In this framework, Shen shows first order of convergence for the guess $\tilde{u}^{m+1}$ in the norm $l^2(0, t_{M+1}; \mathbf{L}^2(\Omega))$. Unfortunately, he is compelled to assume an additional degree of regularity for the solution of (1.1) at time $t = 0$, i.e., $p_t \in L^2(0, T; H^1/\mathbb{R})$. Let us recall that we can not establish a uniform bound for p_t in this topology for times $t \to 0$ in general, owing to the fact that this requires nonlocal compatibility conditions for the data u_0 and f to be satisfied. In particular, it is proved in [16] that this regularity statement can only be valid, if there exists a solution $p_0 \in H^1(\Omega)/\mathbb{R}$ of the overdetermined Neumann problem

$$\Delta p_0 = \operatorname{div}\Big(f(0) - (u_0 \cdot \nabla)u_0\Big), \qquad \text{in } \Omega,$$
$$\nabla p_0|_{\partial\Omega} = [\nu\Delta u_0 + f(0) - (u_0 \cdot \nabla)u_0]|_{\partial\Omega}$$

(1.7)

for the initial pressure p_0. Because of its nonlocal nature, this is virtually uncheckable for given data u_0 and $f(0)$, and a discretization model of (1.1) should be able to cope with an initial breakdown in the solutions regularity without loosing accuracy globally (in time). From this standpoint, Shen's analysis limits the applicability of his convergence results in a restrictive way, not covering problem configurations with incompatibly posed data.

The crucial idea for a rigorous mathematical approach to analyze the scheme of Chorin was proposed by R. Rannacher [30]. In his paper, Chorin's scheme was reinterpreted as a *semi-explicit pressure stabilization scheme,*

$$\frac{1}{k}\{\tilde{u}^{m+1} - \tilde{u}^m\} - \nu\Delta\tilde{u}^{m+1} + (u^m \cdot \nabla)\tilde{u}^{m+1} + \nabla p^m = f^{m+1},$$
$$\mathrm{div}\tilde{u}^{m+1} - k\Delta p^{m+1} = 0, \qquad\qquad\qquad (1.8)$$
$$\tilde{u}^{m+1}|_{\partial\Omega} = 0, \qquad \partial_n p^{m+1}|_{\partial\Omega} = 0.$$

The momentum equation in (1.8) is derived by addition of the first equation in (1.3) and the first equation in (1.4), with the index shifted by -1. The second identity in (1.8) is a quasi-compressibility constraint, which is simply the reformulation of (1.5). Based on this reformulation of the Chorin scheme, the following error estimates will be shown in Chapter 6, *without* requiring any additional amount of regularity property of the solution of (1.1),

$$\max_{0\le m\le M}\left\{\|u(t_{m+1}) - u^{m+1}\| + \tau_{m+1}\|p(t_{m+1}) - p^{m+1}\|_{-1}\right\} \le Ck,$$

$$\max_{0\le m\le M}\left\{\|u(t_{m+1}) - u^{m+1}\|_1 + \sqrt{\tau_{m+1}}\|p(t_{m+1}) - p^{m+1}\|\right\} \le C\sqrt{k}. \quad (1.9)$$

Here, C is a constant that depends on the given data of the problem and we set $\tau_{m+1} := \min\{t_{m+1}, 1\}$. These estimates justify that Chorin's scheme is of first order accuracy for general flows, even pointwise in time. Moreover, these statements with respect to p^{m+1} indicate that this iterate is indeed an approximation of the pressure function in a sufficiently accurate manner, apart from arising (spatial) boundary layers. These global (in space) error bounds will be supplemented by local (in space) statements in Chapter 6. This gives the theoretical justification for the singular perturbation character of the Chorin scheme which is caused by the prescription of an unphysical boundary condition for the pressure function stemming from the projection step. In practical computations, this perturbation is observed as marked boundary layers. We will comment further on this in Chapter 6. Owing to this drawback of the Chorin scheme, one aim of the present study is the construction of a splitting method that does not suffer from boundary layers. This new Chorin-Uzawa scheme and its variants will be proposed and discussed below.

Despite the specified advantages the usefulness of the Chorin method for practical applications is limited, due to its low order of convergence, and we are thus interested in projection schemes of higher order. For that purpose, Van Kan proposed another projection method in 1986, see [45], that combines the advantages of the projection idea with a discretization ansatz of second order (Crank-Nicolson). Then, the resulting scheme is as follows.

1. Start with $u^0 = u_0$ and $p^0 = p(0)$.

2. For $m \geq 0$ and a given triple of functions $\{u^m, \tilde{u}^m, p^m\}$, determine $\tilde{u}^{m+1}$ as the solution of the system:

$$\frac{1}{k}\{\tilde{u}^{m+1} - u^m\} - \frac{1}{2}\nu\Delta\{\tilde{u}^{m+1} + \tilde{u}^m\} + \frac{3}{2}(u^m \cdot \nabla)u^m$$

$$= \overline{f}^{m+1/2} + \frac{1}{2}(u^{m-1} \cdot \nabla)u^{m-1} - \nabla p^m, \qquad (1.10)$$

$$\tilde{u}^{m+1}|_{\partial\Omega} = 0.$$

3. Determine $\{u^{m+1}, p^{m+1}\}$ as the solution of

$$\frac{1}{k}\{u^{m+1} - \tilde{u}^{m+1}\} + \frac{1}{2}\nabla\{p^{m+1} - p^m\} = 0,$$

$$\text{div}\, u^{m+1} = 0, \qquad u^{m+1}|_{\partial\Omega} \cdot n = 0. \qquad (1.11)$$

In the first step $(m = 0)$, the explicit treatment of the convective part is replaced by an implicit version, using trapezoidal rule. — We can proceed as for the Chorin method, reformulating the projection step as a Laplace-Neumann problem for the *pressure correction* $q^{m+1} := p^{m+1} - p^m$,

$$-\Delta q^{m+1} = -\frac{2}{k}\text{div}\,\tilde{u}^{m+1}, \qquad \partial_n q^{m+1}|_{\partial\Omega} = 0, \qquad (1.12)$$

which is followed by the updates

$$p^{m+1} = q^{m+1} + p^m \qquad (1.13)$$

and

$$u^{m+1} = \tilde{u}^{m+1} - \frac{1}{2}k\nabla q^{m+1}. \qquad (1.14)$$

Due to this reformulation, the solution effort corresponds to the one that is caused by Chorin's scheme. Each iteration step consists of solving a problem for $\tilde{u}^{m+1}$, followed by the computation of p^{m+1}. This considerable reduction of expense in the solution process in combination with the improved accuracy of the computed approximation $\{u^{m+1}, p^{m+1}\}$ is the reason for the wide-spread application of Van Kan's scheme in diverse applications.

Despite its relevance for practical computations, numerical studies of Van Kan's scheme are rather rare. We mention the work of Shen [37], where second order of convergence with respect to $\tilde{u}^{m+1}$ in the norm $l^2(0, t_{M+1}; \mathbf{L}^2(\Omega))$ is proved. Especially, the question of *pointwise in time error estimates*, i.e., those for the velocity field in the norm $l^\infty(0, t_{M+1}; \mathbf{L}^2(\Omega))$, remains unanswered according to Shen.

The analysis of the Van Kan scheme and a modification according to Shen is subject of Chapter 7. In that chapter, the following statements of convergence are proved for the iterates of the scheme (1.10), (1.11),

$$
\begin{aligned}
\max_{0 \le m \le M} \left\{ \sqrt{\tau_{m+1}} \| u(t_{m+1}) - \overline{\tilde{u}}^{m+1} \| \right\} &\le Ck^2 \log\frac{1}{k}, \\
\max_{0 \le m \le M} \left\{ \sqrt{\tau_{m+1}} \| p(t_{m+1}) - \overline{\tilde{p}}^{m+1} \| \right\} &\le Ck \log\frac{1}{k},
\end{aligned}
\tag{1.15}
$$

with the averaging notation $\overline{\phi}^{m+1} := \frac{1}{4}\phi^{m+2} + \frac{1}{2}\phi^{m+1} + \frac{1}{4}\phi^m$. — These error statements classify the Van Kan scheme as a time-discretization method of second order of convergence.

Again, the mathematical approach to the Van Kan scheme (1.10)/(1.11) is by reinterpreting it as a *semi-explicit pressure correction scheme*,

$$
\begin{aligned}
\frac{1}{k}\{\tilde{u}^{m+1} - \tilde{u}^m\} &- \frac{1}{2}\nu\Delta\{\tilde{u}^{m+1} + \tilde{u}^m\} + \frac{1}{2}\nabla\{3p^m - p^{m-1}\} \\
&= \overline{f}^{m+1/2} - \frac{1}{2}\{3(u^m \cdot \nabla)u^m - (u^{m-1} \cdot \nabla)u^{m-1}\}, \\
\text{div}\tilde{u}^{m+1} &- k^2\Delta\{\frac{p^{m+1} - p^m}{k}\} = 0, \\
\tilde{u}^{m+1}|_{\partial\Omega} = 0, &\qquad \partial_n\{p^{m+1} - p^m\}|_{\partial\Omega} = 0, \\
u^0 \equiv u_0, &\qquad p^0 \approx p_0.
\end{aligned}
\tag{1.16}
$$

The derivation of these equations is analogous to the corresponding system in the case of Chorin's method. If we compare them with (1.8), we have here a quasi-compressibility constraint with a perturbation term of order $\mathcal{O}(k^2)$ that has evolutionary character now. This is the reason for additional difficulties in our analysis initiated by the limited regularity properties of the solution of (1.1) in the vicinity of the initial time $t = 0$ — apart from the boundary layers as a consequence of the unphysical boundary condition for the pressure function. In other words: General initial data may lead to a reduction of the order of convergence of this scheme. This important aspect will be discussed extensively in Chapter 7 and illustrated by means of computational experiments.

The described problematic nature of incompatibly posed initial data is independent of discretization in time. This is the reason for the analysis of the following continuous versions,

$$\operatorname{div} u^\varepsilon - \varepsilon \Delta p_t^\varepsilon = 0 \tag{1.17}$$

and the new modification

$$\operatorname{div} u^\varepsilon - \varepsilon \Delta \{\tau^r p^\varepsilon\}_t = 0, \qquad \partial_n \{\tau^r p^\varepsilon\}_t|_{\partial\Omega} = 0, \qquad r \geq 2, \tag{1.18}$$

with $\tau = \tau(t) := \min\{t, 1\}$. The study of both perturbations of the incompressibility constraints on the solution is subject of Chapter 4. As the main result, we prove an optimal approximation behavior of convergence for (1.18) for general fluid flows governed by (1.1) without the need of compatibly posed initial data. For further remarks, we refer to the Overview (see Section 7.1) and the results presented in Table 1.

According to the above discussion of the schemes of Chorin and Van Kan these projection schemes suffer from two severe drawbacks:

- *Occurrence of boundary layers:* Owing to the L^2-character of the projection step there result homogeneous Neumann boundary conditions for the pressure iterate.

- *Necessity of a regular solution of (1.1):* In order to assure an optimal convergence behavior of the iterates of the Van Kan scheme, we have to ensure that the solution of (1.1) is sufficiently regular, i.e., $p_t \in L^\infty(0, T; H^1(\Omega)/\mathbb{R})$.

Apart from the analysis of more or less well-known schemes, the aim of this work is the construction and analysis of new projection schemes that do not suffer from the drawbacks listed above any more. In this contents, we will propose a method of first order accuracy which is an improvement over the Chorin method. We will refer to it as the *Chorin-Uzawa scheme*. This scheme differs from the Chorin scheme in the way that *no boundary layers for the pressure approximation are inherent any more.* The idea to construct this scheme relies on a separation of the tasks for the iterate p^{m+1} as an approximation of the actual pressure and its role as the Lagrange multiplier in the projection step. The algorithm has the following structure:

1. Start with the given data $\{u^0, p^0, \tilde{p}^0\}$, such that

$$\|u_0 - u^0\| + \sqrt{k}\|p_0 - p^0\| \le Ck, \qquad \tilde{p}^0 \equiv 0. \tag{1.19}$$

2. Determine $\tilde{u}^{m+1}$ by means of $\{u^m, p^m, \tilde{p}^m\}$,

$$\frac{1}{k}\{\tilde{u}^{m+1} - u^m\} - \nu\Delta\tilde{u}^{m+1} + (u^m \cdot \nabla)\tilde{u}^{m+1}$$
$$+ \nabla\{p^m - \tilde{p}^m\} = f^{m+1}, \tag{1.20}$$
$$\tilde{u}^{m+1}|_{\partial\Omega} = 0.$$

3. Determine u^{m+1} by solving

$$\frac{1}{k}\{u^{m+1} - \tilde{u}^{m+1}\} + \nabla\tilde{p}^{m+1} = 0,$$
$$\operatorname{div} u^{m+1} = 0, \qquad u^{m+1}|_{\partial\Omega} \cdot n = 0. \tag{1.21}$$

4. Compute the actual pressure function,

$$p^{m+1} = p^m - \alpha\nu\operatorname{div}\tilde{u}^{m+1}, \qquad \alpha < 1. \tag{1.22}$$

Note that we have the following splitting in this scheme: whereas p^{m+1} is an approximation of the actual pressure function $p(t_{m+1})$, $\tilde{p}^{m+1}$ takes over the role of the Lagrange multiplier in the projection step. Let us mention that the difference of both quantities appears in (1.20).

The construction of this projection method as well as its analysis rely on the reformulation as another semi-explicit quasi-compressibility method, the so-called *artificial compressibility method,*

$$\frac{1}{k}\{\tilde{u}^{m+1} - u^{m+1}\} - \nu\Delta\tilde{u}^{m+1} + (u^m \cdot \nabla)\tilde{u}^{m+1} + \nabla p^m = f^{m+1},$$
$$\operatorname{div}\tilde{u}^{m+1} + \frac{1}{\alpha\nu}k\frac{p^{m+1} - p^m}{k} = 0,$$
$$\tilde{u}^{m+1}|_{\partial\Omega} = 0. \tag{1.23}$$

Observe that no boundary condition has to be imposed for the pressure function any more. This leads to the conjecture that no boundary layers will arise any more. This will be verified by means of detailed investigations in Chapter 8. — Unfortunately, perturbation effects of the quasi-compressibility constraint stemming from its evolutionary character are still present. This problematic nature will be studied for the continuous analogon of equations,

$$\text{div}u^\varepsilon + \varepsilon p^\varepsilon_t = 0, \tag{1.24}$$

and

$$\text{div}u^\varepsilon + \varepsilon\{\tau^r p^\varepsilon\}_t = 0, \qquad r \geq 2. \tag{1.25}$$

In the course of a comprehensive analysis of the quasi-compressibility constraints arising in numerical applications, we also analyze the *penalty scheme* in Chapter 3,

$$\text{div}u^\varepsilon + \varepsilon p^\varepsilon = 0. \tag{1.26}$$

The results of Hebeker in [15] and Shen in [39] that have been proven for this method will be sharpened in our analysis, see Chapter 3. Especially, we will verify error statements for the pressure function that are pointwise in time,

$$\sup_{[0,T]}\left\{\|u - u^\varepsilon\| + \sqrt{\tau}\|u - u^\varepsilon\|_1 + \tau\|p - p^\varepsilon\|\right\} \leq C\varepsilon. \tag{1.27}$$

With that, we succeed in classifying the penalty method as a first order scheme (in striking norms). The results that are related to the quasi-compressibility constraints (1.24), (1.25) will be presented in Chapter 4.

In order to avoid the problems mentioned above for the application of the Chorin-Uzawa scheme, which stem from the limited regularity features of the solution of (1.1) at times $t \to 0$, we will propose a *modified Chorin-Uzawa scheme* in Chapter 9 that is applicable for general flow 'constellations'. This scheme inherits all advantages of a regular perturbation ansatz, as it is formulated in the framework of the quasi-compressibility methods. Therefore, the following error estimates can be shown for this projection scheme,

$$\max_{0 \leq m \leq M}\left\{\|u(t_{m+1}) - u^{m+1}\|_1 + \|p(t_{m+1}) - p^{m+1}\|\right\} \leq Ck. \tag{1.28}$$

Owing to this, we have succeeded in constructing a *first order projection scheme* that gives velocity and pressure approximations of first order convergence for 'general' fluid flows under consideration.

Let us recall the crucial point of additional regularity requirements with respect to the solution of (1.1) in order to guarantee an optimal performance of the Van Kan scheme. Numerical simulations will be given that show a global reduction of the order of convergence in case of violating this requirement. This will be illustrated in Section 7.4. — As a first idea to circumvent this restrictive requirement, we will propose the *Chorin-/Van Kan scheme* in Section 9.2, which is a projection scheme of higher order giving optimal approximations even for incompatible initial data. The idea of this multi-component scheme is again to combine the "robust" ansatz of Chorin's method with the more accurate scheme of Van Kan for macroscopic times $t > 0$. This leads us to a second order projection scheme which is able to even cope with fluid flows with incompatible initial data in an optimal way.

As a drawback of the Chorin-/Van Kan scheme, this numerical model can only be applied in an optimal way for flows with certain stability properties, see Chapter 7 for further details on this. Unfortunately, the analysis for this numerical scheme cannot be transfered to general flows.

In order to develop a robust scheme of second order accuracy, we introduce the idea of certain time-grids that provide the numerical algorithm with an additional amount of robustness along the initial phase. These time-grid structures do not cause a significant additional amount of computational effort. As one of the main results of the present work, we will propose a slightly modified Van Kan scheme on certain underlying time-grids and prove second order convergence with respect to $\tilde{u}^{m+1}$ for general fluid flows that are governed by the equations (1.1). Corresponding results will be achieved on problem-adapted time-grid structures for the original Chorin-Uzawa scheme. Further, the application of structured time-grids for the latter projection schemes releases us from requiring (sufficiently) accurate initial pressure data.

Finally, we will be concerned with so-called *mixed quasi-compressibility methods* of the type

$$\text{div} u^\varepsilon + \phi_1(\varepsilon_1; p^\varepsilon) + \phi_2(\varepsilon_2; p_t^\varepsilon) = 0, \qquad \varepsilon = (\varepsilon_1, \varepsilon_2), \tag{1.29}$$

together with certain boundary operators that are needed for the well-posedness of the related quasi-compressibility problem. For our purposes, the nonnegative parameters ε_i introduced here will be related to the finite element parameter h and the time discretization parameter k or certain powers of them. As a motivation, one may think of the application of pressure stabilization formulations while applying Van Kan's scheme, together with a bilinear/bilinear (Q1/Q1) discretization in space. Further comments on this will be made in Chapter 5.

Below, the analysis for the Chorin scheme as well as the stationary quasi-compressibility methods are given for the Navier-Stokes equations (1.1). In the remaining cases, we have restricted ourselves to the nonstationary Stokes equations in order to simplify and shorten the corresponding analyses. Nevertheless, let us stress the fact that the analyses can be transfered to the Navier-Stokes equations. In this context, only the Chorin-/Van Kan scheme possesses an exceptional position; we will come back to this in Chapter 9.

The statements of convergence for the proposed schemes will be illustrated by means of numerical test calculations for a model problem. This allows a quantitative comparison between the different schemes. In the interpretation of the subsequent test calculations we focus on elaboration on the perturbation sources that are caused by the diverse quasi-compressibility methods: the errors committed by using a finite element discretization are assumed to be negligible, owing to the selection of sufficiently fine spatial grids.

Test calculations have been performed for boundary conditions for the velocity field that are of Dirichlet type. Here, we will not be concerned with the study of other boundary conditions like for instance those involving both functions, velocity and pressure. For the practically relevant case of mixed boundary conditions, see [19]. We emphasize that additional difficulties arise in the formulation of the projection step, resulting from the necessity of prescribing compatible boundary data.

The computations presented here use the software package FEAT2D [6], which has been developed in the Heidelberg Numerical Mathematics Group. For the finite element discretization, we have chosen a combination of conforming bilinear ansatz spaces for the velocity field and the pressure function, namely Q1/Q1. This pair provides us with high accuracy with respect to the spatial discretization. For questions related to the stability properties of the related schemes, we refer to the discussions in the respective chapters.

Let us finish the introduction with an overview over the organization of

QUASI-COMPESSIBI-LITY METHOD	Penalty	Artificial Compressibility	Damped Artificial Compressibility
$\phi_\Omega(\varepsilon;t,p^\varepsilon) =$	$+\varepsilon p^\varepsilon$	$+\varepsilon d_t p^\varepsilon$	$+\varepsilon(\tau^r p^\varepsilon)_t,\ r \geq 2$
Regularity requirements for the solution of (2.1), additional to (A1), (A2)	none	$\|u_t\|_1 + \|p_t\| \leq C,$ $\forall\, t > 0$	none
Order of Convergence α (in ε^α) for			
u^ε (in $L^\infty(0,T;\mathbf{L}^2)$)	1	$1 - \delta,\ \delta > 0$ arb.	1
u^ε (in $L^\infty(0,T;\mathbf{H}^1)$)	1	$1/2 - \delta,\ \delta > 0$ arb.	1
u^ε (in $L^2(0,T;\mathbf{H}^1)$)	1	$1 - \delta,\ \delta > 0$ arb.	1
p^ε (in $L^\infty(0,T;L_0^2)$)	1	$1/2 - \delta,\ \delta > 0$ arb.	1
p^ε (in $L^\infty(0,T;H^{-1})$)	1	$1/2$	1
THEOREM	3.2	4.1	4.2

Table 1.1: Orders of Convergence for Quasi-Compressibility Methods of type: $\mathrm{div}u^\varepsilon + \Phi_\Omega(\varepsilon;t,p^\varepsilon) = 0$

this work. For that purpose, we refer to Figure 1.1 and Tables 1.1 through 1.4. The subsequent Figure 1.1 demonstrates the conceptual relationships between the proposed projection methods and the corresponding quasi-compressibility methods. The schemes framed with thick lines are new developments that will be dealt with in this work. The results are summarized in Tables 1.1 through 1.4, including the specification of the theorems, in which the results are presented in a more exact way. We stress the fact that all presented results pertaining to convergence are new. Furthermore, the specification of preliminary work to the respective schemes will be subject of the overviews that are the introductory part of every chapter.

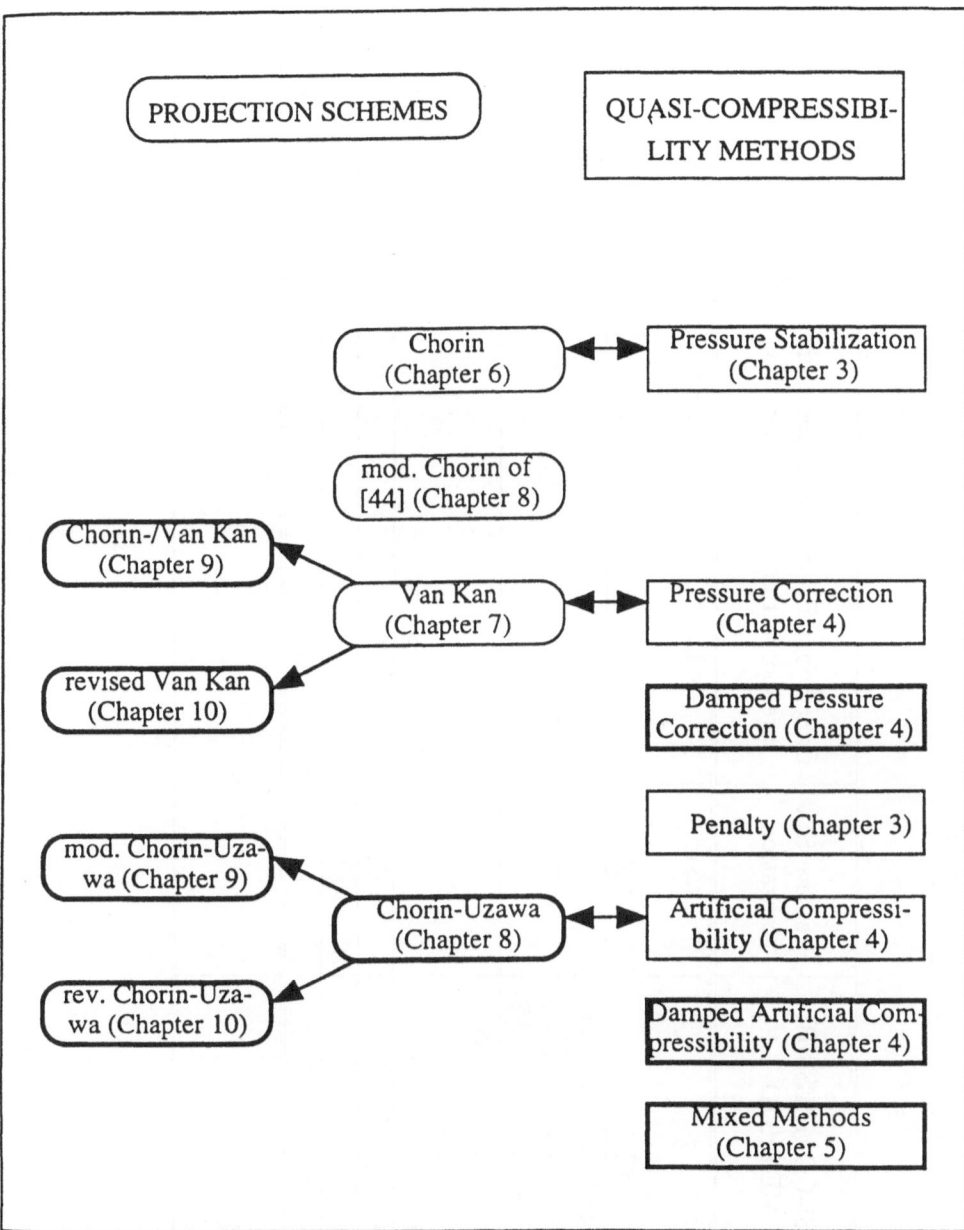

Figure 1.1: Organization of the book: Known (thin borders) and new (thick borders) methods

QUASI-COMPRESSIBILITY METHOD	Pressure Stabilization	Continuous Pressure Correction	Damped Continuous Pressure Correction	Mixed Methods
$\phi_\Omega(\varepsilon; t, p^\varepsilon) =$	$-\varepsilon\Delta p^\varepsilon$	$-\varepsilon\Delta d_t p^\varepsilon$	$-\varepsilon\Delta(\tau^r p^\varepsilon)_t, \ r \geq 2$	$-\varepsilon_1\Delta p^\varepsilon + \varepsilon_2 p^\varepsilon_t,$ $\varepsilon = (\varepsilon_1, \varepsilon_2)$
$\psi_{\partial\Omega}(\varepsilon; t, p^\varepsilon) =$	$\partial_n p^\varepsilon$	$\partial_n p^\varepsilon_t$	$\partial_n(\tau^r p^\varepsilon)_t$	$\partial_n p^\varepsilon$
Regularity requirements for the solution of (2.1), additional to (A1), (A2)	none	$\|u_t\|_2 + \|p_t\|_1 \leq C,$ $\forall t > 0$	none	$\|u_t\|_1 + \|p_t\| \leq C,$ $\forall t > 0$
Order of Convergence α (in ε^α) for				
u^ε (in $L^\infty(0,T; \mathbf{L^2})$)	1	$1 - \delta, \ \delta > 0$ arb.	1	(see Chapter 5)
u^ε (in $L^\infty(0,T; \mathbf{H^1})$)	1/2	1/2	1/2	(see Chapter 5)
p^ε (in $L^\infty(0,T; L^2_0)$)	1/2	1/2	1/2	(see Chapter 5)
p^ε (in $L^\infty(0,T; H^{-1})$)	1	1/2	1	(see Chapter 5)
THEOREM	3.3	4.3	4.4	5.1

Table 1.2: Orders of Convergence for Quasi-Compressibility Methods of type:
$\mathrm{div}\, u^\varepsilon + \Phi_\Omega(\varepsilon; t, p^\varepsilon) = 0, \quad \Psi_{\partial\Omega}(\varepsilon; t, p^\varepsilon)|_{\partial\Omega} = 0$

PROJECTION SCHEME	Chorin	Van Kan	Mod. Chorin	Chorin-Uzawa
Related Quasi-Compressibility Method, with $\phi_\Omega(k; u^{m+1}, p^{m+1}) =$ $\psi_{\partial\Omega}(k; u^{m+1}, p^{m+1}) =$	$-k\Delta p^{m+1}$ $\partial_n p^{m+1}$	$-k^2\Delta d_t p^{m+1}$ $\partial_n d_t p^{m+1}$	$-k\,\mathrm{div}\{-\Delta u^{m+1} + \nabla p^{m+1}\}$ $\{-\nabla\mathrm{div}u^{m+1} + \nabla p^{m+1}\}\cdot n$	$+k d_t p^{m+1}$
Regularity requirements for the solution of (2.1), additional to (A1), (A2)	none	$\|u_t\|_2 + \|p_t\|_1 \le C,$ $\forall t > 0$	none	$\|u_t\|_1 + \|p_t\| \le C,$ $\forall t > 0$
Order of Convergence α (in k^α) for				
u^{m+1} (in $l^\infty(0, t_{M+1}; \mathbf{L}^2)$)	1	$2 - \delta,\ \delta > 0$ arb.	1	$1 - \delta,\ \delta > 0$ arb.
u^{m+1} (in $l^\infty(0, t_{M+1}; \mathbf{H}^1)$)	1/2	1	1/2	$1/2 - \delta,\ \delta > 0$ arb.
u^{m+1} (in $l^2(0, t_{M+1}; \mathbf{H}^1)$)	1/2	1	1/2	$1 - \delta,\ \delta > 0$ arb.
p^{m+1} (in $l^\infty(0, t_{M+1}; L_0^2)$)	1/2	1	1/2	$1/2 - \delta,\ \delta > 0$ arb.
p^{m+1} (in $l^\infty(0, t_{M+1}; H^{-1})$)	1	1	1	$1/2 - \delta,\ \delta > 0$ arb.
THEOREM	6.1	7.1	8.1	8.2

Table 1.3: Orders of Convergence for Projection Schemes

MODIFIED PROJECTION SCHEMES	MULTI-COMPONENT SCHEMES		SCHEMES ON TIME-GRID STRUCTURES	
	Chorin-/Van Kan	modified Chorin-Uzawa	revised Van Kan	revised Chorin-Uzawa
Regularity requirements for the solution of (2.1), additional to (A1), (A2)	none	none	none	none
Order of Convergence α (in k^α) for				
u^ε (in $l^\infty(0, t_{M+1}; \mathbf{L^2})$)	$2-\delta$, $\delta > 0$ arb.	1	$2-\delta$, $\delta > 0$ arb.	$1-\delta$, $\delta > 0$ arb.
u^ε (in $l^\infty(0, t_{M+1}; \mathbf{H^1})$)	1	1	1	$1/2-\delta$, $\delta > 0$ arb.
u^ε (in $l^2(0, t_{M+1}; \mathbf{H^1})$)	1	1	1	$1-\delta$, $\delta > 0$ arb.
p^ε (in $l^\infty(0, t_{M+1}; L^2_0)$)	1	1	1	$1/2-\delta$, $\delta > 0$ arb.
p^ε (in $l^\infty(0, t_{M+1}; H^{-1})$)	1	1	1	$1/2-\delta$, $\delta > 0$ arb.
THEOREM	9.1	9.2	10.3	10.4

Table 1.4: Orders of Convergence for Modified Projection Schemes

Chapter 2

Preliminaries

2.1 Notation

Let $\Omega \subset \mathbb{R}^d$, $d = 2,3$ be a bounded domain. In this contents, we refer to $L^2(\Omega), H^k(\Omega), H_0^k(\Omega)$, k an integer, as the standard Lebesgue and Sobolev spaces, see [9] for further details. These spaces are endowed with the standard scalar products and their induced norms $\|\cdot\|_k$. Further, $H^{-k}(\Omega)$ is the space that is dual to $H^k(\Omega) \cap H_0^1(\Omega)$. We make frequent use of the notation $(\cdot, \cdot) \equiv (\cdot, \cdot)_{L^2}$. L_0^2 is the subspace of $L^2(\Omega)$ consisting of functions with vanishing spatial average, which is isomorphic to $L^2(\Omega)/\mathbb{R}$. Finally, we employ the notation $H^k/\mathbb{R} := H^k \cap L^2/\mathbb{R}$.

The spaces of vector-valued functions will be indicated with boldface letters, for instance $\mathbf{H}_0^1 \equiv (H_0^1)^d$, for $d = 2,3$. In the further analyses, we will not distinguish between the notation of inner products and norms in scalar or vector-valued applications. For convenience, we introduce the spaces

$$\mathbf{J}_0 = \left\{ v \in \mathbf{L}^2, \; \mathrm{div}\, v = 0 \text{ and } v|_{\partial\Omega} \cdot n = 0, \text{ weakly} \right\},$$

and

$$\mathbf{J}_1 = \left\{ v \in \mathbf{H}_0^1, \; \mathrm{div}\, v = 0 \right\},$$

which are completions of the divergence free functions on the domain Ω with respect to the norms $\mathbf{L}^2$ and $\mathbf{H}_0^1$, respectively. Finally, let $P_{\mathbf{J}_0}$ denote the $\mathbf{L}^2$-projection on the space $\mathbf{J}_0$. The Stokes operator will be denoted by $A \equiv -P_{\mathbf{J}_0}\Delta$, and is defined on the domain $D(A) = \mathbf{J}_1 \cap \mathbf{H}^2$.

For the analytical treatment of evolutionary problems, we will employ the following notations: For $1 \leq p < \infty$ and X a Banach space, let $L^p(0, T; X)$

be the space of functions $u \equiv u(x,t)$ such that holds: the evaluation map $t \mapsto \|u(t)\|_X^p$, $t \in [0,T]$ is measurable almost everywhere, and $\int_0^T \|u(s)\|_X^p \, ds < \infty$, for $1 \leq p < \infty$. In the case of $p = \infty$, we demand the property $\sup_{0 \leq s \leq T} \|u(s)\|_X < \infty$ to be satisfied. Correspondingly, we define $C(0,T;X)$ to be the space of functions such that holds: the evaluation map $t \mapsto \|u(t)\|_X$, $t \in [0,T]$ is continuous and we have $\max_{0 \leq s \leq T} \|u(s)\|_X < \infty$. — For time discretizations, we make use of the spaces $l^p(0, t_{M+1}; X)$, $1 \leq p < \infty$, which is the space of functions $\{u^{m+1}\}_{m=0}^M$ with bounded norm $\left(k \sum_{m=0}^M \|u^{m+1}\|_X^p \right)^{1/p}$, for a constant time-step $k = t_{m+1} - t_m$. For time-grids with points that are not equi-distributed and which are described by a grid function k_{m+1}, the modification of this definition is obvious. Finally, $l^\infty(0, t_{M+1}; X)$ is the set of all functions $\{u^{m+1}\}_{m=0}^M$ with finite norm $\max_{0 \leq m \leq M} \|u^{m+1}\|_X$.

2.2 Summary of Results Concerning the Stationary Incompressible Stokes Problem

The importance of the space $\mathbf{J}_0$ will be highlighted in the following Lemma.

Lemma 2.1 *Assume $\Omega \subset \mathbb{R}^d$, $d = 2,3$ to be a bounded domain with Lipschitz boundary. Then, the space $\mathbf{L}^2$ can be decomposed in an orthogonal way:*

$$\mathbf{L}^2 = \mathbf{J}_0 \oplus \mathbf{J}_0^\perp,$$

with $\mathbf{J}_0^\perp = \{u \in \mathbf{L}^2 | \ u = \nabla\phi, \ \phi \in H^1\}$.

For a proof, we refer to [5], for instance.

Lemma 2.2 *The Stokes operator A with domain $D(A) = \{u \in \mathbf{J}_1 | Au \in \mathbf{J}_0\}$ has the following properties:*

i)   It is positive definite and self-adjoint.

ii) The inverse of the Stokes operator, A^{-1}, is a compact operator on $\mathbf{J}_0$.

Again, we refer to [5] for a proof. By means of the Stokes operator A, we can show the following equivalence of norms on the space of weakly divergence free functions.

Lemma 2.3 *For all $\phi \in \mathbf{J}_1$, we have the equivalence of norms*

$$c_1 \|\phi\|_{-1}^2 \leq (A^{-1}\phi, \phi) \leq c_2 \|\phi\|_{-1}^2,$$

with two positive numbers c_1, c_2. Here, the negative norm is defined as follows: $\| \cdot \|_{-1} := \sup_{\phi \in \mathbf{J}_1} \frac{\langle \phi, \cdot \rangle}{\|\nabla \phi\|}$.

Proof:
Assume $u = A^{-1}\phi$, $u|_{\partial\Omega} = 0$ to be the solution of

$$-\Delta u + \nabla p = \phi, \qquad \mathrm{div}\, u = 0.$$

Substitution of these quantities and Lemma 2.1 then give the equations

$$(A^{-1}\phi, \phi) = (u, \phi) = \|\nabla u\|^2 \leq c_2 \|\phi\|_{-1}^2.$$

The last inequality results from a well-known a-priori bound for the solution of the incompressible Stokes equation, compare also Lemma 2.4 below. — The estimation in the inverse direction can be obtained by application of the definition of the negative Sobolev norm,

$$\|\phi\|_{-1} \leq c_1 \sup_{\xi \in \mathbf{J}_1} \frac{\langle \phi, \xi \rangle}{\|\nabla \xi\|} = c_1 \sup_{\xi \in \mathbf{J}_1} \frac{\langle Au, \xi \rangle}{\|\nabla \xi\|} \leq c_1 \|\nabla u\|.$$

$\square$

We finish this section with the presentation of regularity statements for the stationary incompressible Stokes equations.

Lemma 2.4 *Assume $\Omega \subset \mathbb{R}^d$, $d = 2, 3$ to be a bounded domain with boundary of type C^r, $r = \max\{m + 2, 2\}$, and $m \geq -1$ a positive real number. Further, we assume the given data of the generalized Stokes problem*

$$-\Delta u + \nabla p = f, \qquad \mathrm{div}\, u = g, \quad in\ \Omega, \qquad u|_{\partial\Omega} = \phi$$

to possess the following regularity properties: $f \in \mathbf{H}^m$, $g \in H^{m+1}$ and $\phi \in \mathbf{H}^{m+3/2}$. Then, the solution $\{u, p\}$ is in $\mathbf{J}_1 \cap \mathbf{H}^{m+2} \times H^{m+1}/\mathbb{R}$, and the following a-priori statements are valid, with $C = C(m, \Omega)$ a constant,

$$\|u\|_{m+2} + \|p\|_{H^{m+1}/\mathbb{R}} \leq C\{\|f\|_m + \|g\|_{m+1} + \|\phi\|_{m+3/2}\}.$$

A proof of this result can be found in [41].

2.3   A-priori Estimates for the Nonstationary Incompressible Stokes Equations

Let us start with the problem formulation. We are looking for a solution $\{u, p\}$ that satisfies the following nonstationary incompressible Stokes equations,

$$
\begin{aligned}
u_t - \Delta u + \nabla p &= f, &&\text{in } \Omega \times (0, T], \\
\operatorname{div} u &= 0, &&\text{in } \Omega \times (0, T], \\
u(0) &\equiv u_0 \in \mathbf{J}_1 \cap \mathbf{H}^2, &&u|_{\partial\Omega} = 0.
\end{aligned}
\tag{2.1}
$$

The well-posedness of this problem in the tuple of spaces $L^2(0, T; \mathbf{H}_0^1(\Omega)) \cap L^\infty(0, T; \mathbf{L}^2(\Omega)) \times L^2(0, T; L_0^2(\Omega))$ for sufficiently regular given data of the problem is assured, see [16] for instance. Further, the subsequent analyses rely on two additional (standard) assumptions (A1) and (A2), which are as follows:

- *condition (A1), concerning the regularity of the domain:* The unique solution $u \in \mathbf{J}_1$ of the stationary, incompressible Stokes problem with homogeneous boundary data of Dirichlet-type is already in $\mathbf{J}_1 \cap \mathbf{H}^2$, provided $f \in \mathbf{L}^2$, and satisfies the following stability result,

$$
\|u\|_2 \le C\|Au\|.
$$

- *condition (A2), concerning the regularity of the given data:* We suppose the following degrees of regularity for the given data u_0 and f,

$$
u_0 \in \mathbf{J}_1 \cap \mathbf{H}^2, \qquad f, f_t, f_{tt}, f_{ttt} \in C(0, \infty; \mathbf{L}^2(\Omega)).
$$

The assumptions (A1) and (A2) assure the existence and uniqueness of a strong solution of system (2.1) in classical spaces, i.e., $\{u, p\} \in C(0, T; \mathbf{J}_1 \cap \mathbf{H}^2) \times C(0, T; H^1/\mathbb{R})$. Let us mention the fact that (A1) is, e.g., satisfied for convex polygonal domains in two dimensions. — Based on these assumptions, the following results are easy to verify,

Lemma 2.5 *Assume $\{u, p\}$ to be the solution of the incompressible Stokes equations (2.1). Provided (A1) and (A2) are satisfied, the following a-priori statements are valid for the solution,*

i) $\sup_{(0,T]}\{\|\Delta u\| + \|u_t\| + \|\nabla p\|\} \leq C,$

ii) *for tuples* $i \in \{0, 1, 2\}$, $r \in \{1, 2, 3\}$, *with* $i + 2r \leq 7$,

$$\sup_{(0,T]}\{\tau^{r-1+i/2}\|\partial_t^r u\|_i\} + \int_0^T \tau^{2(r-3/2)+i}(s)\|\partial_t^r u(s)\|_i^2 \, ds \leq C,$$

iii) *for values* $r \in \{2, 3\}$,

$$\sup_{(0,T]}\{\tau^{r-3/2}\|\partial_t^r u\|_{-1}\} + \int_0^T \tau^{2r-4}(s)\|\partial_t^r u(s)\|_{-1}^2 \, ds \leq C,$$

iv) $\sup_{(0,T]}\{\tau^r\|\nabla\partial_t^r p\|\} + \int_0^T \tau^{2r+1}(s)\|\nabla\partial_t^{r+1}p(s)\|^2 \, ds \leq C$, $r \in \{0, 1\}$.

Remark 2.1 *We skip the proof of these statements, referring to [16]. Result iii) relies on an isomorphism property of the Stokes operator. In this case, test functions of type* $A^{-1}\partial_t^r u$, $r \geq 0$ *will be employed to verify the result, cf. [33].*

In our further analyses we will frequently make use of the discrete version of Gronwall's inequality. In order to be comprehensive, we will recall it here.

Lemma 2.6 *Let* $\{w_m\}_{m\geq 0}$ *be a sequence of non-negative numbers, satisfying the following inequality,*

$$w_M + \sum_{m=0}^{M} c_m \leq \sum_{m=0}^{M} a_m w_m + b_M, \qquad \text{for all } M \geq 0, \tag{2.2}$$

with nonnegative numbers a_M, b_M, c_M. *The sequence* $\{b_M\}_{M\geq 0}$ *is assumed to be non-decreasing, and we assume* $a_M < 1$, *for all* $M \geq 0$. *Then, if we employ the abbreviative notation* $\sigma_M = (1 - a_M)^{-1}$, *we obtain*

$$w_M + \sum_{m=0}^{M} c_m \leq b_M \exp\left(\sum_{m=0}^{M} \sigma_m a_m\right), \qquad \text{for all } M \geq 0$$

(implicit version of the discrete Gronwall inequality).

In case the sum on the right hand side in (2.2) is only up to $M - 1$, *we arrive at the following simpler form for arbitrary values* $a_m \geq 0$,

$$w_M + \sum_{m=0}^{M} c_m \leq b_M \exp\left(\sum_{m=0}^{M-1} a_m\right), \qquad \text{for all } M \geq 0$$

(explicit version of the discrete Gronwall inequality).

In the following analyses, we make use of C as a generic constant, depending on the domain Ω, the right hand side f and its time derivatives, and the initial data u_0. Further, the constant depends on the time interval indicated by T and t_{M+1}. In addition, we use the notations $\tau \equiv \tau(t) := \min\{t, 1\}$ and $\tau_m := \min\{t_m, 1\}$ and the notation $t_{m+1/2} := \frac{1}{2}\{t_{m+1} + t_m\}$. At last, we make frequent use of the notation for algebraic combinations of iterates,

- $d_t\phi^{m+1} := \frac{1}{k}\{\phi^{m+1} - \phi^m\}$,

- $\overline{\phi}^{m+1/2} := \frac{1}{2}\{\phi^{m+1} + \phi^m\}$ and $\overline{\overline{\phi}}^m := \frac{1}{2}\{\overline{\phi}^{m+1/2} + \overline{\phi}^{m-1/2}\}$.

More specialized notations will be presented in the chapters as needed.

2.4   A Numerical Test Problem

In order to compare the different projection methods that will be dealt with in the subsequent chapters, we consider the following model configuration. Let $(x, y, t) \mapsto \{u(x, y, t), p(x, y, t)\}$ be the solution of the incompressible Stokes equations, with given viscosity parameter ν,

$$u_t - \nu\Delta u + \nabla p = f,$$
$$\text{div} u = 0, \qquad\qquad\qquad\qquad\qquad (2.3)$$
$$u(0) \equiv u_0 \in \mathbf{J}_1 \cap \mathbf{H}^2,$$

on $\mathbf{D} := \{(x, y, t) \in \mathbb{R}^3_+ \mid (x, y, t) \subset Q_T := \Omega \times [0, 2]\}$, with

- $\Omega \equiv [0, 1] \times [0, 1]$, and

- the given functions

$$u_1 : (x, y, t) \mapsto u_1(x, y, t) = x^2(1 - x)^2(2y - 6y^2 + 4y^3),$$
$$u_2 : (x, y, t) \mapsto u_2(x, y, t) = -y^2(1 - y)^2(2x - 6x^2 + 4x^3),$$
$$p \equiv p_\gamma : (x, y, t) \mapsto p_\gamma(x, y, t) = (x^2 + y^2 - \frac{2}{3})t^\gamma, \qquad \gamma \in [0, 1],$$
$$\nu \in \{10, 1, 10^{-1}, 10^{-2}\}.$$

All test calculations use a structured spatial grid with mesh size $h = 1/64$ and can be parameterized by the triple $\{T, \gamma, \nu\}$. Here, T denotes the time, at which the error is evaluated, γ corresponds to the given time-weight in the pressure function p_γ, and ν qualifies the degree of viscosity of the fluid flow.

The computations that have been performed for the various projection schemes to be introduced in the subsequent chapters are based on a Q1/Q1 finite element discretization on a uniform grid that can be parameterized by h. Further, two technical hints will be helpful in order to compare the numerical results qualitatively.

- In the first row of all tables different functions are listed, for which the errors are calculated in the $L^2(\Omega)$-norm at a certain time $T > 0$. The row labeled "order" presents the averaged rates of convergence.

- The isolines depicted in the iso-scale figures represent error values that are staggered in an equi-distant way: the related error gradients are large in regions with a strong concentration of isolines. This makes it possible to get an impression of the global structure of the error function.

Chapter 3

Stationary Quasi-Compressibility Methods: The Penalty Method and the Pressure Stabilization Method

3.1 Overview and Results

The objective of this chapter is the mathematical classification of the *penalty method* and the *pressure stabilization method*. To be more specific, we suggest a new analytical approach to the penalty method and the pressure stabilization method that distinguishes and quantifies the different error mechanisms inherent to them. As a result, we obtain optimal (pointwise in time) error statements as well as sharp a-priori bounds for the solution of the pressure stabilization method, that have not been available before. The latter results are, for instance, necessary in order to derive optimal results in the finite element context.

As we already pointed out, the numerical solution process of (1.1) is made more difficult by the incompressibility constraint that causes a saddle-point character of the system and therefore enforces a stable pair of discrete spaces for determining u and p in the finite element context. In general, stable mixed numerical approaches lead to a large numerical effort in the calculation process. The quasi-compressibility methods that will be analyzed here are to overcome these drawbacks of the standard discretizations once by getting

rid of the pressure (the penalty scheme), and once by stabilizing favorable
ansatz pairs for velocity and pressure approximation (pressure stabilization
method). For further discussions on the relevance of these methods in numerical schemes, we refer to [2], for instance.

We will start by proposing the two quasi-compressibility methods that will
be investigated in this chapter. To this end, we make use of a perturbation
parameter $\varepsilon > 0$. Then the penalty method is formulated as follows: Find a
pair $\{u^\varepsilon, p^\varepsilon\}$, with $p^\varepsilon \in L^2(0, T; L^2_0)$, that satisfies the system

$$u_t^\varepsilon - \nu \Delta u^\varepsilon + \tilde{B}(u^\varepsilon, u^\varepsilon) + \nabla p^\varepsilon = f,$$
$$\mathrm{div} u^\varepsilon + \varepsilon p^\varepsilon = 0, \qquad u^\varepsilon|_{\partial\Omega} = 0, \tag{3.1}$$

with $u^\varepsilon(0) = u_0 \in \mathbf{J_1} \cap \mathbf{H^2}$.

Secondly, the pressure stabilization method to be studied below reads as
follows: Find the solution pair $\{u^\varepsilon, p^\varepsilon\}$, with $p^\varepsilon \in L^2(0, T; H^1/\mathbb{R})$, of the
equations

$$u_t^\varepsilon - \nu \Delta u^\varepsilon + \tilde{B}(u^\varepsilon, u^\varepsilon) + \nabla p^\varepsilon = f,$$
$$\mathrm{div} u^\varepsilon - \varepsilon \Delta p^\varepsilon = 0, \qquad \partial_n p^\varepsilon|_{\partial\Omega} = 0, \qquad u^\varepsilon|_{\partial\Omega} = 0, \tag{3.2}$$

with $u^\varepsilon(0) = u_0 \in \mathbf{J_1} \cap \mathbf{H^2}$. — Both schemes possess a spatial perturbation
character, which is once regular, once singular. The latter term is related to
the prescription of homogeneous Neumann boundary data for the pressure
in (3.2); of course, this condition is not necessarily satisfied by the solution p
of (1.1). — As we will see in subsequent analyses, solutions of (3.1) converge
to the corresponding one in (1.1) with optimal order. In this context, the
prescription of the homogeneous boundary condition of Neumann-type in
(3.2) causes a highly "anisotropic" behavior of convergence in space for the
pressure p^ε as well as for higher order derivatives of the velocity field u^ε for
parameter values $\varepsilon \to 0$, resulting in a loss of global convergence order in
strong norms.

The notation $\tilde{B}(\cdot, \cdot)$ in (3.1) and (3.2) is used as a modification of the
original nonlinearity in (1.1) in order to ensure the stability of the resulting
present systems. We would like to distinguish between two variants. One is
due to Temam, see [41],

$$\tilde{B}_1(u, v) = (u \cdot \nabla)v + \frac{1}{2}(\mathrm{div} u)v, \tag{3.3}$$

whereas another ansatz has been investigated by the author in [27], which is more common even in the case of projection methods,

$$\tilde{B}_2(u, v) = (P_{\mathbf{J}_0} u \cdot \nabla) v. \tag{3.4}$$

Further details for these choices will be given below.

Remark 3.1 *For general flows governed by the incompressible Navier-Stokes equations (1.1) and the perturbed versions (3.1) and (3.2), we are only given a unique strong solution $\{u, p\}$ locally in time. For our further considerations, we will assume well-posedness in the latter sense of the above systems for a given period of time, $[0, T]$, referring to this as* Assumption (A3).

The analysis for both systems, (3.1) and (3.2), rely on striking a-priori statements for the solution $\{u, p\}$ of (1.1). The following results can be found in [16],

$$\sup_{(0,T]}\left\{\tau^{r-1+i/2}\|\partial_t^r u\|_i\right\} + \int_0^T \tau^{2r-1+i}(s)\|\partial_t^{r+1} u(s)\|_i^2 \, ds \le C, \tag{3.5}$$

for values $i \in \{0, 1, 2\}$, $r \in \{1, 2, 3\}$, with $i + 2r \le 7$ and $r - 1 + \frac{i}{2} \ge 0$, and

$$\sup_{(0,T]}\left\{\tau^r\|\nabla\partial_t^r p\|\right\} + \int_0^T \tau^{2r+1}(s)\|\nabla\partial_t^{r+1} p(s)\|^2 \, ds \le C. \tag{3.6}$$

These a-priori statements for the solution of (1.1) establish the framework of our subsequent investigations with respect to convergence of the solutions of (3.1) and (3.2), and we will make frequent use of them without explicit mentioning them.

Let us remark that the well-posedness of the penalized system (3.1) on the pair of spaces $L^\infty(0, T; \mathbf{L}^2) \cap L^2(0, T; \mathbf{H}_0^1) \times L^2(0, T; L_0^2)$ for a time T, which depends on the given data, is assured. Moreover, the additional regularity property $u^\varepsilon \in L^\infty(0, T; \mathbf{H}^2)$ can be proven, provided the given data are regular, i.e. $u_0 \in \mathbf{H}^2$ and $f, f_t \in L^\infty(0, T; \mathbf{L}^2)$. Moreover, striking a-priori bounds that employ ε-independent constants C can easily be verified in corresponding norms, compare [39]. Corresponding, sharp a-priori statements for the pressure stabilization scheme (3.2) are not so obvious to obtain, due to the singular perturbation character. Therefore, they are given in the following theorem that will be proven in this chapter.

Theorem 3.1 *Let $\{u^{\varepsilon}, p^{\varepsilon}\} \in L^{\infty}(0, T; \mathbf{L}^2) \cap L^2(0, T; \mathbf{H}_0^1) \times L^2(0, T; H^1/\mathbb{R})$ be the pair of solutions of the pressure-stabilized scheme, with $\tilde{B} \equiv \tilde{B}_1$ or $\tilde{B} \equiv \tilde{B}_2$. Further, let (A1), (A2), and (A3) be satisfied. Then, the following bounds are valid for a constant C that is independent of $\varepsilon > 0$,*

    i) $\sup_{[0,T]} \|u^{\varepsilon}\|_1 + \int_0^T \|u^{\varepsilon}(s)\|_2^2 \, ds \leq C,$

    ii) $\sup_{[0,T]} \{ \tau \|u_t^{\varepsilon}\| + \tau \|u^{\varepsilon}\|_2 + \|p^{\varepsilon}\|_1 \} \leq C.$

Remark 3.2 *The verification of these regularity results requires sharp error estimates for the differences $\{u - u^{\varepsilon}\}$, $\{p - p^{\varepsilon}\}$ in different norms, pointwise in time. This guarantees a transfer of corresponding known results for the incompressible Navier-Stokes equations (1.1), see also [16].*

Most of the earlier analyses concerning quasi-compressibility methods are restricted to the steady state case. We begin by reviewing earlier studies of the penalty method in the nonstationary case. There are investigations related to the nonstationary Navier-Stokes equations, see [15], which lead to sub-optimal error estimates for the velocity in the norm $L^{\infty}(0, T; \mathbf{L}^2)$. Recently, these results have been drastically improved by Shen, see [39], who derived the following estimates for the solution of (3.1),

$$\sup_{[0,T]} \left\{ \sqrt{\tau}\|u - u^{\varepsilon}\| + \tau\|u - u^{\varepsilon}\|_1 \right\} +$$
$$+ \left(\int_0^T s^2 \|p(s) - p^{\varepsilon}(s)\|^2 \, ds \right)^{1/2} \leq C\varepsilon. \tag{3.7}$$

Nevertheless, the important question of optimal pointwise error estimates for the pressure in time was still unanswered and will be positively answered in the following theorem. Additionally, it will be shown that the estimate (3.7) can be improved with respect to the time-weights that do not admit any statements regarding the error behavior for initial times $t \to 0$.

Theorem 3.2 *Let $\{u^{\varepsilon}, p^{\varepsilon}\}$ be the solution of the penalty scheme (3.1), and $\{u, p\}$, with $p \in L^{\infty}(0, T; L_0^2)$, the solution of (1.1). Let the assumptions (A1), (A2), (A3) concerning the given data be valid, and $\tilde{B} \equiv \tilde{B}_1$ or $\tilde{B} \equiv \tilde{B}_2$. Then, we obtain the following error estimates with a constant $C = C(u_0, f, \Omega, T, \nu)$ that depends on the given data of the problem,*

$$\sup_{[0,T]} \left\{ \|u - u^{\varepsilon}\| + \sqrt{\tau}\|u - u^{\varepsilon}\|_1 + \tau\|p - p^{\varepsilon}\| \right\} \leq C\varepsilon.$$

The proof of this theorem will be given in Section 3.2.

Section 3.3 is devoted to the analysis of the pressure stabilization method (3.2). Subjects are the proof of Theorem 3.1 and the verification of error estimates. To the author's knowledge, there are no analyses leading to optimal error statements for the unsteady Stokes- or Navier-Stokes case. We collect the main results related to the error analysis of (3.2) in another theorem, that will be proven in the specified section.

Theorem 3.3 *Let $\{u^\varepsilon, p^\varepsilon\}$ be the solution of the pressure stabilization scheme (3.2), and $\{u, p\}$, with $p \in L^\infty(0, T; L_0^2)$, the solution of (1.1). Suppose $\tilde{B} \equiv \tilde{B}_1$ or $\tilde{B} \equiv \tilde{B}_2$. Further, let the assumptions (A1), (A2), (A3) for the given data to be given below be valid. Then we obtain the following error estimates, with a constant C that depends on the given data of the problem,*

i) $\quad \sup_{[0,T]}\left\{\|u - u^\varepsilon\| + \tau\|p - p^\varepsilon\|_{-1}\right\} \leq C\varepsilon,$

ii) $\quad \sup_{[0,T]}\left\{\|u - u^\varepsilon\|_1 + \sqrt{\tau}\|p - p^\varepsilon\|\right\} \leq C\sqrt{\varepsilon}.$

Further, the following estimates for the pressure function are valid on interior subdomains $\Omega' \subset\subset \Omega$ [1]

iii) $\quad \sup_{[0,T]} \sqrt{\tau}\left\{\|u - u^\varepsilon\|_{1;\Omega'} + \|p - p^\varepsilon\|_{\Omega'}\right\} \leq \tilde{C}\varepsilon,$

iv) $\quad \sup_{[0,T]}\left\{\sqrt{\tau}\|p - p^\varepsilon\|_{1;\Omega'}\right\} \leq \tilde{C}\sqrt{\varepsilon},$

with a constant $\tilde{C}$ that is now additionally dependent on $\mathrm{dist}(\Omega', \partial\Omega)$.

The proof of the last two theorems is based on a decoupled study of the evolutionary error effects in (3.1) and (3.2) and the error source that stems from the quasi-compressibility constraint. This allows a subtle error analysis giving the optimal estimates in the above theorems. Moreover, a comparison of the results for the penalty and the pressure stabilization method shows that the errors for the velocity, measured in the norm $L^\infty(0, T; \mathbf{L}^2)$, are of the same magnitude. Further, estimates in higher norms as well as results for the pressure are even worse in the case of the pressure stabilization method, which is due to the homogeneous boundary conditions that are given in (3.2). Therefore, optimal error estimates for p^ε can only be achieved in negative norms.

---

[1] "$\subset\subset$" stands for compact imbedding

3.2   Analysis of the Penalty Method

3.2.1   The Stokes Case

The objective of this and the following subsection is the presentation of a proof for Theorem 3.2. We start with an analysis of this method for the Stokes case in order to investigate the influence of the perturbed incompressibility constraint onto the solution behavior. In the subsequent section we will study the amplification effects that are caused by the nonlinearity for the Navier-Stokes equations (3.1). Therefore, we start with the analysis of the problem: Find solutions $\{\hat{u}^\varepsilon, \hat{p}^\varepsilon\} \in L^\infty(0,T; \mathbf{L}^2(\Omega)) \cap L^2(0,T; \mathbf{H}_0^1(\Omega)) \times L^2(0,T; L_0^2(\Omega))$ of the equations

$$
\begin{aligned}
&\hat{u}_t^\varepsilon - \nu \Delta \hat{u}^\varepsilon + \nabla \hat{p}^\varepsilon = F, \\
&\operatorname{div}\hat{u}^\varepsilon + \varepsilon \hat{p}^\varepsilon = 0, \qquad \hat{u}^\varepsilon(0) = u_0,
\end{aligned}
\tag{3.8}
$$

with $F = f - (u \cdot \nabla)u \in L^\infty(0,T;\mathbf{L}^2)$. The well-posedness of the pressure function p^ε results from the Stokes theorem, owing to the prescription of the homogeneous Dirichlet data for the velocity field.

In order to quantify the impact of the quasi-compressibility in (3.8), in-dependent of the nonstationary effects that result from the first identity, let us introduce another auxiliary problem that will be investigated at first. — We look for the solution $\{U^\varepsilon, P^\varepsilon\} \in \mathbf{H}_0^1(\Omega) \times L_0^2(\Omega)$ of the following system,

$$
\begin{aligned}
&-\nu \Delta U^\varepsilon + \nabla P^\varepsilon = F - u_t, \\
&\operatorname{div}U^\varepsilon + \varepsilon P^\varepsilon = 0, \qquad U^\varepsilon|_{\partial\Omega} = 0.
\end{aligned}
\tag{3.9}
$$

Note, that the following arguments can be used due to the fact that the right hand side of the first identity in (3.9) is in $L^\infty(0,T;\mathbf{L}^2)$. We introduce the following notation for the error functions $E \equiv u - U^\varepsilon$ and $\Pi \equiv p - P^\varepsilon$. Then, the error functions $\{E, \Pi\}$ satisfy the following system of equations,

$$
\begin{aligned}
&-\nu \Delta E + \nabla \Pi = 0, \\
&\operatorname{div}E + \varepsilon \Pi = \varepsilon p, \qquad E|_{\partial\Omega} = 0.
\end{aligned}
\tag{3.10}
$$

We obtain these identities through subtraction of the corresponding equalities in (1.1) and (3.9). — The error that is introduced by (3.9) can now easily be determined, using arguments that are already known from the steady state case of the penalized Stokes equations, see e.g. [2],

$$\|\Pi\| + \|E\|_1 \leq C\|\nabla E\| \leq C\sqrt{\varepsilon}|(\Pi, p)|^{1/2}$$
$$\leq \frac{1}{2\alpha}\|\Pi\| + C\alpha\varepsilon\|p\|, \qquad \alpha > 0.$$

Choosing α sufficiently large permits absorption of the first term on the right hand side of this inequality on the left side. Therefore, we arrive at the following estimates that quantify the influence of the penalization in the steady state case,

$$\sup_{[0,T]}\left\{\|u - U^\varepsilon\|_1 + \|p - P^\varepsilon\|\right\} \leq C\varepsilon. \qquad (3.11)$$

For the next part of the error analysis for (3.8) that is related to the determination of error accumulation in time we need estimates for the error part E, differentiated in time. The following estimates for this quantity are even stronger than necessary in the subsequent analysis and can be derived in the same matter, for all $r \in \{1, 2, 3\}$,

$$\tau^{2r-1}(T)\|\nabla \partial_t^r E(T)\|^2 + \int_0^T \tau^{2(r-1)}(s)\|\nabla \partial_t^r E(s)\|^2 \, ds \leq C\varepsilon^2.$$
$$(3.12)$$

In order to succeed in transferring the error statements in (3.11) for the auxiliary problem (3.9) to the problem (3.8), we can restrict our analysis of system (3.8) to the investigation of the following error identities,

$$e_t^\varepsilon - \nu\Delta e^\varepsilon + \nabla\eta^\varepsilon = E_t,$$
$$\text{div} e^\varepsilon + \varepsilon\eta^\varepsilon = 0, \qquad\qquad\qquad\qquad (3.13)$$
$$e^\varepsilon(0) = \{\hat{u}^\varepsilon - U^\varepsilon\}(0), \qquad \|e^\varepsilon(0)\|_1 \leq B\varepsilon,$$

with error functions $e^\varepsilon \equiv \hat{u}^\varepsilon - U^\varepsilon$ and $\eta^\varepsilon \equiv \hat{p}^\varepsilon - P^\varepsilon$. Note that the bound for the initial error stems from the corresponding statement in (3.8) and estimate (3.11). — Employing the time derivative ∂_t^r onto the first identity in (3.13) and testing with $\tau^{2r}\partial_t^r e^\varepsilon$, $r \in \{0, 1, 2\}$, finally integration over the time interval $[0, T]$ gives the next formula (3.14),

$$
\begin{aligned}
\tau^{2r}(T)\|\partial_t^r e^\varepsilon(T)\|^2 + \nu \int_0^T & \tau^{2r}(s)\|\nabla\partial_t^r e^\varepsilon(s)\|^2 \, ds \\
& + \varepsilon \int_0^T \tau^{2r}(s)\|\partial_t^r \eta^\varepsilon(s)\|^2 \, ds \\
& \le C\Big\{ \int_0^T \tau^{2r}(s)\|\partial_t^{r+1} E(s)\|^2 \, ds \\
& + \int_0^T \tau^{2r}(s)\|\partial_t^r e^\varepsilon(s)\|^2 \, ds \Big\},
\end{aligned}
\tag{3.14}
$$

On the other hand, if we employ ∂_t^{r-1} onto the first equation in (3.13) and test it with $\tau^{2r-1}\partial_t^r e^\varepsilon$, $r \in \{1, 2, 3\}$, final integration over the time-interval $[0, T]$ leads to

$$
\begin{aligned}
\int_0^T \tau^{2r-1}(s)\|\partial_t^r e^\varepsilon(s)\|^2 \, ds & + \nu\tau^{2r-1}(T)\|\nabla\partial_t^{r-1} e^\varepsilon(T)\|^2 \\
& + \varepsilon\tau^{2r-1}(T)\|\partial_t^{r-1}\eta^\varepsilon(T)\|^2 \\
& \le C\Big\{ \int_0^T \tau^{2r-1}(s)\|\partial_t^r E(s)\|^2 \, ds \\
& + \int_0^T \tau^{2(r-1)}(s)\|\nabla\partial_t^{r-1} e^\varepsilon(s)\|^2 \, ds \\
& + \varepsilon \int_0^T \tau^{2(r-1)}(s)\|\partial_t^{r-1}\eta^\varepsilon(s)\|^2 \, ds \Big\}.
\end{aligned}
\tag{3.15}
$$

The results (3.14) and (3.15) in combination with (3.12) represent the basis in order to apply an inductive argument that leads to the final estimate, with $i \in \{0, 1\}$, and $r \in \{0, 1, 2\}$,

$$
\begin{aligned}
\tau^{2r+i}(T)\|\partial_t^r e^\varepsilon(T)\|_i^2 & + \int_0^T \tau^{2r-1+i}(s)\|\partial_t^r e^\varepsilon(s)\|_i^2 \, ds \\
& + \varepsilon\Big\{ \tau^{2r+1}(T)\|\partial_t^r \eta^\varepsilon(T)\|^2 + \int_0^T \tau^{2r}(s)\|\partial_t^r \eta^\varepsilon(s)\|^2 \, ds \Big\} \le C\varepsilon^2.
\end{aligned}
\tag{3.16}
$$

Thus, together with the bounds (3.11) we succeeded in proving the following error estimates for the Stokes case (3.8),

$$
\sup_{[0,T]}\big\{ \|u - \hat{u}^\varepsilon\| + \sqrt{\tau}\|u - \hat{u}^\varepsilon\|_1 \big\} \le C\varepsilon.
\tag{3.17}
$$

We complement these statements for the velocity with results that are valid for the pressure $\widehat{p}^\varepsilon$. Therefore, we start from the first identity in (3.13), using a stability result for the divergence operator. We arrive at

$$\|\eta^\varepsilon\| \leq C\{\|E_t\| + \nu\|\nabla e^\varepsilon\| + \|e_t^\varepsilon\|\}. \tag{3.18}$$

In order to bound the first two terms on the right hand side of (3.18) optimally - see (3.12) and (3.15) respectively — we have to multiply (3.18) with the time-weight $\sqrt{\tau}$. This has to be sharpened to τ, due to the last term in (3.18). This completes the proof of the error bound for the pressure,

$$\sup_{[0,T]}\{\tau\|p - \widehat{p}^\varepsilon\|\} \leq C\varepsilon. \tag{3.19}$$

Remark 3.3 *These error estimates for the solution of (3.8) have been derived for exact initial data for the velocity. This assumption can easily be extended to perturbed initial data that satisfy the inequality*

$$\|e(0)\| + \sqrt{\varepsilon}\sqrt{\nu}\|\nabla e(0)\| \leq C\varepsilon. \tag{3.20}$$

To sum up the contents of this section, we derived optimal convergence results (3.17) and (3.19) for the penalized Stokes problem (3.8). They are collected in the following lemma.

Lemma 3.1 *Let $\{\widehat{u}^\varepsilon, \widehat{p}^\varepsilon\}$ be the solution of the penalty scheme (3.8), and $\{u, p\}$, with $p \in L^\infty(0, T; L_0^2)$ the solution of (1.1). Suppose $\tilde{B} \equiv \tilde{B}_1$ or $\tilde{B} \equiv \tilde{B}_2$. Further, let the assumptions (A1), (A2), (A3) on the given data be valid. Then, we obtain the following error estimates, with a constant C that depends on the given data of the problem,*

$$\sup_{[0,T]}\{\|u - \widehat{u}^\varepsilon\| + \sqrt{\tau}\|u - \widehat{u}^\varepsilon\|_1 + \tau\|p - \widehat{p}^\varepsilon\|\} \leq C\varepsilon.$$

These results will now be generalized for the Navier-Stokes case (3.1) in the subsequent subsection.

3.2.2 The Navier-Stokes Equations

As already mentioned, the objective of this section is to quantify the error amplification mechanisms of the nonlinearity in (3.1) in the penalty scheme. This will be done by means of another perturbation argument. — Before we continue our analysis let us come back to the different choices $\tilde{B}_1$ and $\tilde{B}_2$ that have been introduced in the formulae (3.3) and (3.4).

Let us emphasize the fact that certain modifications of the original non-linearity in (1.1) are necessary to assure the well-posedness of the system (3.1) (and (3.2)). In this respect, the modification $\tilde{B}_1$ secures to control the effects of the nonlinearity for velocities that are merely in $\mathbf{H}_0^1$. To be more specific, this causes the related trilinear form

$$\tilde{b}(\phi, u, v) = \frac{1}{2}\big\{b(\phi, u, v) - b(\phi, v, u)\big\}, \qquad \forall \phi, u, v \in \mathbf{H}_0^1 \qquad (3.21)$$

to be again skew-symmetric with respect to the two last arguments, i.e.:

$$\tilde{b}(\phi, u, u) = 0, \qquad \forall \phi, u \in \mathbf{H}_0^1,$$

on the enlarged domain $(\mathbf{H}_0^1)^3$, a fact that is of fundamental relevance for the analytical investigations. Of course, from the computational point of view this stabilization technique increases the complexity of the methods (3.1) and (3.2), owing to the additional term in (3.21).

As a second approach of stabilization, the following modification of the nonlinear term is employed in the context of projection methods,

$$\hat{b}(\phi, \cdot, \cdot) = b(P_{\mathbf{J}_0}\phi, \cdot, \cdot), \qquad \forall \phi \in \mathbf{H}_0^1. \qquad (3.22)$$

The projected incompressible velocity field is now able to control the influence of the nonlinearity in (3.1) (and (3.2)). Note that the same skew-symmetry rule is valid which is essential for the following investigations. In the subsequent studies we simply take $\tilde{B}$ which can be chosen to be $\tilde{B}_1$ or $\tilde{B}_2$.

After these introductory remarks we can complete the proof of Theorem 3.2, using the results from the previous subsection. This permits us to quantify the difference between the solutions of (3.1) and (3.8). Subtracting the corresponding identities and employing the notations $\xi := u^\varepsilon - \hat{u}^\varepsilon$ and $\chi := p^\varepsilon - \hat{p}^\varepsilon$ leads to

$$\begin{aligned} \xi_t - \nu\Delta\xi + \nabla\chi &= -\tilde{B}(u, u) + \tilde{B}(u^\varepsilon, u^\varepsilon), \\ \mathrm{div}\xi + \varepsilon\chi &= 0, \qquad \xi(0) = 0, \end{aligned} \qquad (3.23)$$

because of the expansion property $B(\cdot, \cdot)|_{(\mathbf{J}_1)^2} \equiv \tilde{B}(\cdot, \cdot)|_{(\mathbf{J}_1)^2}$. The right hand side of the first equation can be reformulated to be

$$\begin{aligned} \tilde{B}(u, u) - \tilde{B}(u^\varepsilon, u^\varepsilon) &= -\tilde{B}(\xi, u) - \tilde{B}(u^\varepsilon, \xi) \\ &\quad + \tilde{B}(u^\varepsilon, u - \hat{u}^\varepsilon) + \tilde{B}(u - \hat{u}^\varepsilon, u). \end{aligned} \qquad (3.24)$$

Owing to this, we can proceed by testing the first identity in (3.23) with ξ and finally integrating over the time interval $[0, T]$. Thanks to the skew-symmetry rule above and after some elementary calculations, we are led to

$$\|\xi(T)\|^2 + \nu \int_0^T \|\nabla \xi(s)\|^2 \, ds + \varepsilon \int_0^T \|\chi(s)\|^2 \, ds$$

$$\leq \|\xi(0)\|^2 + C \int_0^T \|\Delta u(s)\|^2 \big\{ \|\xi(s)\|^2 + \|\{u - \widehat{u}^\varepsilon\}(s)\|^2 \big\} \, ds \tag{3.25}$$

$$+ \int_0^T \Big(\tilde{B}(u^\varepsilon, u - \widehat{u}^\varepsilon)(s), \xi(s) \Big) \, ds.$$

The second term on the right hand side stems from the first and the last term in (3.24). Note that no additional term is arising from the second one on the right hand side in (3.24), which is due to the skew-symmetry rule (3.21). Thus, it remains to bound the last integral in (3.25). To do so, we make use of the identity

$$\tilde{B}(u^\varepsilon, u - \widehat{u}^\varepsilon) = \tilde{B}(u^\varepsilon - u, u - \widehat{u}^\varepsilon) + \tilde{B}(u, u - \widehat{u}^\varepsilon). \tag{3.26}$$

The latter result can now be inserted into the last integral in (3.25). Then, the expression stemming from the first term in (3.26) can be bounded as follows:

$$\int_0^T \Big(\tilde{B}(u - u^\varepsilon, u - \widehat{u}^\varepsilon)(s), \xi(s) \Big) \, ds$$

$$\leq \int_0^T \big\{ \|\nabla \xi(s)\| + \|\nabla \{u - \widehat{u}^\varepsilon\}(s)\| \times \tag{3.27}$$

$$\times \|\nabla \{u - \widehat{u}^\varepsilon\}(s)\| \|\xi(s)\|^{1/2} \|\nabla \xi(s)\|^{1/2} \big\} \, ds.$$

On the other side, we can make use of the skew-symmetry rule (3.21) to control the second contribution in (3.26),

$$\int_0^T \Big(\tilde{B}(u, u - \widehat{u}^\varepsilon)(s), \xi(s) \Big) \, ds \leq \int_0^T \Big| \Big(\tilde{B}(u, \xi)(s), \{u - \widehat{u}^\varepsilon\}(s) \Big) \Big| \, ds$$

$$\leq C \int_0^T \|\{u - \widehat{u}^\varepsilon\}(s)\|_1^2 \, ds. \tag{3.28}$$

In order to bound the integral term here, let us recall the results (3.11) and (3.16). If we insert these estimates (3.27) and (3.28) in (3.25), we finally end up with

$$\|\xi(T)\|^2 + \nu \int_0^T \|\nabla\xi(s)\|^2 \, ds + \varepsilon \int_0^T \|\chi(s)\|^2 \, ds$$
$$\leq C\Big\{\varepsilon^2 + \int_0^T \|\xi(s)\|^2 \, ds\Big\}, \tag{3.29}$$

due to the results in the previous section. Now, we can apply Gronwall's lemma, and arrive at the estimate

$$\|\xi(T)\|^2 + \nu \int_0^T \|\nabla\xi(s)\|^2 \, ds + \varepsilon \int_0^T \|\chi(s)\|^2 \, ds \leq C\varepsilon^2. \tag{3.30}$$

In order to derive error estimates for the velocity in higher (spatial) norms or rather for the pressure we again have to employ certain time-weights. Therefore, we start with the verification of another result for the velocity error $\tau\xi_t$, measured in the norm $L^\infty(0, T; \mathbf{L}^2)$. To do so, we differentiate the first equation in (3.23) in time and afterwards test with $\tau^2\xi_t$. This leads to

$$\tau^2(T)\|\xi_t(T)\|^2 + \nu \int_0^T \tau^2(s)\|\nabla\xi_t(s)\|^2 \, ds + \varepsilon \int_0^T \tau^2(s)\|\chi_t(s)\|^2 \, ds$$
$$\leq \int_0^T \tau^2(s)(NLT_1, \xi_t(s)) \, ds + 2 \int_0^T \tau(s)\|\xi_t(s)\|^2 \, ds. \tag{3.31}$$

We use the abbreviation NLT_1 which denotes the differentiated right hand side of (3.24),

$$NLT_1 = -\tilde{B}(\xi_t, u) - \tilde{B}(\xi, u_t) - \tilde{B}(u_t^\varepsilon, \xi) - \tilde{B}(u^\varepsilon, \xi_t)$$
$$+ \tilde{B}(u_t^\varepsilon, u - \hat{u}^\varepsilon) + \tilde{B}(u^\varepsilon, u_t - \hat{u}_t^\varepsilon) \tag{3.32}$$
$$+ \tilde{B}(u_t - \hat{u}_t^\varepsilon, u) + \tilde{B}(u - \hat{u}^\varepsilon, u_t).$$

The treatment of the corresponding terms in (3.31) is now as follows. The first two terms in (3.32) can be estimated in a standard way, using the results of the previous subsection. In order to control the integral expression in (3.31) that stems from the third term in (3.32) we use the identity

$$\tilde{B}(u_t^\varepsilon, \xi) = \tilde{B}(u_t^\varepsilon - u_t, \xi) + \tilde{B}(u_t, \xi). \tag{3.33}$$

The first term on the right hand side of (3.33) will be treated as follows,

$$\Big(\tilde{B}(u_t^\varepsilon - u_t, \xi), \xi_t\Big) \leq C\|u_t^\varepsilon - u_t\|_{L^4}\|\nabla\xi\|\|\nabla\xi_t\|$$
$$\leq C\alpha\Big\{\|\hat{u}_t^\varepsilon - u_t\|\|\hat{u}_t^\varepsilon - u_t\|_1 + \|\nabla\xi\|^2\|\nabla\xi_t\|^2\Big\}\|\nabla\xi\|^2 \tag{3.34}$$
$$+ \frac{1}{4\alpha}\|\nabla\xi_t\|^2,$$

with $\alpha > 0$. If we insert this inequality in (3.31), choosing α sufficiently large, the last term can be absorbed on the left hand side. Thanks to (3.30), the part remaining on the right can be controlled by $C\varepsilon^2$, by additionally using the result $u^\varepsilon \in L^\infty(0, T; \mathbf{H}_0^1)$ and Gronwall's lemma, respectively. — The treatment of the remaining terms in (3.32) and (3.33) can now be done in an analogous fashion. Here, we confine ourselves to analyze the sixth term on the right hand side of (3.32). For that, we use the identity

$$\tilde{B}(u^\varepsilon, u_t - \widehat{u}_t^\varepsilon) = \tilde{B}(\xi, u_t - \widehat{u}_t^\varepsilon) + \tilde{B}(\widehat{u}_t^\varepsilon, u - \widehat{u}_t^\varepsilon). \tag{3.35}$$

The bound

$$\sup_{0 \leq s \leq T} \left\{ \|\nabla \widehat{u}^\varepsilon(s)\| + \sqrt{\tau(s)} \|\nabla \widehat{u}_t^\varepsilon(s)\| \right\} \leq C,$$

from the analyses in Subsection 3.2.1, together with the error bounds (3.12), (3.16) provide the basis for an optimal treatment of the terms in (3.35), together with (3.30).

If we collect these considerations in (3.31), we end up with the inequality

$$\tau^2(T)\|\xi_t(T)\|^2 + \nu \int_0^T \tau^2(s)\|\nabla \xi_t(s)\|^2 \, ds$$

$$+ \varepsilon \int_0^T \tau^2(s)\|\chi_t(s)\|^2 \, ds \leq C\varepsilon^2 + 2 \int_0^T \tau(s)\|\xi_t(s)\|^2 \, ds. \tag{3.36}$$

In order to bound the last term in this inequality, we confine ourselves to the preservation of a sketchy argument, which is as follows: the first equation of (3.23) will be tested with $\tau \xi_t$ and will be finally integrated over the time interval $[0, T]$. By using the earlier results (3.30) and the identity (3.24), we can verify the inequality

$$\int_0^T \tau(s)\|\xi_t(s)\|^2 \, ds + \nu\tau(T)\|\nabla\xi(T)\|^2 + \varepsilon\tau(T)\|\chi(T)\|^2$$

$$\leq C\varepsilon^2 + \frac{1}{4}\nu \int_0^T \tau^2(s)\|\nabla\xi_t(s)\|^2 \, ds. \tag{3.37}$$

Therefore, we have the upper bound $C\varepsilon^2$ for the right hand side of (3.36). — This auxiliary result will be useful in the sequel. Now, the application of $\partial_t(\tau\cdot)$ onto the first identity in (3.23) and testing with $\tau\xi_t$, finally integration over $[0, T]$ gives another error estimate,

$$
\begin{aligned}
\tau^2(T)\|\xi_t(T)\|^2 &+ \nu \int_0^T \tau^2(s)\|\nabla\xi_t(s)\|^2 \, ds \\
&+ \nu\tau(T)\|\nabla\xi(T)\|^2 + \varepsilon\tau(T)\|\chi(T)\|^2 \\
&\leq C\Big\{\varepsilon^2 + \nu \int_0^T \|\nabla\xi(s)\|^2 \, ds \\
&+ \varepsilon \int_0^T \|\chi(s)\|^2 \, ds + \int_0^T \tau(s)\big(NLT_2, \xi_t(s)\big) \, ds \Big\}.
\end{aligned}
\tag{3.38}
$$

Again, we summarize the terms coming from the bilinear forms $\widetilde{B}$ in the abbreviative notation NLT_2. Owing to (3.30), the first two integrals on the right hand side are bounded by $C\varepsilon^2$. Further, easy arguments for the last term in (3.38) can be employed in order to get the same bound for it.

Note that the error for the velocity, differentiated in time, needs a stronger time-weight than the error of the pressure in (3.38) which is quite astonishing. This is due to the modest perturbation of the incompressibility constraint.

3.3 Analysis of the Pressure Stabilization Method

The objective of the second part of the actual chapter is the verification of the error estimates that are formulated in Theorem 3.3, as well as the proof of Theorem 3.1. Therefore, we will start with the analysis for the linear Stokes problem.

3.3.1 The Stokes Case

Primarily, we will focus on the statements i) and ii) of Theorem 3.3. The proof of the local error statements iii) and iv) will be given at the end of the next subsection. In the investigation of (3.2), we follow the same strategy proposed in the former analysis of the penalty scheme. We start with the study of a linear auxiliary problem. Subsequently, we continue with a perturbation analysis that controls the further influence of the nonlinearity on the deviation of the solution $\{u^\varepsilon, p^\varepsilon\}$ of (3.2) from $\{u, p\}$.

We begin the analysis of convergence for (3.2) by investigating the following auxiliary problem,

$$-\nu\Delta U^\varepsilon + \nabla P^\varepsilon = f - u_t - \tilde{B}(u, u),$$
$$\text{div}U^\varepsilon - \varepsilon\Delta P^\varepsilon = 0, \qquad \partial_n P^\varepsilon|_{\partial\Omega} = 0, \qquad U^\varepsilon|_{\partial\Omega} = 0. \tag{3.39}$$

Again, $\tilde{B}$ coincides with the original nonlinearity on the domain $(\mathbf{J_1})^2$. — The objective of the study of this system is to quantify the impact of the pressure stabilization at a fixed time T, whereas the propagation of these errors will be investigated in the next step. Using the notations $E := u - U^\varepsilon$ and $\Pi := p - P^\varepsilon$, subtraction of the corresponding equalities in (1.1) and (3.39) leads to the system

$$-\nu\Delta E + \nabla\Pi = 0,$$
$$\text{div}E - \varepsilon\Delta\Pi = -\varepsilon\Delta p, \qquad \partial_n\Pi|_{\partial\Omega} = \partial_n p|_{\partial\Omega}, \qquad E|_{\partial\Omega} = 0. \tag{3.40}$$

The following estimate follows from easy energy arguments,

$$\frac{1}{\nu}\|\Pi\|^2 + \nu\|\nabla E\|^2 + \varepsilon\|\nabla\Pi\|^2 \le C\varepsilon\|\nabla p\|^2, \tag{3.41}$$

which simultaneously provide an ε-independent bound for the pressure ∇P^ε. In order to get sharp results for the error committed in the velocity approximation, that is to be measured in the norm $L^\infty(0, T; \mathbf{L^2})$, we employ a stationary duality argument: Find the solution $\{W, Q\} \in \mathbf{J_1} \times L^2/\mathbb{R}$ of the following incompressible Stokes problem,

$$-\nu\Delta W + \nabla Q = g, \tag{3.42}$$

with a given right hand side g. If we identify $g \equiv E$, and test (3.42) with E, further the corresponding equation in (3.40) with the divergence-free function W, we can add these identities and get

$$\|\Pi\|_{-1} + \|E\| \le C\varepsilon\|\nabla p\|. \tag{3.43}$$

Here, we have applied a stability result of the divergence operator. If we proceed in an analogous way by inserting E_t and E_{tt} as the right hand side of (3.42), we find

$$\int_0^T \tau(s)\|E_t(s)\|^2\, ds + \tau^2(T)\|E_t(T)\|^2$$
$$+ \int_0^T \tau^3(s)\|E_{tt}(s)\|^2\, ds \le C\varepsilon^2. \tag{3.44}$$

The estimate (3.44) describes the evolutionary behavior of the error part E. It is essential for the further analysis that is related to the nonstationary error effects, resulting from the first equation in (3.2). For this purpose, we introduce another auxiliary problem,

$$\hat{u}_t^\varepsilon - \nu\Delta\hat{u}^\varepsilon + \nabla\hat{p}^\varepsilon = f - \tilde{B}(u, u),$$
$$\text{div}\hat{u}^\varepsilon - \varepsilon\Delta\hat{p}^\varepsilon = 0, \qquad \partial_n\hat{p}^\varepsilon|_{\partial\Omega} = 0, \qquad \hat{u}^\varepsilon|_{\partial\Omega} = 0, \tag{3.45}$$

with $\hat{u}^\varepsilon(0) = u_0$. The main results of this section will be presented in a further lemma.

Lemma 3.2 *Let the tuple $\{\hat{u}^\varepsilon, \hat{p}^\varepsilon\}$ be the solution of problem (3.45), whereas $\{u, p\}$, with $p \in L^\infty(0, T; L_0^2(\Omega))$ is the corresponding one of (1.1). Assuming (A1), (A2), (A3) and $\tilde{B}$ to be of type $\tilde{B}_1$ or $\tilde{B}_2$, we arrive at the following error estimates, with a constant C that depends on the given data of the problem,*

   i) $\sup_{[0,T]}\{\|u - \hat{u}^\varepsilon\| + \tau\|p - \hat{p}^\varepsilon\|_{-1}\} \leq C\varepsilon,$

   ii) $\sup_{[0,T]}\{\|u - \hat{u}^\varepsilon\|_1 + \sqrt{\tau}\|p - \hat{p}^\varepsilon\|\} \leq C\sqrt{\varepsilon}.$

Proof:
We start with the proof of the $L^\infty(0, T; \mathbf{L}^2)$-error estimate for the velocity. For the error functions $e^\varepsilon := \hat{u}^\varepsilon - U^\varepsilon$ and $\eta^\varepsilon := \hat{p}^\varepsilon - P^\varepsilon$, there holds,

$$e_t^\varepsilon - \nu\Delta e^\varepsilon + \nabla\eta^\varepsilon = E_t,$$
$$\text{div}e^\varepsilon - \varepsilon\Delta\eta^\varepsilon = 0, \qquad \partial_n\eta^\varepsilon|_{\partial\Omega} = 0, \qquad e^\varepsilon|_{\partial\Omega} = 0, \tag{3.46}$$

with $\|e^\varepsilon(0)\| = \mathcal{O}(\varepsilon)$, thanks to (3.43) and (3.45). — We test the first equation in (3.46) with τe^ε and integrate over the time interval $[0, T]$. Thanks to (3.44), we get

$$\tau(T)\|e^\varepsilon(T)\|^2 + \nu\int_0^T \tau(s)\|\nabla e^\varepsilon(s)\|^2 \, ds$$
$$+ \varepsilon\int_0^T \tau(s)\|\nabla\eta^\varepsilon(s)\|^2 \, ds \leq CT\varepsilon^2 + \int_0^T \|e^\varepsilon(s)\|^2 \, ds. \tag{3.47}$$

Therefore, it remains to control the last term in (3.47). Because of the estimate (3.43), we are allowed to concentrate on the integral $\int_0^T \|e(s)\|^2 \, ds$, with $e := u - \hat{u}^\varepsilon$. The related error equations follow from subtracting the identities in (1.1) and (3.45), with $\eta := p - \hat{p}^\varepsilon$,

$$e_t - \nu \Delta e + \nabla \eta = 0,$$
$$\text{dive} - \varepsilon \Delta \eta = -\varepsilon \Delta p, \qquad \partial_n \eta|_{\partial\Omega} = \partial_n p|_{\partial\Omega}, \qquad e|_{\partial\Omega} = 0, \qquad (3.48)$$

and $e(0) = 0$. To proceed further, we need an a-priori bound for $\nabla \hat{p}^\varepsilon$ in $L^2(0, T; \mathbf{L}^2(\Omega))$ for the auxiliary problem (3.45). This can be easily verified by testing the first equation in (3.48) with e and finally integrating over the interval $[0, T]$,

$$\|e(T)\|^2 + \nu \int_0^T \|\nabla e(s)\|^2 \, ds + \varepsilon \int_0^T \|\nabla \eta(s)\|^2 \, ds$$
$$\leq C\varepsilon \int_0^T \|\nabla p(s)\|^2 \, ds \leq CT\varepsilon. \qquad (3.49)$$

In particular, we obtain the desired a-priori bound for $\nabla \hat{p}^\varepsilon$, which is independent of ε. After these preparations, let us return to the derivation of an upper bound for $\int_0^T \|e(s)\|^2 \, ds$. To do so, we will introduce the following auxiliary problem that is backward in time, for $T > 0$: Find the solution $\{W, Q\} \in L^\infty(0, T; \mathbf{J}_1) \times L^\infty(0, T; H^1/\mathbb{R})$ of the nonstationary, incompressible Stokes equations,

$$W_t + \nu \Delta W - \nabla Q = g, \qquad W(T) = 0. \qquad (3.50)$$

If we set $g \equiv e$ and test (3.50) with e, further the corresponding relation in (3.48) with W, and add both identities, we arrive at

$$\|e\|^2 = d_t(e, W) - (\nabla Q, e) = d_t(e, W) - \varepsilon(\nabla Q, \nabla \hat{p}^\varepsilon). \qquad (3.51)$$

Integration over $[0, T]$, together with an a-priori statement for the solution Q of (3.50) and inequality (3.49) lead to the following estimate,

$$\int_0^T \|e^\varepsilon(s)\|^2 \, ds \leq CT\varepsilon^2.$$

If we insert the last result in (3.47) and divide by T, we end up with

$$\|e^\varepsilon(T)\|^2 + \nu \frac{1}{T} \int_0^T \tau(s) \|\nabla e^\varepsilon(s)\|^2 \, ds$$
$$+ \varepsilon \frac{1}{T} \int_0^T \tau(s) \|\nabla \varepsilon^\varepsilon(s)\|^2 \, ds \leq C\varepsilon^2. \qquad (3.52)$$

This proves the first statement in i). The remaining part of this section is devoted to the verification of optimal results for the error of $\hat{p}^\varepsilon$ and a further estimate for the one related to $\hat{u}^\varepsilon$ that is now measured in the Dirichlet-norm, pointwise in time.

Subsequently, we will benefit from two stability results of the divergence-operator,

$$\|\eta(T)\|_{-i} \le C\{\|e_t(T)\|_{-i-1} + \nu\|e(T)\|_{-i+1}\}, \qquad i \in \{0, 1\}.$$
(3.53)

Therefore, we can restrict our investigations for the pressure $\hat{p}^\varepsilon$ on further analyses on the behavior of the velocity error. In the following, we need time weights in (3.53), due to the facts that there cannot be taken advantage from negative norms $\|e_t(T)\|_{-i-1}$ on the right hand side of (3.53). This is owing to the fact that the quasi-compressibility equation in (3.45) forces the initial pressure $\hat{p}^\varepsilon(0) \in H^1 \cap L_0^2$ to be identically zero. This is of course not satisfied for $p(0)$, in general. — We start with the case $i = 0$, which requires another result for the velocity error e, measured in the norm $L^\infty(0, T; \mathbf{H}^1)$. This can be achieved by testing the first equation in (3.48) with τe_t and integrating over the time interval $[0, T]$, finally dividing by $\tau(T)$,

$$\frac{1}{\tau(T)} \int_0^T \tau(s)\|e_t(s)\|^2 \, ds + \nu\|\nabla e(T)\|^2 + \varepsilon\|\nabla\eta(T)\|^2$$
$$\le C\varepsilon + \varepsilon \frac{1}{\tau(T)} \int_0^T \tau^2(s)\|\nabla p_t(s)\|^2 \, ds \le C\varepsilon.$$
(3.54)

In here, we benefited from (3.49). On one hand, this result gives the outstanding error bound for the velocity field. Now, in order to complete the derivation of the first error result for the pressure $\hat{p}^\varepsilon$, we have to control $\|e_t(T)\|_{-1}$. This requires a corresponding procedure: Differentiating the first equation in (3.48) in time and testing with $\tau^2 e_t$, finally applying the operator $\frac{1}{\tau(T)} \int_0^T \cdot \, ds$ leads to

$$\tau(T)\|e_t(T)\|^2 + \nu\frac{1}{\tau(T)} \int_0^T \tau^2(s)\|\nabla e_t(s)\|^2 \, ds$$
$$+ \varepsilon\frac{1}{\tau(T)} \int_0^T \tau^2(s)\|\nabla\eta_t(s)\|^2$$
(3.55)
$$\le C\varepsilon\Big\{1 + \frac{1}{\tau(T)} \int_0^T \tau^2(s)\|\nabla p_t(s)\|^2\Big\} \le C\varepsilon.$$

Now, we can insert (3.54) and (3.55) in (3.53) and arrive at the second result in *ii*),

$$\sqrt{\tau}\|\eta\| \leq C\sqrt{\varepsilon}. \tag{3.56}$$

In order to verify an improved estimate for the error η that is measured in a negative norm, we have to come back to the splitting $\eta = \eta^\varepsilon + \Pi$, taking benefit from (3.43). Therefore, proceeding as above, but now for equation (3.46), leads us to an estimate that is necessary for the subsequent step,

$$\frac{1}{\tau(T)} \int_0^T \tau^2(s)\|e_t^\varepsilon(s)\|^2 \, ds + \nu\tau(T)\|\nabla e^\varepsilon(T)\|^2 + \varepsilon\tau(T)\|\nabla \eta^\varepsilon(T)\|^2$$

$$\leq C \frac{1}{\tau(T)} \Big\{ \int_0^T \tau^2(s)\|E_t(s)\|^2 \, ds + \int_0^T \tau(s)\|\nabla e^\varepsilon(s)\|^2 \, ds$$

$$+ \varepsilon \int_0^T \tau(s)\|\nabla \eta^\varepsilon(s)\|^2 \, ds \Big\} \leq C\dot{\varepsilon}^2. \tag{3.57}$$

The validity of the last estimate is a consequence of (3.44) and (3.52). Using this result in the next statement that we obtain through differentiation of (3.46) in time, testing with $\tau^3 e_t^\varepsilon$ and afterwards applying $\frac{1}{\tau(T)} \int_0^T \cdot \, ds$, we arrive at

$$\tau(T)^2\|e_t^\varepsilon(T)\|^2 + \nu\frac{1}{\tau(T)} \int_0^T \tau^3(s)\big\{\nu\|\nabla e_t^\varepsilon(s)\|^2 + \varepsilon\|\nabla \eta_t^\varepsilon(s)\|^2\big\} \, ds$$

$$\leq C\frac{1}{\tau(T)} \int_0^T \tau^2(s)\big\{\tau^2(s)\|E_{tt}(s)\|^2 + \|e_t^\varepsilon(s)\|^2\big\} \, ds \leq C\varepsilon^2. \tag{3.58}$$

The last upper bound is valid, owing to (3.44) and (3.57). — If we insert the results (3.52) and (3.58) in (3.53), we get a second error estimate for the pressure that is presented in part *i*) of the above lemma. This completes the proof. □

We will finish this section with another result that is necessary for the subsequent investigations and provides ε-independent a-priori estimates for the solution of (3.45).

Lemma 3.3 *Let $\{\hat{u}^\varepsilon, \hat{p}^\varepsilon\}$ be the solution of (3.45). Suppose that the given data of the problem satisfy (A1), (A2), (A3), and $\tilde{B} \equiv \tilde{B}_1$ or $\tilde{B} \equiv \tilde{B}_2$. Then there exists a constant $C = C(u_0, f, \Omega, T, \nu)$, independent of the parameter $\varepsilon > 0$, such that the following a-priori estimates are valid,*

$$\sup_{[0,T]}\big\{\|\hat{u}_t^\varepsilon\| + \|\Delta \hat{u}^\varepsilon\| + \|\nabla \hat{p}^\varepsilon\|\big\} \leq C.$$

Proof:
The evidence of the uniform bound for $\widehat{p}^\varepsilon$ in $L^\infty(0, T; H^1/\mathbb{R})$ is a consequence of the second equality in (3.48), in combination with part i) of Lemma 3.2. — The a-priori bound for the first term in the lemma can be derived in the following way. We differentiate the first equation in (3.45) in time. Testing with $\widehat{u}_t^\varepsilon$ and integrating over $[0, T]$ then give

$$\|\widehat{u}_t^\varepsilon(T)\|^2 + \nu \int_0^T \|\nabla \widehat{u}_t^\varepsilon(s)\|^2 \, ds + \varepsilon \int_0^T \|\nabla \widehat{p}_t^\varepsilon(s)\|^2 \, ds$$

$$\leq \|\widehat{u}_t^\varepsilon(0)\|^2 + \int_0^T \|f_t(s)\|^2 \, ds$$

$$- \int_0^T \left\{ \left(\tilde{B}(u_t, u)(s), \widehat{u}_t^\varepsilon(s) \right) + \left(\tilde{B}(u, u_t)(s), \widehat{u}_t^\varepsilon(s) \right) \right\} ds. \tag{3.59}$$

The boundedness of the first term on the right hand side follows from the first identity in (3.45), evaluated at time $t = 0$. A proper estimation of the last integral takes benefit from well-known regularity properties of the solution of the incompressible Navier-Stokes equations and the skew-symmetry rule that needs to be applied to the last term in the brackets. — At last, the verification of an ε-independent bound for $\widehat{u}^\varepsilon$ in $L^\infty(0, T; \mathbf{H^2})$ is a consequence of the last result which can be used in the first equation of (3.45).

□

3.3.2   The Navier-Stokes Case

As already mentioned in the beginning of the preceding subsection, the first part of the actual one is devoted to the transfer of the results that have been derived above to the Navier-Stokes equations (3.2), now focusing on the error amplification effects that stem from the nonlinearity. Firstly, relying on Lemma 3.3, we have to complete the proof of Theorem 3.1. The second part of this section is then concerned with the completion of the proof of Theorem 3.3.

We start with the investigation of the error equations that read as follows:

$$\xi_t - \nu \Delta \xi + \nabla \chi = -\tilde{B}(u^\varepsilon, u^\varepsilon) + \tilde{B}(u, u),$$

$$\text{div} \xi - \varepsilon \Delta \chi = 0, \qquad \partial_n \chi|_{\partial\Omega} = 0, \qquad \xi|_{\partial\Omega} = 0, \tag{3.60}$$

with $\xi(0) = 0$. In this framework, we use the notations $\xi := u^\varepsilon - \widehat{u}^\varepsilon$ and $\chi := p^\varepsilon - \widehat{p}^\varepsilon$. Comparing these equations with the ones related to the penalty method, (3.23), we recognize the same structure of system (3.60). This is because of the fact that the velocity (error) terms stemming from certain estimates to treat the nonlinearities on the right hand side can only be controlled via the first two functions in the first equation (3.60), like in (3.23). Further, the error bounds that are collected in Lemma 3.2 are sufficient in the following to succeed in proving optimal error estimates for the functions $\{\xi, \chi\}$. Therefore, we will confine ourselves to a sketch of the "highlights". The first estimate that corresponds to (3.30) reads as follows,

$$\|\xi(T)\|^2 + \nu \int_0^T \|\nabla \xi(s)\|^2 \, ds + \varepsilon \int_0^T \|\nabla \chi(s)\|^2 \, ds \leq C\varepsilon^2. \tag{3.61}$$

This result provides basic a-priori estimates of the solution of (3.2). Reversely, this result, combined with the second equality in (3.60) and the corresponding one in Lemma 3.3 guarantees the following upper bound,

$$\|\nabla p^\varepsilon(T)\| \leq \|\nabla \widehat{p}^\varepsilon(T)\| + \|\nabla \chi(T)\| \leq \frac{1}{\varepsilon}\|\xi(T)\| \leq C. \tag{3.62}$$

Together with Lemma 3.3 we succeeded in proving the last part of *ii*) in Theorem 3.1. In order to go on with the error analysis, we have to verify part *i*) of the same theorem. Owing to Lemma 3.3, this can be done by using the last result, testing the first equation in (3.60) with $-\Delta\xi$. A standard procedure leads to the estimate

$$\|\nabla u^\varepsilon(T)\|^2 + \int_0^T \|\Delta u^\varepsilon(s)\|^2 \, ds \leq \|\nabla \xi(T)\|^2 + \int_0^T \|\Delta \xi(s)\|^2 \, ds$$
$$+ \|\nabla \widehat{u}^\varepsilon(T)\|^2 + \int_0^T \|\Delta \widehat{u}^\varepsilon(s)\|^2 \, ds$$
$$\leq C\Big\{1 + \int_0^T \|\nabla \xi(s)\|^6 \, ds\Big\}. \tag{3.63}$$

This implies the local (in time) existence of a unique solution of (3.2) and proves part *i*) in Theorem 3.1.

This first regularity result is sufficient for further error estimates. Note that here we again have to deal with a term that is identical to the one in (3.27). This result can equally be used in the case of (3.60), owing to the error results that are collected in Lemma 3.2 and statement (3.62). - Analogously, the same arguments that lead to estimate (3.36) and (3.38) can be applied in this case in order to verify the following inequality,

$$\tau^2(T)\|\xi_t(T)\|^2 + \nu \int_0^T \tau^2(s)\|\nabla\xi_t(s)\|^2\, ds + \tau(T)\|\nabla\xi(T)\|^2$$
$$+ \varepsilon\tau(T)\|\nabla\chi(T)\|^2 \le C\varepsilon^2. \tag{3.64}$$

The two estimates (3.61) and (3.64) complete the error statements for the velocity field that are given in Theorem 3.3. Additionally, (3.64) provides a first result for the error that is related to p^ε. We can now gain another statement in the weaker L^2-norm, using the first identity in (3.60) and taking benefit from (3.61) and (3.64). Together with the statements in Lemma 3.2, we finally arrive at

$$\tau(T)\|\chi(T)\| \le C\tau(T)\big\{\|\nabla\xi(T)\| + \|\xi_t(T)\|\big\} \le C\varepsilon. \tag{3.65}$$

The error norms that provide optimal convergence rates for the pressure in (3.64) and (3.65) are stronger than those in (3.41) and (3.43). This is caused by the fact that the error structure to be investigated in (3.60) does not possess locally deviating features — like in the analysis of the errors $\{E, \Pi\}$ of (3.40).

This completes the proof for the parts i) and ii) of Theorem 3.3. — Further, (3.62), (3.63) and (3.64) in combination with Lemma 3.3 provide the basis for the derivation of the remaining regularity result of ii) in Theorem 3.1. We leave the easy verification to the interested reader.

The end of this section is devoted to the verification of the results iii) and iv) in Theorem 3.3. For this purpose, we recapitulate the steps in the last two subsections that investigated the different error contributions in (3.2). The first step, the analysis of (3.39), consists of the quantification of the pressure stabilization. Due to the pronounced boundary layers, optimal error estimates can only be proven in weak norms, compare (3.41) and (3.43). In order to suppress the influence that stems from the wrong boundary condition for the pressure function in (3.39), we have to restrict the analysis to interior domains $\Omega' \subset\subset \Omega$. The corresponding following error estimates are valid in the stationary Stokes case and are due to Rannacher, [30],

$$\|E(T)\|_{1;\Omega'} + \|\Pi(T)\|_{0;\Omega'} + \sqrt{\varepsilon}\|\nabla\Pi(T)\|_{\Omega'}$$
$$\le C(\Omega')\varepsilon\big\{\|\nabla p(T)\| + \|\Delta p(T)\|\big\}, \tag{3.66}$$

by using the error notations that have been introduced in Subsection 3.2. Further, we have chosen the notation $C(\Omega')$ to denote the dependence of the constant C from $\mathrm{dist}(\partial\Omega, \Omega')$. It is conjectured in [30] that it is exponentially decaying for "decreasing" subdomains $\Omega' \subset\subset \Omega$, but the rigorous mathematical justification of this statement is still an open problem.

In the analysis of the subsequent auxiliary problems (3.45) and (3.60), we investigated the error mechanisms that are caused by the evolutionary and nonlinear character of the first equation in (3.2), respectively. These error parts do not suffer from singular perturbations. The consequences are the following results for the solution of (3.45), employing the notation that has been introduced above,

$$\sup_{[0,T]}\left\{\sqrt{\varepsilon}\sqrt{\tau}\|\nabla\eta^\varepsilon\| + \tau\|\eta^\varepsilon\|\right\} \leq C\varepsilon.$$

The statement for the first term is just a restatement of (3.57), whereas the second estimate follows from the first identity in (3.46), employing a stability result for the divergence operator and the auxiliary results (3.44), (3.57), and (3.58). Finally, corresponding arguments work for (3.60), see (3.64) and (3.65). This completes the proof of Theorem 3.3.

Chapter 4

Nonstationary Quasi-Compressibility Methods

4.1 Overview and Results

The subject of this chapter is the investigation of established as well as new nonstationary quasi-compressibility methods. In contrast to the schemes that have been investigated in the previous chapter the features of the present ones are more sophisticated, due to the additional error transport mechanisms that stem from the evolutionary character of the perturbation in the quasi-compressibility constraint, which is thus no algebraic constraint any more. As we will see, the analysis of these schemes provides the basis for understanding the basic error mechanisms that arise in diverse numerical models like augmented Lagrangean formulations or higher order projection schemes (compare the Van Kan scheme in Chapter 7).

The first part of the present chapter is devoted to the study of the following type of quasi-compressibility constraint,

$$\text{div} u^\varepsilon + \varepsilon p_t^\varepsilon = 0, \qquad \|p_0^\varepsilon - p(0)\| = \mathcal{O}(\sqrt{\varepsilon}), \tag{4.1}$$

with $p_0^\varepsilon \in L_0^2$ and $\varepsilon > 0$. From our point of view, it can be interpreted as an "extension" of the stationary penalty formulation that has been treated in the previous chapter. In particular, we concluded that the penalty constraint represents a *regular* perturbation of the incompressibility constraint in the given framework of regularities. In the case of the *artificial compressibility formulation* (4.1) we will see the impact of an additional error source, stemming from the evolutionary term itself. — Another major topic of this chapter is the analysis of the *pressure correction formulation*,

$$\mathrm{div}u^\varepsilon - \varepsilon\Delta p^\varepsilon_t = 0, \quad \partial_n p^\varepsilon_t|_{\partial\Omega} = 0, \quad \|p^\varepsilon_0 - p(0)\|_1 = \mathcal{O}(\sqrt{\varepsilon}). \tag{4.2}$$

The inherent mechanisms of perturbation effects through time and space are of even higher complexity and will be investigated in Section 4.3. In particular, we get a combination of the *singular* perturbation effects that are concentrated along the boundary $\partial\Omega$ (compare the *pressure stabilization formulation*) and additional error contributions stemming from the evolutionary term of the quasi-compressibility constraint.

In order to "switch off" the error source that is caused by the wrong boundary data we investigate the errors for the pressure in a negative norm, i.e., $H^{-1}(\Omega)$- or in the local (in space) L^2-norm. As already mentioned, this perturbing influence that is due to the prescription of the wrong boundary data for the pressure in (4.2) is persistent. But, additionally, the evolutionary character of the quasi-compressibility constraint causes another crucial impact on the asymptotic approximation behavior to the pressure function of the incompressible problem. This leads to the fact, that the error is no longer concentrated only along the boundary but also possesses a *global portion*. As a result, we get no super-convergence results in negative norms any more.

It is remarkable that the above error sources, which influence the quality of the pressure approximation, do not significantly impact the pointwise in time approximation behavior of the velocity-field in the basic $L^\infty(0, T; \mathbf{L}^2(\Omega))$-norm. This observation will be supported with optimal error bounds in the specified topology. — The functions that give information with respect to the approximation features of the presented stationary and nonstationary quasi-compressibility formulations are

- the gradient of the velocity error, i.e., the function $s \mapsto \|\nabla e(s)\|$, and

- the time-derivative of the velocity error, i.e., $s \mapsto \|e_t(s)\|$,

with the error $e := u - u^\varepsilon$ and the solutions of (2.1) and (4.1) or (4.2), respectively. These functions can be related to the error sources that have been outlined above: The gradient of the velocity error is mainly influenced by the singular perturbation effects in space, which have already been investigated in the case of the stationary schemes (compare the pressure stabilization). On the other hand, the approximation behavior of $s \mapsto \|e_t(s)\|$ reflects the global impacts of the (nonstationary) quasi-compressibility method.

In order to be more specific, let us apply this separation of error effects to the case of the artificial compressibility method (4.1). In comparison with the penalty method, it is not possible to ascertain that the error for the pressure converges at rate $\mathcal{O}(\varepsilon)$ in the norm $L^\infty(0, T; L_0^2)$, see below. But, owing to the preceding discussion, this is not so astonishing: The errors acting globally in space-time lead to a reduction of accuracy with respect to the pressure approximation. In the above outline of error-"indicators", the gradient of the averaged velocity error $\left(\int_0^T \|\nabla e(s)\|^2 \, ds \right)^{1/2}$ converges with optimal, i.e., first order in ε. On the other hand, the term $\left(\int_0^T \|e_t(s)\|^2 \, ds \right)^{1/2}$ only permits an upper bound $C\sqrt{\varepsilon}$, with a constant C that describes the space-time features of the actual solution in dependence of the given data. — Again, let us point out the fact that these considerations are not applicable if we are interested in statements for the velocities in the norm $L^\infty(0, T; \mathbf{L}^2)$ — which are untouched from these error phenomena.

As already mentioned, optimal first order statements for the pressure approximations p^ε cannot be obtained in norms $L^\infty(0, T; X)$, with X a Hilbert space, owing to error mechanisms that are present globally in the space-time cylinder. Moreover, as we will see in the subsequent studies, we need

- sufficiently accurate initial data and

- regularity properties of the solution of the incompressible (Navier-) Stokes equations that are superior to the general ones.

Only the validity of these assumptions will lead to a satisfying convergence behavior of the nonstationary quasi-compressibility formulation. To be precise, we need the following regularity results for the solution to be approximated that are indicated in the second item above, taking $i \in \{1, 2\}$,

Postulate B_i: The solution $\{u, p\} \in L^\infty(0, T; \mathbf{J}_1) \times L^\infty(0, T; H^1/\mathbb{R})$ of the incompressible Stokes equations (2.1) is assumed to satisfy the regularity requirements

$$\sup_{0 \leq s \leq T} \left\{ \|u_t(s)\|_i + \|p_t(s)\|_{i-1} \right\} \leq C,$$

uniformly throughout the range $[0, T]$.

From these considerations, we aim at modifying the above artificial compressibility method (the original one needs *Postulate B_1* to be satisfied) and the pressure correction method (the original one needs *Postulate B_2* to be satisfied) such that no additional requirements as presented above are necessary to ensure an optimal convergence behavior. Therefore, we propose the following modifications of nonstationary quasi-compressibility methods given in (4.1), (4.2),

$$\text{div}u^\varepsilon + \varepsilon\{\tau^r p^\varepsilon\}_t = 0, \qquad r \geq 2, \tag{4.3}$$

and

$$\text{div}u^\varepsilon - \varepsilon\Delta\{\tau^r p^\varepsilon\}_t = 0, \qquad \partial_n\{\tau^r p^\varepsilon\}_t|_{\partial\Omega} = 0, \quad r \geq 2. \tag{4.4}$$

These modifications contain an additional amount of stability features that is necessary to ensure the transfer of a necessary quantity of information of the actual solution behavior of $\{u, p\}$ across the initial time-phase. Moreover, we are not obliged to start with sufficiently accurate initial data and need no further regularity requirements like the Postulates $B_i|_{i\in\{1,2\}}$. Therefore, these schemes combine improved stability in the initial time-phase with optimal accuracy properties inherent to the scheme. Based on these statements, it is not too surprising that these modifications give approximations for the pressure that are quite the same as those that have been obtained for the stationary analoga, i.e.,

- The approximation properties of the solution $\{u^\varepsilon, p^\varepsilon\}$ of (4.3) to $\{u, p\}$ as the solution of (2.1) are quite the same as those derived for the penalty method — at least for times T that are strictly bounded away from the beginning.

- The perturbation impact on the pressure function in (4.4) corresponds to the error phenomena that have been observed in the context of the pressure stabilization method. Again, the significant errors for the pressure function are limited to the neighborhood of the boundary $\partial\Omega$, thus reducing to a local perturbation of the actual solution p.

The remainder of the chapter is arranged as follows. Section 4.2 deals with the analysis of the artificial compressibility and the *modified artificial compressibility method*. The investigation is based on the separate treatment of the error sources as they are given above. We will omit proofs if corresponding ideas are needed for the analysis of the pressure correction scheme (4.2). Complete proofs in the latter case will be established in Section 4.3.

The following analyses are carried out for the Stokes equations. Let us stress the fact that they can easily be extended to the Navier-Stokes case, using ideas that have been given in the investigation of stationary quasi-compressibility methods (see Chapter 3).

4.2 The Artificial Compressibility Scheme and its Modification

4.2.1 Approximation Results for the Artificial Compressibility Scheme and its Modification

The artificial compressibility method is another ansatz that can be applied to solve the incompressible Stokes equations. The corresponding equations read as follows, with a solution $\{u^\varepsilon, p^\varepsilon\} \in L^\infty(0,T;\mathbf{L}^2) \cap L^2(0,T;\mathbf{H}_0^1) \times L^\infty(0,T;L_0^2)$,

$$u_t^\varepsilon - \Delta u^\varepsilon + \nabla p^\varepsilon = f,$$
$$\operatorname{div} u^\varepsilon + \varepsilon p_t^\varepsilon = 0, \tag{4.5}$$
$$u_0^\varepsilon = u_0 + \mathcal{O}(\varepsilon) \text{ in } \mathbf{L}^2, \qquad p_0^\varepsilon = p_0 + \mathcal{O}(\varepsilon) \text{ in } L^2,$$

and $p_0^\varepsilon \in L_0^2(\Omega)$. The idea of extending the penalty approach to this non-stationary quasi-compressibility method can be motivated by convergence acceleration strategies that are employed in order to numerically solve the stationary counterpart of the Stokes equations. We mention iterative strategies like the augmented Lagrangean method or the Uzawa iterative method that are wide-spread solution procedures for stationary saddle-point problems. In the present context of the nonstationary Stokes (or Navier-Stokes) problems, this ansatz is related to the quasi-compressibility constraint

$$\operatorname{div} u_\varepsilon^{m+1} + \delta p_\varepsilon^{m+1} = \delta p_\varepsilon^m, \tag{4.6}$$

with $\delta > 0$ a control parameter.

Subject of the present section is the determination of the difference between the solution $\{u, p\}$ of the incompressible Stokes system (2.1) and the solution $\{u^\varepsilon, p^\varepsilon\}$ as the solution of (4.5). In contrast to the penalty scheme that has been treated in the preceding chapter, nonstationary effects cause instabilities that cover the main errors in this scheme. This is the reason for the fact that *no optimal pointwise in time convergence statements can be obtained for this scheme.* In order to prevent these "negative mechanisms" we have to assume the following facts:

- The starting function for the pressure satisfies

$$\|\{u - u^\varepsilon\}(0)\| + \sqrt{\varepsilon}\|\{p - p^\varepsilon\}(0)\| = \mathcal{O}(\varepsilon) \qquad (4.7)$$

  (accurate initial data).

- The following additional "moderate" regularity property for the pressure function of the Stokes problem (2.1) holds, $p_t \in L^\infty(0, T; L_0^2(\Omega))$. This is the contents of *Postulate B_1*.

Corresponding to the pressure correction formulation to be investigated below, this results from the lack of damping properties in system (4.5). The following result for the artificial compressibility method will be proven subsequently.

Theorem 4.1 *Let the tuple $\{u^\varepsilon, p^\varepsilon\}$ be the solution of the artificial compressibility method (4.5), and let the requirements (4.7) be satisfied. Further, $\{u, p\}$ is the solution of system (2.1). Provided the pressure function $s \mapsto p(s)$ of the incompressible Stokes system satisfies the additional regularity property given in Postulate B_1 and the given data satisfy the assumptions (A1) and (A2), there exists a constant C depending on the given data such that the following estimate is valid for $0 < \varepsilon < 1$,*

$$\sup_{[0,T]} \sqrt{\tau}\{\|u - u^\varepsilon\| + \sqrt{\varepsilon}\|p - p^\varepsilon\|\}$$

$$+ \left(\int_0^T \tau(s)\|\nabla\{u - u^\varepsilon\}(s)\|^2 \, ds\right)^{1/2} \leq C\varepsilon(1 + \log\frac{1}{\varepsilon}).$$

Remark 4.1   *1. The logarithm that is present in this result reflects the problematic regularity nature of this quasi-compressibility ansatz. It can be eliminated in case the pressure function satisfies an additional regularity restriction, for instance: $\exists\, r << \frac{1}{2} : \tau^r(s)p_{tt}(s) \in L^2(0, T; L_0^2(\Omega))$.*

2. *The first sharp results for the quasi-compressibility method (4.1) have been obtained by Shen. However, in his analyses he did not treat the acting error phenomena independently, compare [39]. This forces him to require the following sharper regularity assumptions, namely $p_t, p_{tt} \in L^\infty(0, T; L_0^2(\Omega))$.*

The proof of this theorem will be given in Subsection 4.2.2. Owing to the above argumentation of globally arising evolutionary effects that are caused by the evolutionary character of ansatz (4.1) we have proposed a damped version of it, which is (4.3). This ansatz will free us from the requirement of an approximation property for the pressure function at the beginning, compare (4.7), and the additional regularity constraint with respect to the solution of (2.1) to obtain optimal statements of convergence. The stability of the related operator carries over to the solution, especially for the pressure function p^ε, and we obtain optimal pointwise statements with respect to the convergence of this ansatz for the pressure. The convergence estimates that are valid for this ansatz are collected in another theorem.

Theorem 4.2 *Assume that the basic assumptions (A1), (A2) are valid for the tuple $\{u, p\}$ as the solution of (2.1). Further, let $\{u^\varepsilon, p^\varepsilon\}$ be the solution of the damped artificial compressibility scheme (4.3). Then, for $r \geq 2$, there exists a constant C_r depending on the given data of the problem and r such that the following error bound holds true,*

$$\sup_{[0,T]}\left\{\|u - u^\varepsilon\| + \tau^{r/2}\{\|u - u^\varepsilon\| + \|p - p^\varepsilon\|\}\right\} \leq C_r\varepsilon.$$

The dependence of the employed constant C_r on r is of moderate size for the whole range of values for r.

Remark 4.2   *1. The verification of these results will not be given explic-itly here. Owing to the remarks that have been made before, we will prove "corresponding" error statements for a modified (damped) pres-sure correction scheme in the subsequent section. Let us stress the fact that the presentation of (simplified) arguments in this case do also work here.*

2. The above bound for the constant C_r results from considerations that will presented in the context of the investigation of the damped pressure correction scheme. In case of the above theorem, the growth of C_r can be controlled as follows, $C_r \leq C(2(r - 2) + 1)^{-1}$, with C independent from r.

4.2.2 Analysis of the Original Artificial Compressibility Scheme

Before we start with the true analysis of convergence for the equations (4.5), let us investigate the influence of perturbed initial data for the velocity- and the pressure function. This will permit neglection of error portions stemming from approximated initial data in the further study. — Assume $\{u^\varepsilon, p^\varepsilon; u^\varepsilon(0), p^\varepsilon(0)\}$ and $\{\tilde{u}^\varepsilon, \tilde{p}^\varepsilon; \tilde{u}^\varepsilon(0), \tilde{p}^\varepsilon(0)\}$ to be two solutions of the equalities (4.5), which are uniquely determined. The initial data are assumed to satisfy

$$\|\{u^\varepsilon - \tilde{u}^\varepsilon\}(0)\| + \sqrt{\varepsilon}\|\{p^\varepsilon - \tilde{p}^\varepsilon\}(0)\| \le C\varepsilon. \tag{4.8}$$

Using the abbreviative notations $\xi \equiv \tilde{u}^\varepsilon - u^\varepsilon$ and $\chi \equiv \tilde{p}^\varepsilon - p^\varepsilon$, we get the following system of equations describing the propagation of the initial errors,

$$\begin{aligned} \xi_t - \Delta\xi + \nabla\chi &= 0, \\ \operatorname{div}\xi + \varepsilon\chi_t &= 0, \qquad \xi|_{\partial\Omega} = 0. \end{aligned} \tag{4.9}$$

Now, together with (4.8) we obtain a statement for the error evolution over the time range,

$$\|\xi(T)\| + \left(\int_0^T \|\nabla\xi(s)\|^2\, ds\right)^{1/2} + \sqrt{\varepsilon}\|\chi(T)\| \le C\varepsilon. \tag{4.10}$$

Owing to the fact that this inequality is of the same form as the one given in Theorem 4.1 that is to be verified in the following, we can assume without loss of generality

$$u^\varepsilon(0) = u_0, \qquad p^\varepsilon(0) = p_0. \tag{4.11}$$

Let us now start with the error analysis of problem (4.5), together with the initial data (4.11), by separating the inherent error mechanisms of the method. In doing so, we will investigate the following auxiliary problem: Find a solution $\{U^\varepsilon, P^\varepsilon\} \in L^2(0, T; \mathbf{H}_0^1(\Omega)) \times L^\infty(0, T; L_0^2(\Omega))$ of the system of equations,

$$\begin{aligned} -\Delta U^\varepsilon + \nabla P^\varepsilon &= f - u_t, \\ \operatorname{div}U^\varepsilon + \varepsilon P_t^\varepsilon &= 0, \qquad P^\varepsilon(0) = p(0). \end{aligned} \tag{4.12}$$

The associated error equations are as follows,

$$-\Delta E + \nabla\Pi = 0,$$
$$\operatorname{div} E + \varepsilon\Pi_t = \varepsilon p_t, \qquad \Pi(0) = 0. \tag{4.13}$$

The first equation in (4.13), combined with the statement for the pressure error at initial time, implies the homogeneity of the initial error for the velocity, $E(0) = 0$. Owing to this result, which can be inserted in the second equation of (4.13), we obtain $P_t^\varepsilon(0) \in L_0^2(\Omega)$, thanks to *Postulate* B_1. Now, the derivation of an a-priori statement for $P_t^\varepsilon(s)$, $s \geq 0$ is essential for the further error analysis. In order to succeed in that, we start with the verification of such an inequality at times $0 \leq s \leq \varepsilon$: If we differentiate the first identity in (4.13) in time and test with τE_t, this leads to the following result, after integration over the time range $[0, \varepsilon]$,

$$\int_0^\varepsilon \tau(s)\|\nabla E_t(s)\|^2 \, ds + \varepsilon\tau(\varepsilon)\|\Pi_t(\varepsilon)\|^2 \leq C\varepsilon^2 \int_0^\varepsilon \tau(s)\|p_{tt}(s)\|^2 \, ds$$
$$+ \int_0^\varepsilon \tau(s)\|\Pi_t(s)\|^2 \, ds + \varepsilon \int_0^\varepsilon \|\Pi_t(s)\|^2 \, ds. \tag{4.14}$$

The first term on the right hand side of (4.14) is bounded by $C\varepsilon^2$. In order to control the second term on the right side of (4.14), we employ Gronwall's inequality. In order to bound the last term, we make use of the second error equation in (4.13) and find

$$\varepsilon\int_0^\varepsilon \|\Pi_t(s)\|^2 \, ds \leq C\varepsilon^2 \sup_{0\leq s\leq\varepsilon} \|p_t(s)\|^2 + \frac{1}{\varepsilon}\int_0^\varepsilon \|\nabla E(s)\|^2 \, ds. \tag{4.15}$$

In order to bound the last term on the right hand side of (4.15), we test the first error equation in (4.13), after differentiation in time, with E. This gives after integration,

$$\|\nabla E(T)\|^2 + \varepsilon\int_0^T \|\Pi_t(s)\|^2 \, ds \leq C\varepsilon T \sup_{0\leq s\leq T} \|p_t(s)\|^2. \tag{4.16}$$

We end up with the upper bound $C\varepsilon^2$ for the last term in (4.15). If we make use of this result in (4.14), we have therefore proven the following regularity statement, $P_t^\varepsilon(\varepsilon) \in L_0^2(\Omega)$, together with an a-priori estimate. This auxiliary result can be further employed to derive corresponding statements for all times of the interval $[0, T]$: Thus, we repeat differentiation of the first equality in (4.13) in time and test with E_t. If we integrate this relation over the restricted time interval $[\varepsilon, T]$, with $T \geq \varepsilon$, we find

$$
\int_\varepsilon^T \|\nabla E_t(T)\|^2 \, ds + \varepsilon \|\Pi_t(T)\|^2 \leq \varepsilon \|\Pi_t(\varepsilon)\|^2
$$
$$
+ C\varepsilon (1 + T + \log\frac{1}{\varepsilon}) \int_\varepsilon^T \tau(s)\|p_{tt}(s)\|^2 \, ds
$$
$$
+ \varepsilon (1 + T + \log\frac{1}{\varepsilon})^{-1} \int_\varepsilon^T \frac{1}{\tau(s)}\|\Pi_t(s)\|^2 \, ds.
$$

(4.17)

Thanks to Gronwall's inequality, we can control the last term on the right hand side of (4.17). The resulting constant is moderate, owing to the pre-standing normalizing factor. Owing to the boundedness of the first term on the right hand side of (4.17) we have thus shown that the following regularity statement is valid, $\left(1 + \log\frac{1}{\varepsilon}\right)^{1/2} P_t^\varepsilon \in L^\infty(\varepsilon, T; L_0^2(\Omega))$, together with an appropriate a-priori bound that is independent of the parameter ε. This stability statement is the basis for the following error estimate,

$$
\|\mathrm{div} E\| \leq C\varepsilon \|P_t^\varepsilon\| \leq C\varepsilon (1 + \log\frac{1}{\varepsilon}).
$$

(4.18)

By means of the first equation in (4.13), we can derive sharp statements for the velocity error, measured in the Dirichlet-norm, and the error for the pressure function,

$$
\|\Pi\|^2 \leq C\|\nabla E\|^2 \leq \alpha\|\mathrm{div} E\|^2 + \frac{1}{4\alpha}\|\Pi\|^2, \qquad \alpha > 0,
$$

or, for α sufficiently large,

$$
\|\Pi\| + \|E\|_1 \leq C\varepsilon (1 + \log\frac{1}{\varepsilon}).
$$

(4.19)

We can derive further estimates in an analogous way for the time derivatives of velocity and pressure error, employing certain time-weights. For the subsequent analyses we need an additional a-priori statement for the function P_{tt}^ε, which can be derived via estimate (4.17): If we differentiate the first equation in (4.13) twice and finally test it with τE_t, integration over the range $[\varepsilon, T]$ finally leads to

$$
\tau(T)\|\nabla E_t(T)\|^2 + \varepsilon \int_\varepsilon^T \tau(s)\|\Pi_{tt}(s)\|^2 \, ds \leq \varepsilon \|\nabla E_t(\varepsilon)\|^2
$$
$$
+ C\varepsilon \int_\varepsilon^T \tau(s)\|p_{tt}(s)\|^2 \, ds + 2 \int_\varepsilon^T \|\nabla E_t(s)\|^2 \, ds. \quad (4.20)
$$

Owing to (4.17), it remains to bound the first term on the right hand side of (4.20). This can be done by using estimate (4.14): if we differentiate the first equation of (4.13) twice in time, subsequently test it with $\tau^2 E_t$ and finally integrate over the interval $[0, \varepsilon]$, we are led to

$$
\begin{aligned}
\tau^2(\varepsilon)\|\nabla E_t(s)\|^2 &+ \varepsilon \int_0^\varepsilon \tau^2(s)\|\Pi_{tt}(s)\|^2 \, ds \\
&\leq C\varepsilon^2 \int_0^\varepsilon \tau(s)\|p_{tt}(s)\|^2 \, ds + 2 \int_0^\varepsilon \tau(s)\|\nabla E_t(s)\|^2 \, ds.
\end{aligned}
\tag{4.21}
$$

Thanks to estimate (4.14), we arrive at: $\|\nabla E_t(\varepsilon)\|^2 \leq C$. This result can now be inserted into (4.20), and we obtain the important result

$$
\int_\varepsilon^T \tau(s)\|P_{tt}^\varepsilon(s)\|^2 \, ds \leq C.
\tag{4.22}
$$

An easy consideration then leads us to the statement of convergence,

$$
\int_\varepsilon^T \tau(s)\|E_t(s)\|^2 \, ds \leq C\varepsilon^2.
\tag{4.23}
$$

The *second part* of the analysis of the artificial compressibility scheme is devoted to the study of the evolutionary character of the first equation in (4.5). Thanks to the statements of convergence (4.19) and (4.23), we have to qualify the behavior of the error functions that are governed by the system

$$
\begin{aligned}
e_t^\varepsilon - \Delta e^\varepsilon + \nabla \eta^\varepsilon &= E_t, & e^\varepsilon|_{\partial\Omega} &= 0, \\
\mathrm{div}\, e^\varepsilon + \varepsilon \eta_t^\varepsilon &= 0, & e^\varepsilon &= 0, & \eta^\varepsilon &= 0,
\end{aligned}
\tag{4.24}
$$

with $e^\varepsilon \equiv u^\varepsilon - U^\varepsilon$ and $\eta^\varepsilon \equiv p^\varepsilon - P^\varepsilon$. We test the first equation in (4.24) with e^ε and multiply the resulting equation with τ. Succeeding integration gives after absorption

$$
\begin{aligned}
\tau(T)\|e^\varepsilon(T)\|^2 &+ \int_\varepsilon^T \tau(s)\|\nabla e^\varepsilon(s)\|^2 \, ds + \varepsilon\tau(T)\|\eta^\varepsilon(T)\|^2 \\
&\leq C \int_\varepsilon^T \tau(s)\|E_t(s)\|^2 \, ds + C \int_\varepsilon^T \left\{\|e^\varepsilon(s)\|^2 + \varepsilon\|\eta^\varepsilon(s)\|^2\right\} ds.
\end{aligned}
\tag{4.25}
$$

In the remaining part of the proof, we only have to control the last two terms in (4.25). This is because of the fact that the first term is controllable by means of (4.23). Here, we confine to a sketch of the proof of these estimates.

Corresponding studies will be carried out for the artificial compressibility scheme in the subsequent section. In order to verify the a-priori bounds

$$(1 + \log\frac{1}{\varepsilon}) \int_0^T \|p_t^\varepsilon(s)\|^2 \, ds + \int_0^T \|p^\varepsilon(s)\|^2 \, ds \leq C(1 + \log\frac{1}{\varepsilon}),$$

(4.26)

we start with the error equations for $e \equiv E - e^\varepsilon$ and $\eta \equiv \Pi - \eta^\varepsilon$, that are

$$e_t - \Delta e + \nabla\eta = 0,$$

$$\mathrm{div}\, e + \varepsilon\eta_t = \varepsilon p_t,$$

$$e|_{\partial\Omega} = 0, \qquad e(0) = 0, \qquad \eta(0) = 0.$$

(4.27)

For further studies, we deal with a dual problem, for $T > 0$: Find the solution $\{w, q\} \in L^\infty(0, T; \mathbf{L}^2) \cap L^2(0, T; \mathbf{H}_0^1) \times L^\infty(0, T; L_0^2)$ satisfying

$$w_t + \Delta w - \nabla\eta = 0,$$

$$\mathrm{div}\, w - \varepsilon q_t = \varepsilon g,$$

$$w|_{\partial\Omega} = 0, \qquad w(T) = 0, \qquad q(T) = 0.$$

(4.28)

We set $g = \eta$ and test the second equation with η, further the second equation in (4.27) with g. After adding these identities and using the corresponding first error equalities in (4.27), (4.28), we end up with

$$\varepsilon\|\eta\|^2 = -\varepsilon d_t(q, \eta) + d_t(e, w) + \varepsilon(p_t, q).$$

(4.29)

Subsequent integration over the time-interval $[0, T]$, combined with a-priori statements for the solution of system (4.28), guarantee the validity of estimate (4.26) for the first term on the left hand side. In order to verify the remaining statement in (4.26), we consider another backward auxiliary problem of type (4.28), for $g = \eta_t$. Using arguments as they will be carried out in Subsection 4.3.2 below (for problem (4.60)), we arrive at an identity that corresponds to (4.65),

$$\varepsilon\|\eta_t\| = d_t\{(\mathrm{div}\, w, \eta) + (\mathrm{div}\, e, q)\}$$
$$- d_t(\nabla e, \nabla w) + \varepsilon d_t(p_t, q) - \varepsilon(p_{tt}, q).$$

(4.30)

Integration over the interval $[0, \varepsilon^2]$ now gives, analogously to the bounds (4.70)/(4.74),

$$\varepsilon \int_{\varepsilon^2}^T \|\eta_t(s)\|^2 \, ds \leq \frac{1}{\varepsilon}\{\|\eta(\varepsilon^2)\|^2 + \|e(\varepsilon^2)\|^2\} + C\varepsilon(1 + \log\frac{1}{\varepsilon}).$$

(4.31)

The first expression can be controlled by $C\varepsilon$; this can be verified through an easy consideration for problem (4.27). A corresponding control of integral $\int_0^{\varepsilon^2} \|\eta_t(s)\|^2 \, ds$ can be achieved by means of an inequality that is valid in the initial phase $[0, \varepsilon^2]$, using the length of this time interval. This proves the validity of (4.26).

We can now benefit from the second a-priori result in (4.26), using a third dual problem (again, we refer to the corresponding analysis below). Using a chain of standard-arguments, we are led to an estimation for the first term in brackets on the right hand side of (4.25) by $C\varepsilon^2(1 + \log\frac{T}{\varepsilon})$.

In order to find an upper bound for the second one on the right hand side of (4.25), we employ another backward problem, which is now dual to the error equalities (4.24). This is of type (4.28), owing to the fact that we have no problems concerning the choice of prescribing boundary conditions for the pressure function (unlike for the artificial compressibility method). If we set $g \equiv \eta^\varepsilon$, we are led to the following result,

$$\varepsilon \int_0^T \|\eta^\varepsilon(s)\|^2 \, ds \leq C\varepsilon^2 + \varepsilon \int_\varepsilon^T \|\eta^\varepsilon(s)\|^2 \, ds \leq C\varepsilon^2(1 + \log\frac{1}{\varepsilon}).$$
(4.32)

The details are essentially the same as those presented in the context of the pressure correction method below (see formulae (4.78)ff), using a decomposition $[0, T] = [0, \varepsilon] \cup [\varepsilon, T]$. — This proves Theorem 4.1.

4.3 The Pressure Correction Scheme and its Modification

4.3.1 Approximation Results for the Pressure Correction Scheme and its Modification

Subject of this section is the investigation of the pressure correction scheme, which reads as follows:

Find the solution $\{u^\varepsilon, p^\varepsilon\} \in L^\infty(0, T; \mathbf{L}^2) \cap L^2(0, T; \mathbf{H}_0^1) \times L^2(0, T; H^1/\mathbb{R})$ that is uniquely determined by the following system of equations,

$$
\begin{aligned}
u_t^\varepsilon - \Delta u^\varepsilon + \nabla p^\varepsilon &= f, && \text{in } \Omega \times (0, T], \\
\text{div} u^\varepsilon - \varepsilon \Delta p_t^\varepsilon &= 0, && \varepsilon > 0, && \text{in } \Omega \times (0, T], \\
u^\varepsilon|_{\partial\Omega} &= 0, && \partial_n p_t^\varepsilon|_{\partial\Omega} = 0, \\
u^\varepsilon(0) &= u_0^\varepsilon, && p^\varepsilon(0) = p_0^\varepsilon.
\end{aligned}
$$
(4.33)

As a motivation, it should be pointed out that the analysis of this scheme is necessary to understand many of the error mechanisms that are inherent to higher order projection schemes. This will be pointed out in a later chapter that is dealing with the Van Kan scheme (see Chapter 7). Further, this *pressure correction formulation* is of importance for stabilization strategies of the incompressible Stokes (or Navier-Stokes) equations in a mixed finite element approach, for both the stationary (compare augmented Lagrangean formulations) and nonstationary equations.

In the introductory part of this chapter, we have already stressed the fact that we have to ensure sufficient accuracy of the initial data $\{u^\varepsilon, p^\varepsilon\}$ and compatibility of the related incompressible flow problem to succeed in deriving optimal approximation results in terms of velocity and pressure in strong norms. To this end, we need the validity of *Postulate B_2*. Then, we have the following result.

Theorem 4.3 *Let $\{u^\varepsilon, p^\varepsilon\}$ be the solution of system (4.33). Provided that the data satisfy the assumptions (A1), (A2) and the solution $\{u, p\}$ of the divergence-free Stokes problem (2.1) satisfies Postulate B_2 and an approximation property for the initial data holds,*

$$\|u_0 - u_0^\varepsilon\| + \sqrt{\varepsilon}\{\|u_0 - u_0^\varepsilon\|_1 + \|p(0) - p_0^\varepsilon\|_1\} \le C\varepsilon, \tag{4.34}$$

the following statements of convergence are valid with a constant C, that depends on the given problem data only,

i) $\sup_{[0,T]}\left\{\sqrt{\tau}\{\|u - u^\varepsilon\| + \sqrt{\varepsilon}\|u - u^\varepsilon\|_1\}\right\} \le C\varepsilon(1 + \log\frac{1}{\varepsilon}),$

ii) $\sup_{[0,T]}\left\{\sqrt{\tau}\|p - p^\varepsilon\|\right\} \le C\sqrt{\varepsilon}(1 + \log\frac{1}{\varepsilon}).$

Without loss of generality, we assume $0 < \varepsilon < 1$.

We already stressed the importance of accurate initial data and compatible flow environments for optimal convergence behavior of (4.33). In order to stabilize this scheme with respect to these drawbacks, we suggest the following modification, which we will subsequently refer to as the *damped pressure correction scheme,*

$$\begin{aligned}
u_t^\varepsilon - \Delta u^\varepsilon + \nabla p^\varepsilon &= f, \\
\mathrm{div} u^\varepsilon - \varepsilon\Delta\{\tau^r p^\varepsilon\}_t &= 0, \qquad r \ge 2, \\
\partial_n\{\tau^r p^\varepsilon\}_t|_{\partial\Omega} &= 0, \qquad u^\varepsilon(0) = u_0^\varepsilon, \qquad p^\varepsilon(0) = p_0^\varepsilon.
\end{aligned} \tag{4.35}$$

This approach strengthens the incompressibility of the related velocity field in the initial time phase, by damping perturbation effects across the initial phase. Thus, the nonstationary error effects are expected to be controlled along the crucial initial time phase, and we suspect the singular perturbation character in spatial behavior to be the remaining one in the system. This will be justified in the subsequent theorem by means of error estimates for the pressure approximation in negative norms that guarantee super-convergence properties. This is of course the analytical background for the fact that the dominant error portions are again localized close to the boundary (boundary layers). We stress the fact that *none* of the additional requirements that were essential in the previous theorem are necessary any longer.

Theorem 4.4 *Let $\{u^\varepsilon, p^\varepsilon\}$ be the solution of the damped pressure correction scheme (4.35) and $\{u, p\}$ be the solution of (2.1). Suppose that (A1) and (A2) are valid and the following approximation property for the initial velocity function is satisfied,*

$$\|u_0 - u_0^\varepsilon\| \leq C\varepsilon.$$

Then, the following statements of convergence are valid for $r \geq 2$,

i) $\sup_{[0,T]}\left\{\|u - u^\varepsilon\| + \sqrt{\varepsilon}\tau^{(r-1)/2}\|u - u^\varepsilon\|_1\right\} \leq C_r\varepsilon,$

ii) $\sup_{[0,T]}\left\{\tau^{r/2}\{\|p - p^\varepsilon\|_{-1} + \sqrt{\varepsilon}\|p - p^\varepsilon\|\}\right\} \leq C_r\varepsilon.$

The dependence of the employed constant C_r on r is of moderate size for the whole range of values for r.

Remark 4.3   1. *The dependence of C_r on the parameter r can be stated in a more explicit way, $C_r \leq C(2(r-2)+1)^{-1}$, with C independent of r.*

2. *The proof of Theorem 4.4 splits into several parts, by investigating the various error contributions in the scheme. Nevertheless, the study in these parts is entirely different from the one for the original pressure correction method (4.33) that possesses completely different stability properties.*

The remainder of this section is organized as follows: We start with the investigation of the pressure correction method (4.33), proving Theorem 4.3. Subject of Subsection 4.3.3 is the verification of Theorem 4.4.

4.3.2 Perturbation Analysis for the Pressure Correction Method

This subsection is devoted to the verification of the error statements presented in Theorem 4.3, with respect to the parameter $\varepsilon > 0$. In order to do so, let us introduce the abbreviative notations for the errors $e := u - u^\varepsilon$ and $\eta := p - p^\varepsilon$. Subtraction of the systems (2.1) and (4.33) then leads to the following equations describing the evolution of the errors,

$$
\begin{aligned}
&e_t - \Delta e + \nabla \eta = 0, \\
&\operatorname{div} e - \varepsilon \Delta \eta_t = -\varepsilon \Delta p_t, \\
&\partial_n \eta_t|_{\partial\Omega} = \partial_n p_t|_{\partial\Omega}, \qquad e(0) = 0, \qquad \eta(0) = 0.
\end{aligned}
\qquad (4.36)
$$

Owing to the prescription of vanishing initial errors, we can now employ an easy perturbation argument that justifies to use homogeneous initial errors instead of those that are truly valid,

$$
\|e(0)\| + \sqrt{\varepsilon}\big\{\|\nabla e(0)\| + \|\nabla \eta(0)\|\big\} \leq C\varepsilon.
$$

We omit the presentation of the details.

The analysis of equation (4.36) causes several difficulties. On one hand, the first as well as the second equation possess an evolutionary operator, requiring initial data for the velocity *and* the pressure function. The quasi-compressibility constraint is *no algebraic constraint* any more. Secondly, the term on the right hand side of the second equation in (4.36) exhibits a singular behavior, for times $t \to 0$. The third difficulty arises from the prescription of homogeneous Neumann data for the pressure function, giving rise to boundary layers in the pressure approximation.

Part I: Introduction and Analysis of an Auxiliary Problem

Corresponding to the procedure of investigating the pressure stabilization scheme we split the difficulties that determine the approximation behavior of (4.33), by introducing auxiliary problems. In the first step, let us investigate the problems stemming from the pressure correction. The second part of the actual analysis is then devoted to the investigation of the error propagation owing to the evolutionary character of the momentum equation in (4.33). —

We consider the following problem that determines the tuple $\{U^\varepsilon, P^\varepsilon\}$ of

$$- \Delta U^\varepsilon + \nabla P^\varepsilon = f - u_t,$$
$$\mathrm{div} U^\varepsilon - \varepsilon \Delta P^\varepsilon_t = 0, \qquad\qquad\qquad (4.37)$$
$$\nabla P^\varepsilon(0) = \nabla p(0), \qquad U^\varepsilon|_{\partial\Omega} = 0, \qquad \partial_n P^\varepsilon_t|_{\partial\Omega} = 0.$$

Using the abbreviative notations $\Pi = p - P^\varepsilon$ and $E = u - U^\varepsilon$, we obtain

$$- \Delta E + \nabla \Pi = 0,$$
$$\mathrm{div} E - \Delta \Pi_t = -\varepsilon \Delta p_t, \qquad\qquad\qquad (4.38)$$
$$E|_{\partial\Omega} = 0, \qquad \partial_n \Pi_t|_{\partial\Omega} = \partial_n p_t|_{\partial\Omega}, \qquad \nabla \Pi(0) = 0.$$

We can now benefit from the requirements that are formulated in Postulate B_2 in order to derive stability estimates for the solution of (4.37).

Remark 4.4   1. *Owing to the injectivity of the Laplace-Dirichlet operator, the first equation in (4.38) implies the homogeneity of the starting data for the velocity error.*

2. *The second equation in (4.38), combined with the statement in Postulate B_2, gives the following statement for the time-derivative of the pressure function of problem (4.37), evaluated at time $t = 0$,*

$$\|\nabla P^\varepsilon_t(0)\| \leq C. \qquad\qquad\qquad (4.39)$$

3. *If we analyze problem (4.38), we do not need the statements formulated in Postulate B_2. This allows the derivation of pointwise error statements, based on the standard requirements (A1), (A2) and the damping properties of system (4.38). But unfortunately, these general assumptions are not sufficient to verify the a-priori property given in Lemma 4.1, ii) below, which is crucial for the further analysis.*

In order to get statements for the gradient of the pressure that is differentiated in time for certain times, we make use of the "generating mechanism" of the following energy statements. They can easily be obtained from (4.38), for $r \geq 0$, with $\delta > 0$,

$$\|\nabla \partial^r_t E\|^2 + \varepsilon d_t \|\nabla \partial^r_t \Pi\|^2 \leq C\delta\varepsilon\|\nabla \partial^{r+1}_t p\|^2 + \frac{1}{4\delta}\varepsilon\|\nabla \partial^r_t \Pi\|^2, \qquad (4.40)$$

and

$$d_t\|\nabla \partial_t^r E\|^2 + \varepsilon\|\nabla \partial_t^{r+1}\Pi\|^2 \leq C\varepsilon\|\nabla \partial_t^{r+1} p\|^2. \tag{4.41}$$

In order to use these inequalities for the case $r \geq 0$, by employing some damping time weights, we have to verify an a-priori bound for the last term on the right hand side of (4.40) for the case $r = 0$. Following the integration over the time interval $[0, T]$, we apply Gronwall's lemma to (4.40) owing to the fact, that we can set $\delta \equiv \delta(T) = T$ as normalizing time-weights in (4.40). This is to avoid an exponential growth of resulting stability constants.

Lemma 4.1 *Suppose (A1), (A2) are satisfied and the solution $\{u, p\}$ of the incompressible Stokes equations (2.1) satisfies Postulate B_2. Then, the following a-priori statements are valid for the solution $\{U^\varepsilon, P^\varepsilon\}$ of problem (4.37), for $r \geq 1$, using a constant C that is depending on the given data of the problem, and $0 < \varepsilon < 1$,*

i) $\sup_{[0,T]}\left\{\|\nabla P^\varepsilon\| + \tau^{r-1/2}\|\nabla \partial_t^r P^\varepsilon\|\right\}$
$\qquad + \int_0^T \tau^{2(r-1)}(s)\|\nabla \partial_t^r P^\varepsilon(s)\|^2 \, ds \leq C.$

Further, the following estimate is valid,

ii) $\int_\varepsilon^T \tau(s)\|\nabla P_{tt}^\varepsilon(s)\|^2 \, ds \leq C(1 + \log\frac{1}{\varepsilon}), \qquad \forall\, T \geq \varepsilon > 0.$

Proof:
In order to verify these a-priori statements for system (4.37), we start from the error equations (4.38) and make use of the standard-regularity results for the solution of the incompressible Stokes problem (2.1) that are presented in Lemma 2.5.

$\quad$ *i)* : At first, simple estimates of convergence for the solution of (4.38) can be obtained using (4.40), and setting $r = 0$. The control of the second term in this relation is possible by means of Gronwall's inequality. This gives us a bound for the function ∇P^ε in the $L^\infty(0, T; \mathbf{L}^2)$-norm. Additionally, we intend to present another possibility to gain a stability statement for ∇P^ε in the norm of $L^2(0, T; \mathbf{L}^2)$ that illustrates the arising problematic natures of approximations for irregular solutions of (2.1). Therefore, we look at the following dual problem, with initial time $T > 0$,

$$-\Delta w + \nabla q = 0,$$
$$\text{div}\, w + \varepsilon\Delta q_t = -\varepsilon\Delta\Pi, \tag{4.42}$$
$$\partial_n q_t|_{\partial\Omega} = -\partial_n \Pi|_{\partial\Omega}, \qquad q(T) = 0.$$

We stress the importance of the boundary conditions in the dual problem. Testing the second identity in (4.42) with Π, correspondingly the second equation in (4.38) with q, adding both identities leads to

$$\varepsilon\|\nabla\Pi\|^2 = -\varepsilon(\nabla q_t, \nabla\Pi) - (w, \nabla\Pi) + (E, \nabla q) - \\ - \varepsilon(\nabla\Pi_t, \nabla q) + \varepsilon(\nabla p_t, \nabla q). \tag{4.43}$$

Integration over the time range $[0, T]$ then gives,

$$\int_0^T \|\nabla\Pi(s)\|^2\, ds \leq C\Big\{1 + \int_0^T \|\nabla p_t(s)\|^2\, ds\Big\}. \tag{4.44}$$

The verification for $r > 0$ is by using the generating mechanism of the inequality pair (4.40), (4.41). This gives the proof of statement i).

ii) : In order to verify this inequality, we start with the first equation in (4.38), differentiated twice in time. Testing of this identity with E_t and multiplication with τ, finally integrating over the time interval $[\varepsilon, T]$ give

$$\tau(T)\|\nabla E_t(T)\|^2 + \varepsilon\int_\varepsilon^T \tau(s)\|\nabla P_{tt}^\varepsilon(s)\|^2\, ds \leq \varepsilon\|\nabla E_t(\varepsilon)\|^2 \\ + C\varepsilon\int_\varepsilon^T \tau(s)\|\nabla p_{tt}(s)\|^2\, ds + \int_\varepsilon^T \|\nabla E_t(s)\|^2\, ds. \tag{4.45}$$

The first term on the right hand side is bounded by $C\varepsilon$. The verification of this statement requires some auxiliary considerations that will be presented now: Let us start with the relations (4.40) and (4.41): If we set $r = 1$ in (4.41) and multiply this inequality with τ^2, finally integrate over the range $[0, \varepsilon]$, we obtain

$$\tau^2(\varepsilon)\|\nabla E_t(\varepsilon)\|^2 + \varepsilon\int_0^\varepsilon \tau^2(s)\|\nabla\Pi_{tt}(s)\|^2\, ds \\ \leq C\varepsilon^2\int_0^\varepsilon \tau(s)\|\nabla p_{tt}(s)\|^2\, ds + 2\int_0^\varepsilon \tau(s)\|\nabla E_t(s)\|^2\, ds. \tag{4.46}$$

The second term on the right hand side of (4.46), we apply statement (4.40), with $r = 1$, put the weight τ and finally integrate over the range $[0, \varepsilon]$,

$$\int_0^\varepsilon \tau(s)\|\nabla E_t(s)\|^2\, ds + \varepsilon\tau(\varepsilon)\|\nabla\Pi_t(\varepsilon)\|^2 \\ \leq C\Big\{\varepsilon^2\int_0^\varepsilon \tau(s)\|\nabla p_{tt}(s)\|^2\, ds + \int_0^\varepsilon \tau(s)\|\nabla\Pi_t(s)\|^2\, ds \tag{4.47} \\ + \varepsilon\int_0^\varepsilon \|\nabla\Pi_t(s)\|^2\, ds\Big\}.$$

Thanks to the second equation in (4.38), we can bound the last term in (4.47) as follows,

$$\varepsilon \int_0^\varepsilon \|\nabla \Pi_t(s)\|^2 \, ds \le C\varepsilon^2 \sup_{[0,\varepsilon]} \|\nabla p_t\|^2 + \frac{1}{\varepsilon} \int_0^\varepsilon \|\nabla E(s)\|^2 \, ds. \tag{4.48}$$

Because of (4.41), for $r = 0$, and the regularity condition in Postulate B_2, we now get the following bound for the last term in (4.48)

$$\|\nabla E(T)\|^2 + \varepsilon \int_0^T \|\nabla \Pi_t(s)\|^2 \, ds \le C\varepsilon T \sup_{0 \le s \le T} \|\nabla p_t(s)\|^2 + \|\nabla E(0)\|^2, \tag{4.49}$$

and therefore

$$\frac{1}{\varepsilon} \int_0^\varepsilon \|\nabla E(s)\|^2 \, ds \le C\varepsilon^2. \tag{4.50}$$

Here, we take benefit from the coincidence of the initial data for the velocity. — Now, we insert error bound (4.50) in (4.48), which can now be employed in (4.47). This gives the upper bound $C\varepsilon^2$ for the first as well as the last term in (4.47). Thanks to the choice of the initial time interval $[0, \varepsilon]$, we can bound the remaining term on the right hand side, using Gronwall's inequality. At the end, we can use (4.46) to arrive at

$$\|\nabla E_t(\varepsilon)\| \le C. \tag{4.51}$$

Let us mention the fact that the boundedness of $\|\nabla E_t(T)\|$ independent of the parameter ε can only be guaranteed for times $T \ge C\varepsilon$ under the assumptions of Postulate B_2.

Let us come back to inequality (4.45). From the latter considerations, we can bound the first term on the right hand side of (4.45) by $C\varepsilon$, as already asserted above, and we are left with the analysis of the last term in (4.45). This can be done, starting from (4.40), with $r = 1$, and integrating over the time interval $[\varepsilon, T]$. Successively, this gives bounds for the resulting terms,

$$\int_\varepsilon^T \|\nabla E_t(s)\|^2 \, ds + \varepsilon \|\nabla \Pi_t(T)\|^2$$

$$\le C\varepsilon \|\nabla \Pi_t(\varepsilon)\|^2 + \varepsilon \int_\varepsilon^T \left(\nabla p_{tt}(s), \nabla \Pi_t(s) \right) ds$$

$$\le C\varepsilon + C\varepsilon \int_\varepsilon^T \tau(s) \|\nabla p_{tt}(s)\|^2 \, ds + \varepsilon \int_\varepsilon^T \frac{1}{\tau(s)} \|\nabla \Pi_t(s)\|^2 \, ds. \tag{4.52}$$

To verify the controllability of the last term on the right hand side of (4.52), we have to multiply (4.41) with τ^{-1}, for $r = 0$, and to integrate over the range of time $[\varepsilon, T]$,

$$\int_\varepsilon^T \frac{1}{\tau^2(s)} \|\nabla E(s)\|^2 \, ds + \frac{1}{\tau(T)} \|\nabla E(T)\|^2$$

$$+ \varepsilon \int_\varepsilon^T \frac{1}{\tau(s)} \|\nabla \Pi_t(s)\|^2 \, ds \le \frac{1}{\tau(\varepsilon)} \|\nabla E(\varepsilon)\|^2 + C\varepsilon \log \frac{T}{\varepsilon}. \qquad (4.53)$$

By means of (4.49), we can bound the right hand side by $C(1 + \log \frac{T}{\varepsilon})\varepsilon$. — This gives the same bound for the right hand side of (4.52), and part ii) is proved. $\qquad \Box$

Now, we continue with the analysis of convergence for problem (4.37), based on the regularity statements presented in Postulate B_2. The next considerations make use of the stability statements for the pressure function and its differentiations that are collected in Lemma 4.1. We employ a dual problem, backward in time and formulated on the space of divergence-free velocity functions. This will imply optimal statements of convergence for the velocity U^ε in the $L^\infty(0, T; \mathbf{L}^2)$-norm. The results are presented in another lemma.

Lemma 4.2 *Suppose that Postulate B_2 is valid, and further (A1), (A2). Let $\{u, p\}$ be the solution of (2.1). Then, we have the following statements that describe the perturbation effects of the system (4.37) on the solution $\{U^\varepsilon, P^\varepsilon\}$, with a constant C that is only depending on the given data of the problem, for $0 < \epsilon < 1$,*

$i)$ $\sup_{[0,T]} \sqrt{\tau} \|u - U^\varepsilon\| + \left(\int_0^T \tau^2(s) \|\{u - U^\varepsilon\}_t(s)\|^2 \, ds \right)^{1/2} \le C\varepsilon,$

$ii)$ $\sup_{[0,T]} \sqrt{\tau} \{ \|u - U^\varepsilon\|_1 + \|p - P^\varepsilon\| \} \le C\sqrt{\varepsilon},$

$iii)$ $(1 + \log\frac{1}{\varepsilon}) \left(\int_0^T \|\{u - U^\varepsilon\}(s)\|^2 \, ds \right)^{1/2}$

$\qquad + \left(\int_\varepsilon^T \tau(s) \|\{u - U^\varepsilon\}_t(s)\|^2 \, ds \right)^{1/2} \le C\varepsilon(1 + \log\frac{1}{\varepsilon}).$

Proof:

The statements for the pressure error in $ii)$ can be shown, using two stability properties of the divergence operator from the first error equation in (4.38). The perturbation statements for the velocity, as they are given in $i)$, $ii)$ and $iii)$, are results that can be obtained on the basis of stationary duality arguments for divergence-free functions,

$$
\begin{aligned}
-\Delta w + \nabla q &= g, \\
\mathrm{div}\, w = 0, \qquad w|_{\partial\Omega} &= 0.
\end{aligned}
\qquad (4.54)
$$

Thanks to the a-priori statements in Lemma 4.1 we can proceed, setting $g \equiv E$ and $g \equiv E_t$ to prove the above statements. $\qquad\square$

At the end of this part, we focus on the necessity of the rather restrictive regularity requirement $p_t \in L^\infty(0,T;H^1/\mathbb{R})$ instead of $p_t \in L^2(0,T;H^1/\mathbb{R})$, which is presented in Postulate B_2 and provides the basis for the previous analysis. Therefore, we come back to the statements of convergence, given in Lemma 4.2. They show the fact that even irregular solutions of the incompressible Stokes equations can be approximated after a certain initial time, which is due to an inherent damping mechanism. More important for the following, the *"adaption phase" for fluid flows of regularity $p_t \in L^2(0,T;H^1/\mathbb{R})$ is normally much longer than the corresponding time interval for pressure functions of regularity $p_t \in L^\infty(0,T;H^1/\mathbb{R})$*. This contents gives rise to the usage of stronger time-weights for results of convergence. As we have already hinted at in the beginning of this section, the evolutionary character of the momentum equation puts further influence on the ability of approximation in (4.33). Now, provided the perturbation effects stemming from the quasi-compressibility constraint are not vanishing at a sufficiently fast rate, this leads to an amplification mechanism arising from this second error source of the scheme, the *evolutionary character of the first equation*, which cannot be controlled any more. Therefore, our presented scenario of regularities (see Postulate B_2) is even adapted to cope with the inherent error mechanisms leading to optimal convergence statements.

The subsequent subsection is devoted to the study of the effects stemming from the evolutionary momentum equation.

Part II: Continuation of the Error Analysis for Problem (4.33)

Owing to the results that have been obtained in the previous subsection for problem (4.37), it is sufficient to compare the systems (4.33) and (4.37). Thus, introducing the notations $e^\varepsilon := u^\varepsilon - U^\varepsilon$ and $\eta^\varepsilon := p^\varepsilon - P^\varepsilon$, the error equations read as follows,

$$
\begin{aligned}
e_t^\varepsilon - \Delta e^\varepsilon + \nabla \eta^\varepsilon &= E_t, \\
\text{div} e^\varepsilon - \varepsilon \Delta \eta_t^\varepsilon &= 0, \\
e^\varepsilon(0) = 0, \qquad \eta^\varepsilon(0) = 0, &\qquad \partial_n \eta_t^\varepsilon|_{\partial\Omega} = 0.
\end{aligned}
\tag{4.55}
$$

The homogeneity of the initial data for both, velocity and pressure function, is a result from the formulation of the auxiliary problem (4.37) and the original error equations (4.36).

In order to verify error estimates for system (4.55), we need parameter-independent a-priori statements for the pressure gradient of problem (4.33) and its time-derivative. They are presented in the next theorem.

Theorem 4.5 *Let the tuple $\{u^\varepsilon, p^\varepsilon\}$ be the solution of the system of equations (4.33). Provided, (A1), (A2) are satisfied and furthermore Postulate B_2 for the solution $\{u, p\}$ of (2.1), we obtain the following sharp bounds, with a constant C that is only dependent on the given data of the problem, for $0 < \epsilon < 1$,*

i) $\int_0^T \|\nabla p^\varepsilon(s)\|^2 \, ds \leq C,$

ii) $\int_0^T \|\nabla p_t^\varepsilon(s)\|^2 \, ds \leq C(1 + \log\frac{1}{\epsilon}).$

Proof:
In order to proof these stability properties, we make use of the original error equations (4.36) and related dual problems.

i) Consider a backward problem with solution $\{w, q\}$, parameterized in time $T > 0$, and dual to problem (4.36),

$$
\begin{aligned}
w_t + \Delta w - \nabla q &= 0, \\
\text{div} w + \varepsilon \Delta q_t &= -\varepsilon \Delta \eta, \\
\partial_n q_t|_{\partial\Omega} = -\partial_n \eta|_{\partial\Omega}, \qquad w(T) = 0, &\qquad q(T) = 0.
\end{aligned}
\tag{4.56}
$$

Corresponding to the dual problem (4.42) in the previous subsection, the boundary data are selected in a problem-compatible way. Now, testing of the second equation in (4.56) with η and integration by parts, further adding the second identity in (4.36), tested with q, we get:

$$
\begin{aligned}
\varepsilon\|\nabla\eta\|^2 &= -\varepsilon(\nabla q_t, \nabla\eta) - (w, \nabla\eta) - \varepsilon(\nabla q, \nabla\eta_t) \\
&\quad + (e, \nabla q) + \varepsilon(\nabla p_t, \nabla q) \\
&= -\varepsilon d_t(\nabla q, \nabla\eta) - (w, \nabla\eta) + (e, \nabla q) + \varepsilon(\nabla p_t, \nabla q).
\end{aligned}
\tag{4.57}
$$

Now, we make use of the first equations of the systems (4.56) and (4.36), each, to be tested with e and w, respectively. Thanks to the identities

$$
\begin{aligned}
(e_t, w) + (\nabla e, \nabla w) + (\nabla\eta, w) &= 0, \\
(w_t, e) - (\nabla e, \nabla w) - (\nabla q, e) &= 0,
\end{aligned}
$$

we obtain after summation,

$$
-(\nabla\eta, w) + (\nabla q, e) = d_t(e, w).
\tag{4.58}
$$

This enables us to bring (4.57) in the following form,

$$
\varepsilon\|\nabla\eta\|^2 = -\varepsilon d_t(\nabla q, \nabla\eta) + d_t(e, w) + \varepsilon(\nabla p_t, \nabla q).
\tag{4.59}
$$

Now, we integrate this equality in time over the interval $[0, T]$. For the first term on the right hand side of (4.59), we employ a stability result for the pressure (as the solution of (4.56)). After integration, the second term can be split via Young's Inequality, one part to be controlled by means of another a-priori stability result for the velocity of the dual problem (4.56), whereas the second part is bounded by $C\varepsilon^2$. In order to govern the last term on the right hand side of (4.59), we can take benefit from the framework of given regularities that are formulated in Postulate B_2.

ii): In order to verify this a-priori bound, we introduce another dual problem which is again related to problem (4.36). We stress the fact, that in this case we essentially need a dual problem formulation to derive the given a-priori bound for p_t^ε in $L^2(0, T; H^1(\Omega)/\mathbb{R})$. We can not expect to succeed in proving a corresponding result for $L^\infty(0, T; H^1(\Omega)/\mathbb{R})$, because of the initial incompatibility of flow and prescribed initial data and their combination with the quasi-compressibility constraint. The auxiliary problem to be investigated now is as follows, for times $T > 0$,

$$w_t + \Delta w - \nabla q = 0,$$
$$\text{div} w + \varepsilon \Delta q_t = -\varepsilon \Delta \eta_t,$$
$$\partial_n q_t|_{\partial\Omega} = \partial_n \eta_t|_{\partial\Omega}, \qquad w(T) = 0, \qquad q(T) = 0. \tag{4.60}$$

As a first step, we test the second equation with η_t and use the compatibility of the boundary data. Further, adding this to the second equation in (4.36) which is tested with q_t, we arrive at

$$\varepsilon\|\nabla\eta_t\|^2 = -\varepsilon(\nabla q_t, \nabla\eta_t) - (w, \nabla\eta_t) + \varepsilon(\nabla q_t, \nabla\eta_t)$$
$$- (e, \nabla q_t) - \varepsilon(\nabla p_t, \nabla q_t). \tag{4.61}$$

The first and the third term cancel, and the remaining terms will be reformulated by using differentiation by parts. This procedure represents the loss of stability, which is caused by the evolutionary character of the quasi-compressibility constraint, for a differentiated pressure function of regularity $p_t \in L^2(0, T; H^1(\Omega)/\mathbb{R}) \backslash L^\infty(0, T; H^1(\Omega)/\mathbb{R})$ of the incompressible Stokes equations. Thus, this procedure of differentiation by parts in time is done for two reasons: Once, in order to make use of the first equations in (4.36) and (4.60), cf. (4.62), secondly, cf. (4.63), because of the fact, that no striking, i.e., parameter-independent, a-priori bounds can be obtained for the pressure function of system (4.60), differentiated in time. — Now, owing to the identities

$$-(w, \nabla\eta_t) - (e, \nabla q_t) = -d_t\big\{(w, \nabla\eta) + (e, \nabla q)\big\}$$
$$+ (w_t, \nabla\eta) + (e_t, \nabla q), \tag{4.62}$$

and

$$\varepsilon(\nabla p_t, \nabla q_t) = \varepsilon d_t(\nabla p_t, \nabla q) - \varepsilon(\nabla p_{tt}, \nabla q), \tag{4.63}$$

and by means of the first equations in (4.36) and (4.60) that give the following statements,

$$(e_t, w_t) + (\nabla e, \nabla w_t) + (\nabla\eta, w_t) = 0,$$
$$-(e_t, w_t) + (\nabla w_t, \nabla e) + (\nabla q, e_t) = 0,$$

we arrive at the following result after adding,

$$(\nabla q, e_t) + (w_t, \nabla\eta) = -d_t(\nabla e, \nabla w). \tag{4.64}$$

This permits us to write instead of (4.61),

$$\varepsilon\|\nabla\eta_t\|^2 = -d_t\big\{(w, \nabla\eta) + (e, \nabla q)\big\} - d_t(\nabla e, \nabla w)$$
$$+ \varepsilon d_t(\nabla p_t, \nabla q) - \varepsilon(\nabla p_{tt}, \nabla q). \tag{4.65}$$

If we would continue by integration over the time interval $[0, T]$, this would only be efficient in the case that we have an a-priori bound of type

$$\int_0^T \|\nabla p_{tt}(s)\|^2 \, ds \leq C.$$

In order to circumvent this additional regularity assumption, we split the integration of time of (4.65) in two intervals. — In order to verify an optimal a-priori bound for the term $\int_0^\varepsilon \|\nabla\eta_t(s)\|^2 \, ds$, we consider system (4.36), testing the first equation with e and finally integrating over the time range $[0, \varepsilon]$. We can now take benefit from the second identity in this system, which leads to the inequality

$$\|e(\varepsilon)\|^2 + \int_0^\varepsilon \|\nabla e(s)\|^2 \, ds + \varepsilon\|\nabla\eta(\varepsilon)\|^2$$
$$\leq C\varepsilon\|\nabla\eta(0)\|^2 + \|e(0)\|^2$$
$$+ C\varepsilon \int_0^\varepsilon \Big\{\alpha\|\nabla p_t(s)\|^2 + \frac{1}{4\alpha}\|\nabla\eta(s)\|^2\Big\} \, ds, \qquad \alpha > 0. \tag{4.66}$$

We obtain, thanks to Postulate B_2,

$$\leq C\varepsilon^2\alpha + C\varepsilon\frac{1}{4\alpha}\int_0^\varepsilon \|\nabla\eta(s)\|^2 \, ds. \tag{4.67}$$

Now, we can apply Gronwall's lemma. If we set $\alpha = \varepsilon$, we finally end up with the upper bound

$$\leq C\varepsilon^2. \tag{4.68}$$

This result allows application of the second identity in (4.36) to obtain a (crude) bound for the desired term,

$$\int_0^\varepsilon \|\nabla\eta_t(s)\|^2 \, ds \leq C. \tag{4.69}$$

Making use of the estimate (4.66)/(4.68) that provides statements for the error in pressure and velocity at time $T = \varepsilon$, it is possible to derive another estimate for the pressure function of (4.36) which is differentiated in time, and averaged over the time interval $[0, T]$. We obtain from (4.65),

$$\varepsilon \int_\varepsilon^T \|\nabla \eta_t(s)\|^2 \, ds \leq \left(w(\varepsilon), \nabla \eta(\varepsilon)\right) + \left(e(\varepsilon), \nabla q(\varepsilon)\right)$$

$$+ \left(\nabla e(\varepsilon), \nabla w(\varepsilon)\right) - \varepsilon\left(\nabla p_t(\varepsilon), \nabla q(\varepsilon)\right) - \varepsilon \int_\varepsilon^T \left(\nabla p_{tt}(s), \nabla q(s)\right) \, ds$$

$$\leq C\Big\{\|\nabla \eta(\varepsilon)\|^2 + \frac{1}{\varepsilon}\|e(\varepsilon)\|^2 + \frac{1}{\varepsilon}T\|\nabla e(\varepsilon)\|^2 + \varepsilon\|\nabla p_t(\varepsilon)\|^2 \qquad (4.70)$$

$$+ \varepsilon \int_\varepsilon^T \frac{1}{\tau(s)} \, ds \int_\varepsilon^T \tau(s)\|\nabla p_{tt}(s)\|^2 \, ds.\Big\}$$

The last inequality needs further explanations. The first two terms on the right hand side of inequality (4.70) arise from application of the Young Inequality and well-known a-priori statements for the solution of system (4.60). The third term on the right hand side of the equality sign uses another a-priori statement for the system (4.60) that we achieve for the velocity gradient, when testing (4.60) with w_t and subsequently integrating over the time,

$$\frac{1}{4T}\varepsilon \int_T^{T-t} \|w_t(s)\|^2 \, ds + \frac{1}{4T}\varepsilon\|\nabla w(T-t)\|^2$$

$$\leq \varepsilon\frac{1}{4T} \int_\varepsilon^T \int_\varepsilon^T \|\nabla \eta_t(s')\|^2 \, ds'ds. \qquad (4.71)$$

The (enlarged) term on the right hand side of the last inequality can be absorbed through the left hand side of (4.70). - In order to control the third term on the right hand side of inequality (4.70) by $C\varepsilon$, we proceed as follows: We test the first relation in (4.36) with e_t and integrate over the time interval $[0, \varepsilon]$. This gives

$$\int_0^\varepsilon \|e_t(s)\|^2 \, ds + \|\nabla e(\varepsilon)\|^2 \leq C \int_0^\varepsilon \|\nabla \eta(s)\|^2 \, ds \leq C\varepsilon^2. \qquad (4.72)$$

The last estimate can be obtained, using result (4.66), by initially integrating over $[0, t]$, with $0 \leq t \leq \varepsilon$, and finally integrating over the interval $[0, \varepsilon]$. This gives

$$\varepsilon\|\nabla \eta(t)\|^2 \leq C\varepsilon^2 \int_0^t \|\nabla p_t(s)\|^2 \, ds + \int_0^t \|\nabla \eta(s)\|^2 \, ds + C\varepsilon^2,$$

and application of Gronwall's inequality leads us to

$$\|\nabla \eta(t)\|^2 \leq C\varepsilon\Big\{1 + \int_0^t \|\nabla p_t(s)\|^2 \, ds\Big\}\exp\Big(\frac{1}{\varepsilon}t\Big). \qquad (4.73)$$

Integration over $0 \leq t \leq \varepsilon$ leads to the desired result that ensures the last estimate in (4.72). Provided with these auxiliary results, we come back to the right hand side of (4.70), thus obtaining the upper bound $C\varepsilon(1 + \log\frac{T}{\varepsilon})$ for the right hand side of this inequality. Therefore, we end up with

$$\int_\varepsilon^T \|\nabla p_t^\varepsilon(s)\|^2 \, ds \leq C(1 + \log\frac{T}{\varepsilon}), \tag{4.74}$$

which completes the proof.   □

Remark 4.5 *The subsequent remarks refer to the proof for part i) of Theorem 4.5.*

1. *In order to derive an a-priori bound for the term presented in i), we may also apply Gronwall's lemma, employing time-normalizing weights. This gives an a-priori bound for p^ε in a norm that is pointwise in time. On the other hand, the above derivation of a regularity statement $p^\varepsilon \in L^2(0, T; H^1(\Omega)/\mathbb{R})$ is more "problem specific" and offers further insight into the problematic nature of regularity.*

2. *Multiplying equation (4.59) with a time-weight before integration gives further integral terms for velocity and pressure that cannot be handled any more.*

Remark 4.6 *The following discussion refers to the proof of part ii) of 4.5.*
   The term that makes a further analysis of convergence impossible without additional assumptions with respect to the regularity of the solution of (2.1) is the first one on the right hand side of (4.65), since it is competing with the last term. Without the additional amount of regularity assured by Postulate B_2, we would have to multiply this equality with a linear time-weight. The further terms arising from such an approach could be handled altogether, provided $p_t \in L^2(0, T; H^1(\Omega)/\mathbb{R})$. Solely the terms that arise from such an approach from the first term on the right hand side of (4.65) after integration by parts cannot be controlled by $C\varepsilon$.
   Finally, in the case of $p_t \in L^2(0, T; H^1(\Omega)/\mathbb{R}) \backslash L^\infty(0, T; H^1(\Omega)/\mathbb{R})$ we can only prove the bound $\sqrt{\varepsilon}(\int_0^T \tau(s)\|\nabla p_t^\varepsilon(s)\|^2 \, ds)^{1/2} \leq C$. The necessary ε-parameter in this a-priori bound implies a global loss of convergence order in our error analysis. For corresponding computational observations in the context of Van Kan's scheme we refer to Section 7.4.

Remark 4.7 *In order to verify the a-priori statement (4.74) for the pressure gradient of problem (4.33) differentiated in time, we established a splitting of the time-interval in $[0, \varepsilon^r]$ and $[\varepsilon^r, T]$, with $r = 1$. Setting $r = \frac{1}{2}$ is not applicable for our investigations (see (4.72)). Nevertheless, the choice $r = \frac{3}{4}$ is already sufficient. We omit the verification of this assertion, because it causes no "significant" impact onto the a-priori bound (4.74).*

After these stability investigations, we can continue with the proper error analysis. In contrast to the latter analysis, we will now come back to the reduced error equations (4.55), which has already been established to be justified. This permits us to take benefit from the homogeneity of the second error equation, with non-vanishing perturbation in the first equation that is of order preserving type. Finally, we will benefit from the homogeneous boundary data (of Neumann type) for the pressure error.

After this preparatory work, it is now possible to quantify the error quantities $\{e^\varepsilon, \eta^\varepsilon\}$ in (4.55). Testing of the first equation in (4.55) with e^ε, multiplication with τ and finally integration over the range $[\varepsilon, T]$ leads to the following inequality, using Lemma 4.2, ii),

$$\tau(T)\|e^\varepsilon(T)\|^2 + \int_\varepsilon^T \tau(s)\|\nabla e^\varepsilon(s)\|^2 \, ds + \varepsilon\tau(T)\|\nabla\eta^\varepsilon(T)\|^2$$

$$\leq C\varepsilon^2\Big\{1 + \log\frac{T}{\varepsilon}\Big\} + \int_\varepsilon^T \|e^\varepsilon(s)\|^2 \, ds + \varepsilon\int_\varepsilon^T \|\nabla\eta^\varepsilon(s)\|^2 \, ds. \quad (4.75)$$

In order to prove optimal error bounds for the two remaining integral terms on the right hand side of this inequality, we make use of duality arguments below. For the first term describing the averaged (in time) error evolution for the velocity function, we employ an auxiliary problem that is dual to the error equations (4.36), formulated for divergence-free functions. A simple analysis leads to the following bound,

$$\int_0^T \|e(s)\|^2 \, ds \leq C\varepsilon^2 \int_0^T \|\nabla p_t^\varepsilon(s)\|^2 \, ds \leq C\varepsilon^2(1 + \log\frac{1}{\varepsilon}). \quad (4.76)$$

Here, we benefited from the stability statement ii) in Theorem 4.5. — In order to control the integral pressure error, we introduce the following backward problem that is *dual to the equations (4.55)*:

For given $T > 0$, find the solution $\{w, q\}$ of

$$w_t + \Delta w - \nabla q = 0,$$
$$\text{div} w + \varepsilon\Delta q_t = -\varepsilon\Delta\eta^\varepsilon, \quad\quad\quad (4.77)$$
$$\partial_n q_t|_{\partial\Omega} = -\partial_n \eta^\varepsilon|_{\partial\Omega}, \quad q(T) = 0, \quad w(T) = 0,$$

for times $0 \le t \le T$. Now, we test the second equation in (4.77) with η^ε, and, correspondingly, the second identity in (4.55) with q. After subtraction of both relations and integration over the time interval $[\varepsilon, T]$ we get

$$
\varepsilon \int_\varepsilon^T \|\nabla \eta^\varepsilon(s)\|^2 \, ds = -\int_\varepsilon^T \left\{ \left(w(s), \nabla \eta^\varepsilon(s)\right) - \left(e^\varepsilon(s), \nabla q(s)\right) \right\} \, ds
$$
$$
+ \varepsilon \left(\nabla q(\varepsilon), \nabla \eta^\varepsilon(\varepsilon) \right). \tag{4.78}
$$

At this point, we cannot integrate over the whole time interval $[0, T]$. Fortunately, this is also not necessary for the derivation of an optimal error statement of the corresponding term in inequality (4.75). This procedure is possible owing to the fact that the boundedness of the pressure gradient of (4.36) or (4.38) in $L^\infty(0, T; H^1/\mathbb{R})$ enables using the measure of the initial time interval $[0, \varepsilon]$: If we test the first equation in (4.36) with e, successively integrate over times $0 \le t \le \varepsilon$ and finally over $[0, \varepsilon]$, we arrive at the desired estimate, using result $i)$ in Lemma 4.1. — In order to control the last term on the right hand side of equality (4.78) by $C\varepsilon^2$, we make use of Young's Inequality, an a-priori bound for the pressure gradient as the solution of problem (4.77) and a statement of convergence for the error in the pressure gradient at time $t = \varepsilon$. The latter comes from (4.66)/(4.68) and (4.38), by testing the related first equation with E and finally integrating over the time interval $[0, \varepsilon]$. We omit the details here. — In order to treat the integral term on the right hand side of (4.78), we employ the first equations of the systems (4.55) and (4.77). If we test them with w and e^ε, respectively, and finally add the resulting identities, we obtain the equality

$$
(w, \nabla \eta^\varepsilon) - (e^\varepsilon, \nabla q) = -d_t(e^\varepsilon, w) + (E_t, w). \tag{4.79}
$$

Integration over the interval $[\varepsilon, T]$ now gives, using the a-priori statements for problem (4.77),

$$
\int_\varepsilon^T \left\{ \left(w(s), \nabla \eta^\varepsilon(s)\right) - \left(e^\varepsilon(s), \nabla q(s)\right) \right\} \, ds
$$
$$
\le C\alpha \left\{ \|e(\varepsilon)\|^2 + \left(1 + \log\frac{T}{\varepsilon}\right) \int_\varepsilon^T \tau(s)\|E_t(s)\|^2 \, ds \right\}
$$
$$
+ \frac{1}{\alpha} \sup_{\varepsilon \le s \le T} \|w(s)\|^2 \tag{4.80}
$$
$$
\le C\alpha \varepsilon^2 \left(1 + \log\frac{T}{\varepsilon}\right) + \frac{C\varepsilon}{\alpha} \int_\varepsilon^T \|\nabla \eta^\varepsilon(s)\|^2 \, ds,
$$

with $\alpha > 0$. If we insert this result in (4.78), we can absorb the last term in (4.80) on the left hand, for α sufficiently large. In the end, we can employ the results (4.76) and (4.78)/(4.80) in (4.75). Recalling the results from Lemma 4.2, this proves Theorem 4.3.

Remark 4.8   *1. The verification of the $L^\infty(0, T; L^2/\mathbb{R})$-error statements for the pressure error, given in Theorem 4.3, is an immediate consequence of Lemma 4.2, ii) and (4.75), using a Poincare inequality. Owing to the evolutionary character of the second equation in (4.55), it is not possible to derive a super-convergence result for the pressure error in a negative norm, $s \mapsto \|\{p^\varepsilon - P^\varepsilon\}(s)\|_{-1}$. We refer to Section 4.1 for further discussions.*

2. We stress the choice of the employed boundary conditions of Neumann type for the pressure functions in the dual problems (4.56), (4.60) and (4.77). This gives "problem-compatibly" posed auxiliary problems leading to the desired statements.

4.3.3   A Modified Pressure Correction Ansatz

Results of Existence and Stability

In the previous section, we considered the perturbation effects stemming from the pressure correction ansatz. Especially, we extensively discussed the necessity of certain regularity conditions (see Postulate B_2) as well as accurate initial data to obtain full approximation ability of the scheme versus the solution of the incompressible counterpart. The present section is devoted to the proposal and analysis of a damped form of the quasi-compressibility constraint. As the result, this scheme will not suffer from the marked drawbacks, while improving the accuracy of the pressure approximation (super-convergence results).

The objective of this section is to verify optimal statements of convergence *without requiring regularity assumptions that exceed the basic assumptions (A1) and (A2)*. Further, we will verify super-convergence results for the pressure functions in negative norms, leading to improved approximation results in compact interior domains $\Omega' \subset\subset \Omega$.

The problem to be considered in the following is as follows: Find the solution $\{u^\varepsilon, p^\varepsilon\}$ of the equations

$$u_t^\varepsilon - \Delta u^\varepsilon + \nabla p^\varepsilon = f,$$
$$\text{div} u^\varepsilon - \varepsilon \Delta \{\tau^r p^\varepsilon\}_t = 0, \qquad r \geq 2,$$
$$\partial_n \{\tau^r p^\varepsilon\}_t|_{\partial\Omega} = 0, \qquad u^\varepsilon = u_0^\varepsilon, \qquad p^\varepsilon(0) = p_0^\varepsilon. \tag{4.81}$$

In order to verify the full approximation property, we only need the property

$$\|u_0 - u_0^\varepsilon\| \leq C\varepsilon.$$

Remark 4.9 *Note, that we do not need sufficiently accurate initial pressure data any more. We only have to assure that there holds $p_0^\varepsilon \in H^1/\mathbb{R}$.*

An elementary consideration provides us with the stability of the operator related to (4.81) on the pair of spaces $L^\infty(0, T; \mathbf{L}^2(\Omega)) \cap L^2(0, T; \mathbf{H}_0^1(\Omega)) \times L^\infty(0, T; H^1(\Omega)/\mathbb{R})$, for $\varepsilon > 0$. Now, we investigate the regularity of the solution, provided we are given the initial regularities,

$$u^\varepsilon(0) \equiv u_0 \in \mathbf{J}_1(\Omega) \cap \mathbf{H}^2(\Omega), \qquad p^\varepsilon(0) \in H^1(\Omega)/\mathbb{R}.$$

Owing to the fact that we are interested in getting stability results for the solution of (4.81) which are independent of the parameter ε, we start from the error equations, resulting from subtraction of the identities in (2.1) and (4.81),

$$e_t - \Delta e + \nabla \eta = 0,$$
$$\text{div} e - \varepsilon \Delta \{\tau^r \eta\}_t = -\varepsilon \Delta \{\tau^r p^\varepsilon\}_t,$$
$$\partial_n \{\tau^r \eta\}_t|_{\partial\Omega} = \partial_n \{\tau^r p\}_t|_{\partial\Omega}, \tag{4.82}$$
$$e(0) = u_0 - u_0^\varepsilon, \qquad \eta(0) = p(0) - p_0^\varepsilon.$$

with $e := u - u^\varepsilon$ and $\eta := p - p^\varepsilon$. If we test the first equation in (4.82) with e and finally integrate over time, we can arrange the resulting relation to

$$\|e(T)\|^2 + \int_0^T \|\nabla e(s)\|^2 \, ds$$
$$+ \varepsilon \tau^r(T)\|\nabla \eta(T)\|^2 + r\varepsilon \int_0^T \tau^{r-1}(s)\|\nabla \eta(s)\|^2 \, ds \tag{4.83}$$
$$\leq C\varepsilon \int_0^T \tau^{1-r}(s)\|\nabla \{\tau^r p\}_t(s)\|^2 \, ds + \|e(0)\|^2.$$

In particular, for values $r \geq 2$, the first term on the right hand side of (4.83) can be bounded by $C\varepsilon$.

Remark 4.10 1. *In comparison to the analysis in subsection 4.3.2 we do not need any (normalized) Gronwall estimate. This is, because on the left hand side of (4.83) there appears an additional term for the pressure gradient, implying improved stability of the system in case of irregular solutions $\{u, p\}$ of the system (2.1). This gives us a statement for the quantity ∇p^ε which is determining the dynamical nature of the system in a problem adapted norm.*

2. *In order to arrive at (4.83), we need no further assertions apart from (A1), (A2).*

3. *Relation (4.83) now gives the following uniform a-priori bound for the pressure function of problem (4.81),*

$$\sup_{0 \le s \le T} \left\{ \tau^{r/2}(s) \| \nabla p^\varepsilon(s) \| \right\} \le C, \tag{4.84}$$

for $r \ge 2$. By means of compactness arguments, this gives the existence and uniqueness of a solution $p^\varepsilon \in L^\infty(0, T; H^1(\Omega)/\mathbb{R})$.

In order to derive a corresponding verification of an a-priori bound for the velocity function, we test the first equation in (4.82) with e_t, multiply with $\bar{\tau}^{r-1} := \min\{\tau, \tau^{r-1}\}$ and finally integrate over the time interval. We can apply inequality (4.83) to conclude

$$\int_0^T \bar{\tau}^{r-1}(s) \| e_t(s) \|^2 \, ds + \bar{\tau}^{r-1}(T) \| \nabla e(T) \|^2$$

$$\le C \int_0^T \bar{\tau}^{r-1}(s) \| \nabla \eta(s) \|^2 \, ds + C \int_0^T \| \nabla e(s) \|^2 \, ds \le C. \tag{4.85}$$

Remark 4.11 1. *We can obtain a bound that corresponds to inequality (4.85) by multiplying the first equality in (4.82) with the time-weight τ^r before differentiating in time and finally testing this identity with e_t.*

2. *The subsequent analyses of system (4.82) can also be applied to the implicit semi-discretization in time of Adams-Moulton type. We will not discuss this point any further.*

Proof of Theorem 4.4

The contents of this subsection is the proof of Theorem 4.4. Here, we are interested in putting weight on the derivation of super-convergence results for the pressure function p^ε in negative norms. — In order to get optimal error statements for (4.82), we start with the investigation of a more easy auxiliary problem of the following type, possessing one "algebraic equation". — Let $\{U^\varepsilon, P^\varepsilon\}$ be the solution of the problem

$$
\begin{aligned}
&- \Delta U^\varepsilon + \nabla P^\varepsilon = f - u_t, \\
&\mathrm{div} U^\varepsilon - \varepsilon \Delta \{\tau^r P^\varepsilon\}_t = 0, \qquad r \geq 2, \\
&U^\varepsilon(0) = u_0, \qquad P^\varepsilon(0) = p(0), \\
&U^\varepsilon|_{\partial\Omega} = 0, \qquad \partial_n \{\tau^r P^\varepsilon\}_t|_{\partial\Omega} = 0.
\end{aligned}
\tag{4.86}
$$

The choice of accurate initial data for the pressure function provides us with good ones for the velocity, owing to the algebraic constraint in (4.86), and gives rise to good approximations for the solution $\{u, p\}$ of the incompressible Stokes equations (2.1), even for small times $t \in [0, 1]$. In the further analysis, we will benefit from the special choice of the initial function for the pressure function.

The relevant error equations result from subtraction of the systems (4.86) and (2.1). They read as follows, using the notations $E := u - U^\varepsilon$ and $\Pi := p - P^\varepsilon$,

$$
\begin{aligned}
&- \Delta E + \nabla \Pi = 0, \\
&\mathrm{div} E - \varepsilon \Delta \{\tau^r \Pi\}_t = -\Delta \{\tau^r p\}_t, \\
&E|_{\partial\Omega} = 0, \qquad \partial_n \{\tau^r \Pi\}_t|_{\partial\Omega} = \partial_n \{\tau^r p\}_t|_{\partial\Omega}, \qquad \nabla \Pi(0) = 0.
\end{aligned}
\tag{4.87}
$$

We start with a stability analysis of the system of equations (4.87), gathering a couple of a-priori statements for the solution of (4.86) in the following lemma.

Lemma 4.3 *Let (A1), (A2) be satisfied. Then, the following a-priori bounds are valid for the solution $\{U^\varepsilon, P^\varepsilon\}$ of problem (4.86), provided $r \geq 2$, with a constant C that is independent of ε but depends on the given data of the problem and the parameter r,*

i) $\sup_{[0,T]} \left\{ r \tau^{r-1} \|\nabla U^\varepsilon\|^2 + \|\nabla \{\tau^r P^\varepsilon\}_t\|^2 \right\} + \int_0^T \tau^r(s) \|\nabla U_t^\varepsilon(s)\|^2 \, ds \leq C,$

ii) $\sup_{[0,T]} \left\{ \tau^{r/2} \|\nabla \{\tau^r P^\varepsilon\}_{tt}\| \right\} + \int_0^T \|\nabla \{\tau^r P^\varepsilon\}_{tt}(s)\|^2 \, ds \leq C,$

iii) $\int_0^T \tau^{r+1}(s) \|\nabla \{\tau^r P^\varepsilon\}_{ttt}(s)\|^2 \, ds \leq C.$

Remark 4.12 *We can give an upper bound for the constant C employed in the latter lemma with respect to its dependence on the parameter r, $C \leq C_0(2(r-2)+1)^{-1}$. The verification of this statement is a consequence of the following results (4.94) and (4.95). Let us point out that the restriction to the values $r \geq 2$ is necessary to cope with the nonstationary character of the momentum equation, see problem (4.110) in the subsequent proof.*

Proof:
We start from system (4.87) in order to verify the statements of this lemma, in combination with the standard regularity properties of the solution of the incompressible Stokes equations, compare (A1) and (A2).

$i)$: The verification of this statement is based on two estimates, that equally serve as estimates of convergence for the gradient of the velocity: We test the first equation in (4.87) with E and get

$$\|\nabla E\|^2 + \varepsilon r \tau^{r-1}\|\nabla\Pi\|^2 + \frac{1}{2}\varepsilon \tau^r d_t\|\nabla\Pi\|^2$$
$$\leq \varepsilon r(\tau^{r-1}\nabla\Pi, \nabla p) + \varepsilon(\tau^r \nabla\Pi, \nabla p_t). \qquad (4.88)$$

In order to obtain this inequality, we have benefited from the stabilizing effect of the time weights that appear in (4.87). Now, we multiply this relation with τ^{r-3}. By means of Young's Inequality we immediately obtain after integration,

$$\int_0^T \tau^{r-3}(s)\|\nabla E(s)\|^2\, ds + \varepsilon \int_0^T \tau^{2(r-2)}(s)\|\nabla\Pi(s)\|^2\, ds$$
$$+ \frac{1}{2}\varepsilon\tau^{2r-3}(T)\|\nabla\Pi(T)\|^2$$
$$\leq \frac{1}{2}\varepsilon\tau^{2r-3}(0)\|\nabla\Pi(0)\|^2 + C\varepsilon \int_0^T \tau^{2(r-1)}(s)\|\nabla p_t(s)\|^2\, ds \qquad (4.89)$$
$$+ C\varepsilon \int_0^T \tau^{2(r-2)}(s)\|\nabla p(s)\|^2\, ds.$$

For values $r \geq 2$, the right hand side can be bounded by $C\varepsilon$. — For the further studies, we need additional bounds at the starting time for the velocity and the pressure function, together with their time-differentiated versions, multiplied with time weights. At first, we can use the second identity in (4.87) to arrive at the subsequent result,

$$\varepsilon\|\nabla\{\tau^r\Pi\}_t(0)\|^2 \leq \varepsilon\|\nabla\{\tau^r p\}_t(0)\|^2 + \frac{1}{\varepsilon}\|E(0)\|^2 = 0. \qquad (4.90)$$

This statement is already valid for parameter values $r > 1$. Another estimation for the error at initial time can be extracted from the first equation in (4.87), which is to be tested with $\tau^{r-2}E$:

$$2\tau^{r-2}(s)\|\nabla E(s)\|^2 + \varepsilon\tau^{2r-3}(s)\|\nabla\Pi(s)\|^2$$
$$+ \varepsilon d_t\left(\tau^{2(r-1)}(s)\|\nabla\Pi(s)\|^2\right) \qquad\qquad (4.91)$$
$$\leq \varepsilon r\tau^{2r-3}(s)\|\nabla p(s)\|^2 + \varepsilon r\tau^{2r-1}(s)\|\nabla p_t(s)\|^2.$$

The evaluation at time $s = 0$ gives, owing to the positivity of the last term on the left hand side of this result,

$$\|\tau^{r/2-1}\nabla E(0)\|^2 = 0, \qquad r > 3/2. \qquad\qquad (4.92)$$

This statement can be employed in the following. — We start from the identity

$$-\Delta\{\tau^r E\}_t + \nabla\{\tau^r\Pi\}_t = 0. \qquad\qquad (4.93)$$

Testing with E_t gives

$$\tau^r\|\nabla E_t\|^2 + r\tau^{r-1}d_t\|\nabla E\|^2 + \varepsilon d_t\|\nabla\{\tau^r\Pi\}_t\|^2$$
$$\leq C\frac{1}{T}\varepsilon\|\nabla\{\tau^r\Pi\}_t\|^2 + CT\varepsilon\|\nabla\{\tau^r p\}_{tt}\|^2. \qquad\qquad (4.94)$$

Thanks to the result (4.90), we can conclude,

$$\int_0^T \tau^r(s)\|\nabla E_t(s)\|^2\,ds + r\tau^{r-1}(T)\|\nabla E(T)\|^2$$
$$+ \varepsilon\|\nabla\{\tau^r\Pi\}_t(T)\|^2 \leq C\varepsilon. \qquad\qquad (4.95)$$

The latter bound holds for $r \geq 2$. Further, the application of Gronwall's inequality is unproblematic, because of the normalizing factor $\frac{1}{T}$ in front of the first term on the right hand side of (4.94). This proves the first part of the lemma.

In order to get another estimate for the error at time $s = 0$, we start from (4.93), testing this identity with $\tau^{-1}E$,

$$\tau^{r-1}(s)d_t\|\nabla E(s)\|^2 + 2r\tau^{r-2}(s)\|\nabla E(s)\|^2 + \varepsilon\frac{1}{\tau(s)}\|\nabla\{\tau^r\Pi\}_t(s)\|^2$$
$$\leq \frac{1}{2}\varepsilon\frac{1}{\tau(s)}\|\nabla\{\tau^r p\}_t(s)\|^2. \qquad\qquad (4.96)$$

Now, we set $s = 0$ which gives the result: $\|\frac{1}{\sqrt{\tau}}\nabla\{\tau^r\Pi\}_t\|(0) = 0$. This identity can now be employed in (4.94), leading us to another result,

$$\|\tau^{r/2}\nabla E_t\|(0) + \|\frac{1}{\sqrt{\tau}}\nabla\{\tau^r\Pi\}_t\|(0) = 0. \tag{4.97}$$

Here, we benefited from the fact that the terms arising on the left hand side of (4.94) are all positive at time $s = 0$. — This result enables us to get a striking result for the velocity error estimation in the energy norm: Formally differentiating the first equation in (4.87) in time and subsequently testing with E gives,

$$\frac{1}{2}d_t\|\nabla E\|^2 + \varepsilon\tau^r\|\nabla\Pi_t\|^2 + \frac{1}{2}\varepsilon r\tau^{r-1}d_t\|\nabla\Pi\|^2$$
$$\leq C\varepsilon\|\tau^{r-1}\nabla p + \tau^r\nabla p_t\|^2. \tag{4.98}$$

Before integrating, we multiply this relation with a time-weight τ^{r-2}, and we obtain

$$\tau^{r-2}(T)\|\nabla E(T)\|^2 + \int_0^T \tau^{2(r-1)}(s)\|\Pi_t(s)\|^2 \, ds$$
$$+ \frac{r}{2}\varepsilon\tau^{2r-3}(T)\|\nabla\Pi(T)\|^2$$
$$\leq \|\tau^{r/2-1}\nabla E\|^2(0) + C\varepsilon\{\|\tau^{r-3/2}\nabla\Pi\|^2(0) \tag{4.99}$$
$$+ \int_0^T \tau^{3r-4}(s)\|\nabla\{p + \tau p_t\}(s)\|^2 \, ds + \int_0^T \tau^{r-3}(s)\|\nabla E(s)\|^2 \, ds\}.$$

For values $r \geq 2$, the right hand side of this inequality can be controlled by $C\varepsilon$, thanks to (4.89).

$ii)$: In order to verify this part of the lemma, we make use of identity (4.97). Therefore, we differentiate identity (4.93) another time with respect to the time, and finally test with E_t. An elementary calculation leads us to

$$2r\tau^{r-1}\|\nabla E_t\|^2 + \frac{1}{2}\tau^r d_t\|\nabla E_t\|^2 + r(r-1)\tau^{r-2}d_t\|\nabla E\|^2$$
$$+ \varepsilon\|\nabla\{\tau^r\Pi\}_{tt}\|^2 \leq C\varepsilon\|\nabla\{\tau^r p\}_{tt}\|^2. \tag{4.100}$$

Owing to the previous considerations, integration gives

$$\int_0^T \tau^{r-1}(s)\|\nabla E_t(s)\|^2 \, ds + \frac{1}{2}\tau^r(T)\|\nabla E_t(T)\|^2$$

$$+ r(r-1)\tau^{r-2}(T)\|\nabla E(T)\|^2 + \varepsilon \int_0^T \|\nabla\{\tau^r\Pi\}_{tt}(s)\|^2 \, ds$$

$$\leq C \int_0^T \tau^{r-3}(s)\|\nabla E(s)\|^2 \, ds + C\varepsilon. \tag{4.101}$$

The initial terms that correspond to the second and third term on the left hand side of this inequality vanish, owing to the statements (4.92) and (4.97). Thanks to (4.89), we can bound the first term on the right hand side of the last inequality through $C\varepsilon$, in case $r \geq 2$.

iii) : In order to verify the last part of the lemma, we start with some preliminary considerations. At first, we will use the identity

$$\left\{\tau^r\{\nabla\tau^r\Pi\}_t\right\}_{tt} = r(r-1)\tau^{r-2}\{\nabla\tau^r\Pi\}_t + 2r\tau^{r-1}\{\nabla\tau^r\Pi\}_{tt}$$

$$+ \tau^r\{\nabla\tau^r\Pi\}_{ttt}. \tag{4.102}$$

Now, differentiation of (4.93) in time and finally testing this identity with $\{\tau^r E\}_{tt}$ gives the following result

$$\|\nabla\{\tau^r E\}_{tt}\|^2 + r\varepsilon\tau^{r-1}\|\nabla\{\tau^r\Pi\}_{tt}\|^2 + \frac{1}{2}\varepsilon d_t\left\{\tau^r\|\nabla\{\tau^r\Pi\}_{tt}\|^2\right\}$$

$$+ \frac{1}{2}r(r-1)\varepsilon d_t\left\{\tau^{r-2}\|\nabla\{\tau^r\Pi\}_t\|^2\right\}$$

$$\leq C\varepsilon\tau^{r-3}\|\nabla\{\tau^r\Pi\}_t\|^2 + C\varepsilon\tau^{1-r}\|\{\tau^r\{\nabla\tau^r p\}_t\}_{tt}\|^2. \tag{4.103}$$

Subsequently, integration over the time-interval $[0, s]$ implicates further terms at time $s = 0$ on the right hand side, stemming from the third and fourth terms on the left. The first one can be treated using statement (4.97), in combination with the case $r \geq 2$, and the second equality in (4.87). This gives the result

$$\|\nabla\tau^{r/2}\{\tau^r\Pi\}_{tt}\|(0) = 0. \tag{4.104}$$

The controllability of the second term in question follows analogously from (4.92),

$$\|\nabla\tau^{r/2-1}\{\tau^r\Pi\}_t\|(0) = 0. \tag{4.105}$$

Further, the boundedness of the resulting first integral term on the right hand side of (4.103) can be shown by means of elementary calculations and the estimate (4.89) and (4.97). Finally, standard regularity statements give the boundedness of the last term on the right hand side of (4.103) after integration, for $r \geq 2$. Therefore, we can write,

$$\int_0^T \|\nabla\{\tau^r E\}_{tt}(s)\|^2 \, ds + r\varepsilon \int_0^T \tau^{r-1}(s)\|\nabla\{\tau^r\Pi\}_{tt}(s)\|^2 \, ds$$
$$+ \varepsilon\tau^r(T)\|\nabla\{\tau^r\Pi\}_{tt}(T)\|^2$$
$$+ r(r-1)\varepsilon\tau^{r-2}(T)\|\nabla\{\tau^r\Pi\}_t(T)\|^2 \leq C\varepsilon. \tag{4.106}$$

For further investigations, we need another a-priori bound for the time-derivatives of the solution of system (4.87). Therefore, we start from (4.103), multiplying this inequality with a linear time factor. Thanks to (4.104) and (4.105) — they guarantee the positivity of the third and the fourth term on the left hand side of this inequality at time $s = 0$ — we arrive at the statement

$$\|\nabla\tau^{1/2}\{\tau^r E\}_{tt}\|(0) = 0. \tag{4.107}$$

This auxiliary result helps us to verify the inequality of part $iii)$ in the lemma. We twice differentiate in time equation (4.93), and finally test it with $\{\tau^r E\}_{tt}$. After multiplication with τ and final integration, we arrive at the following estimate, owing to equation (4.102),

$$d_t\{\tau\|\nabla\{\tau^r E\}_{tt}\|^2\} + \varepsilon\tau^{r+1}\|\nabla\{\tau^r\Pi\}_{ttt}\|^2$$
$$\leq \|\nabla\{\tau^r E\}_{tt}\|^2 + C\varepsilon\{\tau^{r-1}\|\nabla\{\tau^r\Pi\}_t\|^2 \tag{4.108}$$
$$+ \tau^r\|\nabla\{\tau^r\Pi\}_{tt}\|^2\} + C\varepsilon\tau^{1-r}\|\{\tau^r\nabla\{\tau^r p\}_t\}_{tt}\|^2.$$

Now, we can employ the results (4.106), (4.107), (4.95) and (4.101), when integrating the latter inequality. For the restriction $r \geq 2$, we can find a constant C such that

$$\tau(T)\|\nabla\{\tau^r E\}_{tt}(T)\|^2 + \frac{1}{2}\varepsilon\int_0^T \tau^{r+1}(s)\|\nabla\{\tau^r\Pi\}_{ttt}(s)\|^2 \, ds \leq C\varepsilon. \tag{4.109}$$

This completes the proof of the above lemma. □

Based on this stability analysis, it is now possible to verify striking error bounds for problem (4.81), using stationary duality arguments. We omit the explicit verification of the the following results.

Lemma 4.4 *Suppose $\{u, p\}$ to be the solution of (2.1). Further, let (A1), (A2) be satisfied. Then, the following results are valid for the solution of problem (4.81), for $r \geq 2$,*

i) $\sup_{[0,T]}\{\|u - U^\varepsilon\| + \tau^{r/2}\|\{u - U^\varepsilon\}_t\|\}$

$$+ \left(\int_0^T \|\{u - U^\varepsilon\}_t(s)\|^2 \, ds\right)^{1/2}$$

$$+ \left(\int_0^T \tau^{r+1}(s)\|\{u - U^\varepsilon\}_{tt}(s)\|^2 \, ds\right)^{1/2} \leq C\varepsilon,$$

ii) $\sup_{[0,T]}\{\|p - P^\varepsilon\|_{-1} + \sqrt{\varepsilon}\tau^{r/2-1}\{\|p - P^\varepsilon\| + \|u - U^\varepsilon\|_1\}\} \leq C\varepsilon.$

The constant C used in these inequalities possesses the properties mentioned in Lemma 4.3.

Remark 4.13 *The verification of statement i) for the second term on the left hand side of this estimate is by means of estimate (4.106) that gives the following a-priori estimate,*

$$\sup_{0 \leq s \leq T} \{\tau^{r/2}(s)\|\nabla\{\tau^r P^\varepsilon\}_{tt}(s)\|\} \leq C.$$

Note, that we have to employ another time weight, in case we look for a bound in a "non-implicit term structure",

$$\sup_{0 \leq s \leq T} \{\tau^{(3r+1)/2}(s)\|\nabla P_{tt}^\varepsilon(s)\|\} \leq C.$$

In order to complete the proof of the error estimates for problem (4.81) as they are given in Theorem 4.4, it is sufficient to analyze the following system of equations, resulting from subtraction of (4.81) and (4.86). If we apply the notation $e^\varepsilon := u^\varepsilon - U^\varepsilon$ and $\eta^\varepsilon := p^\varepsilon - P^\varepsilon$, we are led to the system

$$e_t^\varepsilon - \Delta e^\varepsilon + \nabla\eta^\varepsilon = E_t,$$
$$\text{div}\, e^\varepsilon - \varepsilon\Delta\{\tau^r\eta^\varepsilon\}_t = 0,$$
$$\partial_n\{\tau^r\eta^\varepsilon\}_t|_{\partial\Omega} = 0,$$
$$e^\varepsilon(0) = e(0), \qquad \|\nabla\eta^\varepsilon(0)\| = \mathcal{O}(1).$$

(4.110)

Thanks to the results of convergence given in Lemma 4.4, concerning the time-derivative of the velocity-field error E_t, we succeed in testing the first equation of the latter system with e^ε and final integration. We find

$$\|e^\varepsilon(T)\|^2 + \int_0^T \|\nabla e^\varepsilon(s)\|^2 \, ds + \frac{1}{2}\varepsilon\tau^r(T)\|\nabla\eta^\varepsilon(T)\|^2$$

$$+ \frac{r}{2}\varepsilon \int_0^T \tau^{r-1}(s)\|\nabla\eta^\varepsilon(s)\|^2 \, ds \qquad (4.111)$$

$$\leq \|e(0)\|^2 + C \int_0^T \|E_t(s)\|^2 \, ds \leq C\varepsilon^2.$$

We stress that no additional terms arise on the right hand side of the last inequality that are integral terms of the velocity or the pressure-gradient error over the range $[0, T]$. This is a consequence of the improved stability at initial times, thanks to the employed time-weights.

In order to verify further optimal statements for the error in the pressure function in (4.110), we need further pointwise error statements for the velocity error function and its derivatives in time. Therefore, we multiply the first equation in (4.110) with τ^r, afterwards differentiate in time and finally test this identity with e_t^ε. Elementary calculations lead us to

$$\frac{1}{2}d_t\{\tau^r\|e_t^\varepsilon\|^2\} + \frac{1}{2}r\tau^{r-1}\|e_t^\varepsilon\|^2 + \tau^r\|\nabla e_t^\varepsilon\|^2$$

$$+ \frac{r}{2}d_t\{\tau^{r-1}\|\nabla e^\varepsilon\|^2\} + \frac{1}{2}\varepsilon d_t\|\nabla\{\tau^r\eta^\varepsilon\}_t\|^2 \qquad (4.112)$$

$$\leq r(r-1)\tau^{r-2}\|\nabla e^\varepsilon\|^2 + C\tau^{r-1}\{\|E_t\|^2 + \tau^2\|E_{tt}\|^2\}.$$

Subsequent integration leads to an initial term that is related to the first one in the latter inequality. It can be controlled by means of the first equation in (4.110), by taking into account result (4.97). The same comments can be made for the last term on the left hand side of (4.112), by taking benefit from the second equation in (4.110). We emphasize, that we need the parameter restriction $r \geq 2$ here to be able to control the first term on the right hand side of the last inequality, by employing result (4.111). Finally, we recall Lemma 4.4, i) to bound the remaining expression in brackets on the right hand side of (4.112). This leads us to

$$\tau^r(T)\|e_t^\varepsilon(T)\|^2 + \frac{r}{2}\int_0^T \tau^{r-1}(s)\|e_t^\varepsilon(s)\|^2 \, ds + \int_0^T \tau^r(s)\|\nabla e_t^\varepsilon(s)\|^2 \, ds$$

$$+ \tau^{r-1}(T)\|\nabla e^\varepsilon(T)\|^2 + \varepsilon\|\nabla\{\tau^r\eta^\varepsilon\}_t(T)\|^2 \leq C\varepsilon^2.$$

Now, we can combine this result together with Lemma 4.4, i) for the velocity error E_t to get

$$\tau^{r/2}(T)\|\eta^{\varepsilon}(T)\|_{-1} \leq C\varepsilon.$$

This proves Theorem 4.4.

Chapter 5

Mixed Quasi-Compressibility Methods

5.1   Overview

In the Chapters 3 and 4, we have been concerned with stationary and nonstationary quasi-compressibility methods in order to approximate the solutions of the incompressible (Navier-) Stokes equations. The drawback of nonstationary methods is that the corresponding finite element discretization schemes are only stable under certain restrictions $F(\varepsilon, h, k) \geq 0$, for tuples of ansatz spaces that do not satisfy the *LBB constraint*. This prevents the applicability of the numerical scheme for the simulation of quickly varying space and time features of the solution. Therefore, we are naturally led to the investigation of so-called *mixed quasi-compressibility methods*. These schemes are composed of both stationary as well as nonstationary perturbation parts in the resulting quasi-compressibility scheme. In doing so, these constraints are as follows:

$$\text{div} u^\varepsilon + \phi_1(\varepsilon_1; p^\varepsilon) + \phi_2(\varepsilon_2; p_t^\varepsilon) = 0, \tag{5.1}$$

together with related trace operators $\big(\psi_1(p^\varepsilon), \psi_2(p_t^\varepsilon)\big)$. Here, we have used the notation $\varepsilon = (\varepsilon_1, \varepsilon_2)$. As already mentioned, the following schemes are practically relevant in the framework of pressure stabilizations for tuples of ansatz spaces that do not satisfy the *LBB constraint*,

$$\text{div} u^\varepsilon - \varepsilon_1 \Delta p^\varepsilon + \phi(\varepsilon_2; p_t^\varepsilon) = 0, \qquad \partial_n p^\varepsilon|_{\partial\Omega} = 0, \tag{5.2}$$

	ε_1	ε_2	$\phi(\varepsilon_2; p_t^\varepsilon) =$	$\psi(\varepsilon_2; p_t^\varepsilon) =$	projection method
i)	h^2	k	kp_t^ε		Chorin-Uzawa
ii)	h^2	k^2	$-k^2\Delta p_t^\varepsilon$	$\partial_n p_t^\varepsilon$	Van Kan
iii)	h^2	k^2	$-k^2\Delta(\tau^r p^\varepsilon)_t$	$\partial_n(\tau^r p^\varepsilon)_t$	

Table 5.1: Correspondences of Mixed Methods and Projection Schemes

with the associated trace condition $\psi(\varepsilon_2; p_t^\varepsilon) = 0$ on the boundary $\partial\Omega$. If we set $\varepsilon_1 = \mathcal{O}(h^2)$ and $\varepsilon_2 = \mathcal{O}(k)$, the reference to pressure stabilization becomes visible. In particular, the importance of the following studies even for projection methods will be stressed in subsequent chapters. As a preview, correspondences between mixed quasi-compressibility methods and certain projection schemes are listed in Table 5.1.

The subject of this section is to provide an error analysis for the scheme that is given for case i) in Table 5.1. The selection of this one is only to fix the ideas that are necessary for an analysis of mixed quasi-compressibility methods. Therefore, we omit the formulation and proof of related statements valid for the case ii). Further, we have discussed the impact of nonstationary perturbations of the incompressibility constraint to a large extend in the previous chapters, emphasizing the global reduction of accuracy in the case of incompatible flow constellations or inaccurate initial data for the pressure. From that studies, we know in principle how to modify the nonstationary perturbation by means of damping time-weights in order to slow down the initial dynamical solution features in a sufficient way. Therefore, the analysis of case iii) ($r \geq 2$) can be done based on the material that is presented in the remainder of this section and Chapter 4, and it involves no new ideas. Because of this, it is omitted.

Coming back to the discussion of general mixed methods, there arises the question of whether there are stability conditions $F(\varepsilon) \geq 0$ necessary in order to guarantee a stable balancing of the nonstationary and stationary parts in (5.2). As a result, the stability properties with respect to the velocity field approximation remain untouched, no matter how $\varepsilon \in (\mathbb{R}^+)^2$ is chosen. On the other hand, the different scaling of the pressure approximation with respect to time and space behavior in the mixed constraint requires an adjustment of the choice of parameters, depending on which properties of the approximation goodness one is interested in. To be more specific, if one is interested in, say, good accuracy of p^ε in the topology $L^\infty(0, T; L_0^2(\Omega))$, we have to observe the following condition that limits the choice for the parameter tuple ε, $\varepsilon_1 <<$ ε_2. Note that this is no severe restriction on the flexibility in the scheme: the corresponding condition for related numerical schemes is that the space discretization parameter h (or even a parameter function h_K, with K an element of the underlying triangulation of the domain $\Omega \in \mathbb{R}^n$) is bounded from above by powers of the (global) time discretization parameter k.

On the other hand, suppose that the structure of the solution is determined by time-effects rather than spatial ones, we have to reduce the width of our time-grid, keeping the spatial mesh unchanged. This situation corresponds to the case $\varepsilon_1 >> \varepsilon_2$. As we will see in future chapters dealing with the projection schemes of Chorin-Uzawa and Van Kan, the consequence will be that the selected Q1/Q1 finite element ansatz spaces do not provide a stable spatial discretization of the original projection schemes any more.

The subject of the present chapter is to illustrate that mixed quasi-compressibility methods *are* capable of approximating time-space solution structures of different scales over the domain $Q_T := \Omega \times (0, T]$ in a stable manner, especially for finite element tuples that do *not* satisfy the *LBB condition*. In order to avoid technical difficulties that arise if we carry out the analysis for the Navier-Stokes equations we restrict ourselves to the Stokes case. This seems to be justified because we have learned how to principally deal with the nonlinearity in the error analysis in previous chapters. Corresponding considerations can be used here in principle.

Finally, let us recall the relevance of the succeeding investigations in this section for projection methods of higher order. For instance, by further studies that will be carried out in the second part of the book for the so-called Chorin-Uzawa and the Van Kan scheme, it will become clear how to modify them to have a stable time-space discretization model that is even applicable to finite element pairs that do not satisfy the *LBB condition*.

Remark 5.1 *1. Again, we mention that the stability properties of the pressure function p^ε depend on a parameter restriction $F(\varepsilon) \geq 0$. As a matter of fact, for a large amount of "eccentricity", i.e., for cases $\max\{\varepsilon_1/\varepsilon_2, \varepsilon_2/\varepsilon_1\} >> 1$, we have no correlated time-space stability properties for the solution of scheme (5.1) any more, and the stability property can only be formulated in terms of time or space behavior. This will be reflected in the subsequent theorem, presenting error estimates* in different norms for different problem constellations (i.e., $\varepsilon_1/\varepsilon_2 \gg 1$ and $\varepsilon_1/\varepsilon_2 \ll 1$). Contrary to this, the approximation quality of the velocity in the $L^\infty(0,T;\mathbf{L}^2)$-norm remains untouched from this problematic nature.

2. Apart from the mixture of stationary and nonstationary perturbations in the quasi-compressibility method, it is also interesting to study systems with a relaxed incompressibility constraint of type

$$\mathrm{div}u^\varepsilon + \phi(\varepsilon; p_t^\varepsilon) = 0,$$

with $\phi(\varepsilon; p_t^\varepsilon) = \Phi(\phi_i(\varepsilon_i; p_t^\varepsilon))$ and ε a vector. In Section 10.3, we will be concerned with a special one, namely

$$\mathrm{div}u^\varepsilon + \varepsilon_1 p_t^\varepsilon - \varepsilon_2\Delta p_t^\varepsilon = 0, \qquad \partial_n p_t^\varepsilon|_{\partial\Omega} = 0,$$

for $\varepsilon \equiv (\varepsilon_1, \varepsilon_2) \equiv (k, k^2)$. For further discussions of this scheme, we refer to the revised Chorin-Uzawa scheme in Section 10.3.

5.2   Error Analysis for the Case i)

The objective of this subsection is the investigation of the following system of equations with respect to the error behavior of its solution $\{u^\varepsilon, p^\varepsilon\}$, depending on $\varepsilon = (\varepsilon_1, \varepsilon_2)$,

$$\begin{aligned}
u_t^\varepsilon - \Delta u^\varepsilon + \nabla p^\varepsilon &= f, \\
\mathrm{div}u^\varepsilon - \varepsilon_1\Delta p^\varepsilon + \varepsilon_2 p_t^\varepsilon = 0, \qquad \partial_n p^\varepsilon|_{\partial\Omega} &= 0, \\
u^\varepsilon(0) = u_0, \qquad p^\varepsilon(0) &= p^0.
\end{aligned} \tag{5.3}$$

The next theorem implies that the application of this mixed quasi-compressibility method (5.3) is justified and ensures optimal error behavior for the velocity field, without demanding certain stability restrictions on the choices of $\varepsilon = (\varepsilon_1, \varepsilon_2)$. On the other side, the approximation features for the pressure heavily rely on the eccentricity, owing to a change of the stability behavior of the solution p^ε. Owing to the evolutionary part of the perturbation, we are again forced to consider the restricted class of fluid flow problems satisfying Postulate B_1. Let us mention that — provided this condition is not satisfied for the actual flow — this causes a global loss of convergence rate with respect to the parameter ε_2. — Nevertheless, the previously introduced idea to circumvent a global breakdown in the order of convergence by using a damped version (see case iii)) can be applied to generalize the schemes to general flows.

Theorem 5.1 *Let $\{u^\varepsilon, p^\varepsilon\}$ be the solution of (5.3), whereas the tuple $\{u, p\}$ is determined to be the solution of the incompressible Stokes equations (2.1). Suppose (A1), (A2) to be satisfied. Further, we assume the solution $\{u, p\}$ to satisfy Postulate B_1. Suppose the given initial data functions to be sufficiently accurate, i.e.,*

$$\|u^0 - u_0\| + \sqrt{\varepsilon_2}\|p^0 - p(t_0)\| \leq C|\varepsilon|.$$

Then, for $0 < \varepsilon_1, \varepsilon_2 < 1$, the following error estimates are valid, with a parameter function $\mathcal{H}(\varepsilon) := \sqrt{|\varepsilon|}(\sqrt{|\varepsilon|} + \sqrt{\varepsilon_2}\log\frac{1}{\varepsilon_2}) < 1$ and a constant C that is only depending on the given problem data,

i) $\quad \sup_{0 \leq s \leq T} \|u(s) - u^\varepsilon(s)\| \leq C(1 + \log\frac{1}{\mathcal{H}(\varepsilon)})\mathcal{H}(\varepsilon),$

ii) $\quad \sqrt{\varepsilon_1} \sup_{0 \leq s \leq T} \|p(s) - p^\varepsilon(s)\| + \sqrt{\varepsilon_2} \int_0^T \|p(s) - p^\varepsilon(s)\|^2 \, ds$

$$\leq C(1 + \log\frac{1}{\mathcal{H}(\varepsilon)})\mathcal{H}(\varepsilon),$$

Proof:
In order to verify the statement given in this theorem we proceed in a fashion which is oriented at former proof strategies and has been successfully applied in the treatment of the quasi-compressibility methods discussed before. Thus, in order to control the errors that arise if we use scheme (5.3), we split the diverse parts of error sources, investigating certain auxiliary problems.

1st step: Discussion of a stationary auxiliary problem
The first auxiliary problem reads as follows: Find the solution $\{U^\varepsilon, P^\varepsilon\}$ of the following equations:

$$- \Delta U^\varepsilon + \nabla P^\varepsilon = -u_t + f,$$
$$\text{div} U^\varepsilon - \varepsilon_1 \Delta P^\varepsilon + \varepsilon_2 P_t^\varepsilon = 0, \qquad \partial_n P^\varepsilon|_{\partial\Omega} = 0, \qquad (5.4)$$
$$P^\varepsilon(0) \equiv p(0).$$

Note that the mixed quasi-compressibility constraint remains unchanged, without separating the nonstationary from the stationary part in separated studies. It is essential for an effective study that the coupling of the different perturbation sources is preserved. Further, we mention that the terms on the right hand side of the first equation in (5.4) are pointwise in time in $L^2(\Omega)$, which is again substantial for the succeeding considerations. Using the error functions $E = u - U^\varepsilon, \Pi = p - P^\varepsilon$, the identities to be investigated are:

$$- \Delta E + \nabla \Pi = 0,$$
$$\operatorname{div} E - \varepsilon_1 \Delta \Pi + \varepsilon_2 \Pi_t = -\varepsilon_1 \Delta p + \varepsilon_2 p_t, \tag{5.5}$$
$$\partial_n \Pi|_{\partial\Omega} = \partial_n p|_{\partial\Omega}, \qquad \Pi(0) \equiv 0.$$

1) Stability analysis for system (5.4): Testing the first identity in (5.5) with E and subsequent integration over the time interval $[0, T]$ leads us to the following estimate, owing to the a-priori result $\sup_{0 \le s \le T} \|p_t\| \le C$,

$$\int_0^T \|\nabla E(s)\|^2 \, ds + \varepsilon_1 \int_0^T \|\nabla \Pi(s)\|^2 \, ds$$
$$+ \varepsilon_2 \|\Pi(T)\|^2 \le CT\{\varepsilon_1 + \varepsilon_2\}. \tag{5.6}$$

We see from this result that the stability properties in space and time of the pressure P^ε are depending on the choice of the parameter vector ε. Nevertheless, this will *not* influence the property of convergence for the velocity field in a way that we need a stability restriction $F(\varepsilon) \ge 0$, which will be shown now. Therefore, we start from the differentiated form,

$$- \Delta E_t + \nabla \Pi_t = 0, \tag{5.7}$$

and test with E. — For the following, note that $E(0) \equiv 0$ is a consequence of the first algebraic identity in (5.5). We arrive at

$$\|\nabla E(T)\|^2 + \varepsilon_1\|\nabla \Pi(T)\|^2 + \varepsilon_2 \int_0^T \|\Pi_t(s)\|^2 \, ds \le CT\{\varepsilon_1 + \varepsilon_2\}. \tag{5.8}$$

For the further considerations concerning the study of the differentiated functions, we have to do a splitting of the interval of integration in $[0, T] = [0, \varepsilon_2] \cup [\varepsilon_2, T]$, in order to avoid the requirement of additional regularity requirements to the solution of (2.1).

a) *considerations on* $[0, \varepsilon_2]$. Testing of (5.7) with τE_t and integration over the interval $[0, \varepsilon_2]$ gives

$$
\int_0^{\varepsilon_2} \tau(s) \|\nabla E_t(s)\|^2 \, ds + \varepsilon_1 \int_0^{\varepsilon_2} \tau(s) \|\nabla \Pi_t(s)\|^2 \, ds + \varepsilon_2 \tau(\varepsilon_2) \|\Pi_t(\varepsilon_2)\|^2
$$

$$
\leq C\varepsilon_2^2 \int_0^{\varepsilon_2} \tau(s) \|p_{tt}(s)\|^2 \, ds + \int_0^{\varepsilon_2} \tau(s) \|\Pi_t(s)\|^2 \, ds
$$

$$
+ \varepsilon_2 \int_0^{\varepsilon_2} \|\Pi_t(s)\|^2 \, ds + C\varepsilon_1 \int_0^{\varepsilon_2} \tau(s) \|\nabla p_t(s)\|^2 \, ds \qquad (5.9)
$$

$$
\leq C|\varepsilon|^2 + \int_0^{\varepsilon_2} \tau(s) \|\Pi_t(s)\|^2 \, ds + C\varepsilon_2 \int_0^{\varepsilon_2} \|\Pi t(s)\|^2 \, ds,
$$

due to well-known a-priori bounds for the solution of the incompressible Stokes equations. — We can bound the right hand side of this inequality by means of Gronwall's lemma, getting

$$
\leq C\{|\varepsilon|^2 + \varepsilon_2 \int_0^{\varepsilon_2} \|\Pi_t(s)\|^2 \, ds\}. \qquad (5.10)
$$

Thus, it remains to control the last term in (5.10). This can be done by using relation (5.8). Therefore, we arrive at

$$
\int_0^{\varepsilon_2} \tau(s) \|\nabla E_t(s)\|^2 \, ds + \varepsilon_1 \int_0^{\varepsilon_2} \tau(s) \|\nabla \Pi_t(s)\|^2 \, ds
$$

$$
+ \varepsilon_2^2 \|\Pi_t(\varepsilon_2)\|^2 \leq C|\varepsilon|^2. \qquad (5.11)
$$

b) *considerations on* $[\varepsilon_2, T]$. In this part, we intend to verify statements that correspond to those in *a)*, but now on the time interval $[\varepsilon_2, T]$. Again, we start from the following relation that is easily obtained from (5.7),

$$
\|\nabla E_t(s)\|^2 + \frac{1}{2}\varepsilon_1 \|\nabla \Pi_t(s)\|^2 + \varepsilon_2 d_t \|\Pi_t(s)\|^2
$$

$$
\leq \frac{\alpha_s \varepsilon_2}{2} \|p_{tt}(s)\|^2 + \frac{\varepsilon_2}{2\alpha_s} \|\Pi_t(s)\|^2 + C\varepsilon_1 \|\nabla p_t(s)\|^2, \qquad (5.12)
$$

with a positive parameter function $\alpha_s, \forall \, s \in [0, T]$ that needs to be fixed. Therefore, let us recall the fact that the last term in this inequality is bounded after integration over the whole time interval $[0, T]$, which is owing to Postulate B_1. Now, we can choose $\alpha_s = (1 + T + \log\frac{1}{\varepsilon_2})$ and integrate over the time interval $[\varepsilon_2, T]$. Together with Gronwall's lemma, we get

$$\int_{\varepsilon_2}^{T} \|\nabla E_t(s)\|^2 \, ds + \varepsilon_1 \int_{\varepsilon_2}^{T} \|\nabla \Pi_t(s)\|^2 \, ds + \varepsilon_2 \|\Pi_t(T)\|^2$$

$$\leq \varepsilon_2 \|\Pi_t(\varepsilon_2)\|^2 + C\varepsilon_2(1 + \log\frac{T}{\varepsilon_2}) + C\varepsilon_1$$

$$\leq C\Big\{\varepsilon_2 + \varepsilon_1 + \frac{\varepsilon_1^2}{\varepsilon_2} + \varepsilon_2(1 + \log\frac{T}{\varepsilon_2})\Big\} \tag{5.13}$$

$$= C\Big\{(1 + \frac{\varepsilon_1}{\varepsilon_2})\varepsilon_1 + (1 + \log\frac{T}{\varepsilon_2})\varepsilon_2\Big\}.$$

Again, we recognize from this inequality the varying stability properties of the pressure function P^ε with respect to time and space behavior. In order to guarantee a parameter-independent stability behavior for the pressure, we have to impose certain stability requirements that relate the given parameters ε_1 and ε_2.

From (5.12), with $\alpha_s \equiv 1, \forall \ s \in [0, T]$, we get another estimate that will be useful for the subsequent error analysis, permitting an upper bound without a factor that limits the stability of the involved pressure functions. Therefore, multiplication of the relation (5.12) with the time weight $\tau(s)$ and subsequent integration over the time interval $[\varepsilon_2, T]$ give

$$\int_{\varepsilon_2}^{T} \tau(s)\Big\{\|\nabla E_t(s)\|^2 + \varepsilon_1 \|\nabla \Pi_t(s)\|^2\Big\} \, ds + \varepsilon_2 \tau(T) \|\Pi_t(T)\|^2 \leq C|\varepsilon|. \tag{5.14}$$

Here, we have employed the estimates (5.8) and (5.11) and related a-priori bounds for the solution of the incompressible Stokes problem. We leave the easy details to the reader.

2) *Error bounds for* $\{E, \Pi\}$: Provided with these a-priori statements, we can now perform the error analysis to derive an upper sharp bound for the error $E(s), s \in [0, T]$, measured in the norm $L^\infty(0, T; L^2(\Omega))$. To do so, let us formulate the following *dual problem*: For given g, find the solution $\{w, q\} \in H_0^1(\Omega) \times L_0^2(\Omega)$ of

$$-\Delta w + \nabla q = g, \quad \text{div} w = 0, \quad \text{in } \Omega. \tag{5.15}$$

Now, setting $g \equiv E(T)$ and testing with $E(T)$, we arrive at

$$\|E(T)\|^2 \leq C\{\varepsilon_2^2 \|\Pi_t(T)\|^2 + \varepsilon_1^2 \|\nabla\Pi(T)\|^2\}$$

$$\leq C\{\varepsilon_1\varepsilon_2 + \varepsilon_1^2 + \varepsilon_2^2(1 + \log\frac{1}{\varepsilon_2})\} \leq C\{|\varepsilon|^2 + \varepsilon_2^2\log\frac{1}{\varepsilon_2}\}. \quad (5.16)$$

This proves an optimal error estimate for the velocity, without an additional assumption that limits the choice of the parameters $\varepsilon_i, i \in \{1, 2\}$.

We can now easily verify error statements for the pressure function P^ε, using the corresponding results above for the velocity field, and the following stability properties of the div-operator,

$$\|\chi\|_{-i} \leq C \sup_{\phi \in H_0^1 \cap H^{i+1}} \frac{(\nabla\chi, \phi)}{\|\phi\|_{i+1}}, \qquad \forall \chi \in L_0^2, \quad (5.17)$$

with $i \in \{0, 1\}$. Using the first identity in (5.5), we arrive at error statements for the pressure function,

$$\|\Pi(T)\|_{-1} \leq C\{|\varepsilon| + \varepsilon_2\log\frac{1}{\varepsilon_2}\} \qquad \text{and} \qquad \|\Pi(T)\| \leq C\sqrt{|\varepsilon|}. \quad (5.18)$$

The error estimates (5.8), (5.16) and (5.18) indicate an optimal error behavior for the solution of the auxiliary problem (5.4). This justifies a restriction of the corresponding analysis for the equation (5.3): we can concentrate on estimating the gap between the solutions of (5.3) and (5.4). But, as we will see subsequently, we need further information concerning the propagation of the error $t \mapsto \{E(t), \Pi(t)\}$, for $t \in [\varepsilon_2, T]$ before. This will be provided in the next part.

Error bounds for E_t: Again, we have to provide sharp stability statements for the time-derivatives of the pressure function, P_t^ε. Differentiating (5.7) in time, afterwards testing with $\tau(s)E_t(s)$ and finally integrating over $[\varepsilon_2, T]$ gives,

$$\tau(T)\|\nabla E_t(T)\|^2 + \varepsilon_1\tau(T)\|\nabla\Pi_t(T)\|^2 + \varepsilon_2\int_{\varepsilon_2}^T \tau(s)\|\Pi_{tt}(s)\|^2 \, ds$$

$$\leq \varepsilon_2\|\nabla E_t(\varepsilon_2)\|^2 + \varepsilon_1\varepsilon_2\|\nabla\Pi_t(\varepsilon_2)\|^2 + \int_{\varepsilon_2}^T \|\nabla E_t(s)\|^2 \, ds$$

$$+ \varepsilon_1\int_{\varepsilon_2}^T \|\nabla\Pi_t(s)\|^2 \, ds + \varepsilon_2\int_{\varepsilon_2}^T \tau(s)\|p_{tt}(s)\|^2 \, ds \quad (5.19)$$

$$+ \varepsilon_1\int_{\varepsilon_2}^T \tau(s)\big(\nabla p_t(s), \nabla\Pi_{tt}(s)\big) \, ds.$$

The last integral term can be transformed via integration by parts,

$$
\int_{\varepsilon_2}^{T} \tau(s)\left(\nabla p_t(s), \nabla \Pi_{tt}(s)\right) \, ds = \tau(T)\left(\nabla p_t(T), \nabla \Pi_t(T)\right)
$$
$$
- \varepsilon_2\left(\nabla p_t(\varepsilon_2), \nabla \Pi_t(\varepsilon_2)\right) - \int_{\varepsilon_2}^{T} \left(\nabla p_t(s), \nabla \Pi_t(s)\right) \, ds \qquad (5.20)
$$
$$
- \int_{\varepsilon_2}^{T} \tau(s)\left(\nabla p_{tt}(s), \nabla \Pi_t(s)\right) \, ds.
$$

Using Cauchy's inequality, corresponding a-priori estimates for the incompressible Stokes equations and result (5.13), we can proceed estimating the right hand side of (5.19) further,

$$
\leq \varepsilon_2 \|\nabla E_t(\varepsilon_2)\|^2 + \varepsilon_1\varepsilon_2 \|\nabla \Pi_t(\varepsilon_2)\|^2 + C\left\{(1 + \frac{\varepsilon_1}{\varepsilon_2})\varepsilon_1 + (1 + \log\frac{T}{\varepsilon_2})\varepsilon_2\right\}
$$
$$
+ C\left\{\varepsilon_2 + \varepsilon_1\alpha \int_{\varepsilon_2}^{T} \left\{\|\nabla p_t(s)\|^2 + \tau^2(s)\|\nabla p_{tt}(s)\|^2\right\} \, ds\right\}
$$
$$
+ \varepsilon_1 \frac{1}{\alpha} \int_{\varepsilon_2}^{T} \|\nabla \Pi_t(s)\|^2 \, ds + \varepsilon_1 T \|\nabla p_t(T)\|^2, \qquad (5.21)
$$

and we choose $\alpha = 1 + T + \log\frac{1}{\varepsilon_2}$. Now, we can apply Gronwall's lemma which gives

$$
\tau(T)\|\nabla E_t(T)\|^2 + \varepsilon_2 \int_{\varepsilon_2}^{T} \tau(s)\|\Pi_{tt}(s)\|^2 \, ds + \varepsilon_1 \tau(T)\|\nabla \Pi_t(T)\|^2
$$
$$
\leq C\left\{(1 + \frac{\varepsilon_1}{\varepsilon_2} + \log\frac{1}{\varepsilon_2})\varepsilon_1 + (1 + \log\frac{1}{\varepsilon_2})\varepsilon_2\right\}
$$
$$
+ \varepsilon_1\varepsilon_2 \|\nabla \Pi_t(\varepsilon_2)\|^2 + \varepsilon_2 \|\nabla E_t(\varepsilon_2)\|^2. \qquad (5.22)
$$

As a consequence of the mixing of the stationary and nonstationary perturbations appearing in the quasi-compressibility constraint, the approximation features of the functions differentiated in time of $\{U^\varepsilon, P^\varepsilon\}$ are dependent on a stability restriction $F(\varepsilon) \geq 0$. — In order to complete this step it remains to bound the two last terms in inequality (5.22), which will be done next.

Two times differentiating in time the first equation in (5.5) and testing with $\tau^2(s)E_t(s)$, afterwards integrating over the interval $[0, \varepsilon_2]$, gives

$$\tau^2(\varepsilon_2)\|\nabla E_t(\varepsilon_2)\|^2 + \varepsilon_2 \int_0^{\varepsilon_2} \tau^2(s)\|\Pi_{tt}(s)\|^2 \, ds + \varepsilon_1\tau^2(\varepsilon_2)\|\nabla\Pi_t(\varepsilon_2)\|^2$$

$$\leq C\Big\{ \int_0^{\varepsilon_2} \tau(s)\|\nabla E_t(s)\|^2 \, ds + \varepsilon_1 \int_0^{\varepsilon_2} \tau(s)\|\nabla\Pi_t(s)\|^2 \, ds \Big\}$$

$$+ \varepsilon_1 \int_0^{\varepsilon_2} \tau^2(s)\big(\nabla p_t(s), \nabla\Pi_{tt}(s)\big) \, ds \tag{5.23}$$

$$+ \varepsilon_2 \int_0^{\varepsilon_2} \tau^2(s)\big(p_{tt}(s), \Pi_{tt}(s)\big) \, ds.$$

The third integral on the right side can be controlled, using integration by parts at first. We obtain

$$\varepsilon_1 \int_0^{\varepsilon_2} \tau^2(s)\big(\nabla p_t(s), \nabla\Pi_{tt}(s)\big) \, ds \leq \varepsilon_1\varepsilon_2^2 \big|\big(\nabla p_t(\varepsilon_2), \nabla\Pi_t(\varepsilon_2)\big)\big|$$

$$- 2\varepsilon_1 \int_0^{\varepsilon_2} \tau(s)\big(\nabla p_t(s), \nabla\Pi_t(s)\big) \, ds$$

$$- \varepsilon_1 \int_0^{\varepsilon_2} \tau^2(s)\big(\nabla p_{tt}(s), \nabla\Pi_t(s)\big) \, ds. \tag{5.24}$$

We proceed further, using diverse a-priori bounds for the solution of the unperturbed Stokes problem. Thanks to (5.11) and the length of the interval of integration, we can continue in (5.24),

$$\leq C(1+\alpha)|\varepsilon|^2 + \frac{1}{\alpha}\varepsilon_1\varepsilon_2^2\|\nabla\Pi_t(\varepsilon_2)\|^2, \tag{5.25}$$

with $\alpha > 0$ to be chosen arbitrarily. If we return back to inequality (5.23), using the statement (5.25) with α sufficiently large, we finally arrive at the result

$$\tau^2(\varepsilon_2)\|\nabla E_t(\varepsilon_2)\|^2 + \varepsilon_2 \int_0^{\varepsilon_2} \tau^2(s)\|\Pi_{tt}(s)\|^2 \, ds$$

$$+ \varepsilon_1\tau^2(\varepsilon_2)\|\nabla\Pi_t(\varepsilon_2)\|^2 \leq C|\varepsilon|^2. \tag{5.26}$$

Now, this auxiliary result enables us to give an upper bound for the right hand side of inequality (5.22). Therefore, we end up with

$$\tau(T)\|\nabla E_t(T)\|^2 + \varepsilon_2 \int_{\varepsilon_2}^T \tau(s)\|\Pi_{tt}(s)\|^2 \, ds + \varepsilon_1\tau(T)\|\nabla\Pi_t(T)\|^2$$

$$\leq C\Big\{ (1 + \frac{\varepsilon_1}{\varepsilon_2} + \log\frac{1}{\varepsilon_2})\varepsilon_1 + (1 + \log\frac{1}{\varepsilon_2})\varepsilon_2 \Big\}. \tag{5.27}$$

Owing to this additional stability result we can employ another duality argument similar to (5.15), but now for the error function that is differentiated in time. Then we obtain, thanks to the statements (5.27) and (5.14),

$$
\int_{\varepsilon_2}^T \tau(s)\|E_t(s)\|^2 \, ds \le C\Big\{\varepsilon_2^2 \int_{\varepsilon_2}^T \tau(s)\|\Pi_{tt}(s)\|^2 \, ds
$$
$$
+ \varepsilon_1^2 \int_{\varepsilon_2}^T \tau(s)\|\nabla\Pi_t(s)\|^2 \, ds\Big\}\le C\mathcal{H}^2(\varepsilon). \tag{5.28}
$$

Like in the error statement (5.16), we emphasize the fact that there is *no* stability factor taking account of the parameters $\{\varepsilon_1, \varepsilon_2\}$. This seems to be surprising if we remember the auxiliary estimates for the pressure error. So, owing to the results (5.16) and (5.28), choosing parameters $\{\varepsilon_1, \varepsilon_2\}$ of different magnitudes does only effect the accuracy of scheme (5.4) but not the stability of the convergence behavior with respect to the velocity.

As has already been anticipated above, the remainder of this proof is devoted to the study of the interaction of the quasi-compressibility constraint in (5.3) and the evolutionary character of the first identity in the same system of equations.

2nd step: Discussion of the propagation effects in (5.3)
The proof is complete when having under control the evolutionary effects that stem from the momentum equation. Therefore, we have to investigate the following system,

$$
\begin{aligned}
&e_t^\varepsilon - \Delta e^\varepsilon + \nabla \eta^\varepsilon = E_t, \\
&\operatorname{div} e^\varepsilon - \varepsilon_1 \Delta \eta^\varepsilon + \varepsilon_2 \eta_t^\varepsilon = 0, \\
&e^\varepsilon(0) \equiv E(0), \qquad \eta^\varepsilon(0) \equiv 0, \\
&e^\varepsilon|_{\partial\Omega} = 0, \qquad \partial_n \eta^\varepsilon|_{\partial\Omega} = 0,
\end{aligned} \tag{5.29}
$$

with error functions $e^\varepsilon := u^\varepsilon - U^\varepsilon$ and $\eta^\varepsilon := p^\varepsilon - P^\varepsilon$. — In order to start the error analysis for this system, we test the first identity with e^ε. Owing to (5.28), we are allowed to integrate over the interval $[\mathcal{H}(\varepsilon), T]$,

$$
\|e^\varepsilon(T)\|^2 + \int_{\mathcal{H}(\varepsilon)}^T \|\nabla e^\varepsilon(s)\|^2 \, ds + \varepsilon_1 \int_{\mathcal{H}(\varepsilon)}^T \|\nabla\eta^\varepsilon(s)\|^2 \, ds + \varepsilon_2\|\eta^\varepsilon(T)\|^2
$$
$$
\le C\Big\{\mathcal{H}(\varepsilon)\{\varepsilon_1 + \varepsilon_2\} + |\varepsilon|^2 + \varepsilon_2^2 \log\frac{1}{\varepsilon_2}\Big\}
$$
$$
+ C\alpha \int_{\mathcal{H}(\varepsilon)}^T \tau(s)\|E_t(s)\|^2 \, ds + \frac{1}{\alpha}\int_{\mathcal{H}(\varepsilon)}^T \frac{1}{\tau(s)}\|e^\varepsilon(s)\|^2 \, ds. \tag{5.30}
$$

The first term on the right hand side of the inequality sign is owing to the errors at time $t = \mathcal{H}(\varepsilon)$. Now, we can apply a normalized version of Gronwall's lemma, setting $\alpha = \log(\frac{T}{\mathcal{H}(\varepsilon)})$. This completes the proof. $\square$

Chapter 6

The Projection Scheme of Chorin

6.1 Overview and Results

In some respect, Chorin's scheme is the most basic projection method in use. It has been proposed at the end of the 1960's, in order to decouple the calculation of the actual velocity field and the pressure function at each iteration step. This reduces the computational effort drastically and, additionally, introduces a stabilization of finite element pairs, that will be discussed in the following. The algorithm of Chorin is defined as follows, see [3]:

1. Start with $u^0 \equiv u_0$.

2. For $m \geq 0$, given u^m, first compute $\tilde{u}^{m+1}$ as the solution of the linear equation

$$\frac{1}{k}\{\tilde{u}^{m+1} - u^m\} - \nu\Delta\tilde{u}^{m+1} + (u^m \cdot \nabla)\tilde{u}^{m+1} = f^{m+1},$$

$$\tilde{u}^{m+1}|_{\partial\Omega} = 0. \tag{6.1}$$

3. Given $\tilde{u}^{m+1}$, compute $\{u^{m+1}, p^{m+1}\}$ by solving

$$\frac{1}{k}\{u^{m+1} - \tilde{u}^{m+1}\} + \nabla p^{m+1} = 0,$$

$$\operatorname{div} u^{m+1} = 0, \qquad u^{m+1}|_{\partial\Omega} \cdot n = 0. \tag{6.2}$$

Therefore, for times $t_{m+1} \equiv (m+1)k > 0$, we get approximate solutions $\{u^{m+1}, p^{m+1}\} \approx \{u(t_{m+1}), p(t_{m+1})\}$. For computational purposes, the system (6.2) can be reformulated as a Poisson equation for the pressure by applying the div-operator to the first equation in (6.2) and keeping in mind the boundary conditions for both velocity functions involved. Therefore, instead of solving the coupled system (6.2), this is equivalent to first solving a problem for the pressure,

$$- \Delta p^{m+1} = -\frac{1}{k} \mathrm{div}\, \tilde{u}^{m+1}, \qquad \partial_n p^{m+1}|_{\partial\Omega} = 0, \tag{6.3}$$

and then updating the velocity

$$u^{m+1} = \tilde{u}^{m+1} - k\nabla p^{m+1}. \tag{6.4}$$

We emphasize the *homogeneous* boundary condition of Neumann type for the computed pressure which is generally not satisfied by the exact pressure function. There are lots of publications concerning the question of whether Chorin's method is first order accurate or not. In particular, the question has been posed of whether this scheme leads to approximations p^{m+1} that have a "physical" meaning. For a brief survey on the history of achieving optimal convergence results for this time-discretization scheme, see Shen [33]. The first striking results in this direction were given by Shen, [33], [34], where he shows the following error estimates,

$$\left(k \sum_{m=0}^{M} \| u(t_{m+1}) - u^{m+1} \|^2 \right)^{1/2}$$
$$+ \sqrt{k} \left(k \sum_{m=0}^{M} \| p(t_{m+1}) - p^{m+1} \|^2 \right)^{1/2} \leq Ck, \tag{6.5}$$

provided that an additional regularity property for the solution of the incompressible Navier-Stokes equations holds true, i.e., $p_t \in L^2(0, T; H^1(\Omega)/\mathbb{R})$. Let us recall that this requirement is not satisfied for general practical flow problems, because of the lack of compatibility of the given data at times $t \to 0$. We refer to [16] for a comprehensive discussion of this breakdown of regularity for the solution of (1.1).

The objective of this chapter is to present a proof that shows optimal error estimates for Chorin's method without using the regularity assumptions needed by Shen for $t \to 0$. Moreover, we will improve the averaged error estimates (6.5) to pointwise statements in order to ensure first order accuracy in $l^\infty(0, t_{M+1}; \mathbf{L}^2(\Omega))$ for the velocity. Finally, we want to show that p^{m+1} is indeed an approximation of the pressure $p(t_{m+1})$ in a relevant sense. This statement is verified by error estimates in the norms $l^\infty(0, t_{M+1}; L^2(\Omega)/\mathbb{R})$ and $l^\infty(0, t_{M+1}; H^{-1}(\Omega))$, leading to half order and first order convergence, respectively. The analysis of Chorin's method is supplemented with local error estimates in the norms $l^\infty(0, t_{M+1}; L^2(\Omega'))$ and $l^\infty(0, t_{M+1}; H^1(\Omega'))$ on compact interior subdomains $\Omega' \subset\subset \Omega$. These statements are to demonstrate that the error effects caused by the unphysical homogeneous boundary condition for the pressure are limited to a region close to the boundary.

The subsequent theorem deals with *global* error statements for the solution of the Chorin method. They classify this method as a discretization scheme which is of first order accuracy, pointwise in time.

Theorem 6.1 *Let $\{\tilde{u}^{m+1}, p^{m+1}\}$ be the (semi-)discrete solution of Chorin's method (6.1), (6.2), whereas $\{u(t_{m+1}), p(t_{m+1})\}$ is the solution of the Navier-Stokes equations (1.1) at time $0 < t_{m+1} \leq T \equiv t_{M+1}$. Under the basic assumptions (A1), (A2) and (A3) on the given data, and for sufficiently small time-steps $k \leq k_0(T)$, there exists a constant C, which only depends on the given data of the problem, such that the following estimates hold,*

i) $\max_{0 \leq m \leq M}\left\{\|u(t_{m+1}) - \tilde{u}^{m+1}\| + \tau_{m+1}\|p(t_{m+1}) - p^{m+1}\|_{-1}\right\} \leq Ck,$

ii) $\max_{0 \leq m \leq M}\left\{\|u(t_{m+1}) - \tilde{u}^{m+1}\|_1 + \sqrt{\tau_{m+1}}\|p(t_{m+1}) - p^{m+1}\|\right\} \leq C\sqrt{k}.$

Let us stress, that first order convergence results are only possible for the pressure approximation in a negative norm. This reflects the phenomenon of numerical boundary layers caused by Chorin's method, as it is also observed in numerical experiments; see below. In the next theorem, we present *local* error estimates for the pressure. With these statements, the function p^{m+1} is guaranteed to be a good approximation of the actual pressure $p(t_{m+1})$ in the interior of the domain Ω.

Theorem 6.2 *Under the same conditions as in Theorem 6.1 the following local error estimates are valid on interior subdomains $\Omega' \subset\subset \Omega$,*

i) $\max_{0 \leq m \leq M}\left\{\sqrt{\tau_{m+1}}\|u(t_{m+1}) - \tilde{u}^{m+1}\|_{1;\Omega'}\right\} \leq \tilde{C}k,$

ii) $\max_{0 \leq m \leq M} \sqrt{\tau_{m+1}}\left\{\|p(t_{m+1}) - p^{m+1}\|_{\Omega'}\right.$
$\left. + \sqrt{k}\|p(t_{m+1}) - p^{m+1}\|_{1;\Omega'}\right\} \leq \tilde{C}k.$

In comparison with the constant C used above, the constant $\tilde{C}$ depends additionally on the distance $d(x) = \mathrm{dist}(\partial\Omega, \Omega')$ in a reciprocal way.

Remark 6.1   *1. If we compare the results of the Theorems 6.1 and 6.2, we observe that the local super-convergence for the velocity function in the Dirichlet-norm needs an initial time-adjustment to be overcome. This result reflects the application of the "wrong" initial function $p^0 \equiv 0$ (in the asymptotic limit). For further considerations in that respect, we refer to the subsequent analysis.*

   2. Taking benefit from improved stability properties of particular fluid flows under consideration, the constants $C, \tilde{C}$ can be uniformly bounded in time. A corresponding analysis for the case of the implicit Euler scheme, applied to the incompressible Navier-Stokes equations, is presented in [17].

   3. The error estimates in Theorems 6.1 and 6.2 are given for the auxiliary function $\tilde{u}^{m+1}$ that is computed in the first part of each iteration. The same estimates can be derived for the projected velocity function u^{m+1} by using certain standard stability properties of the operator $P_{\mathbf{J}_0}$ in the spaces $\mathbf{L}^2$ and $\mathbf{H}_0^1$.

   4. The interior error estimates presented in Theorem 6.2 do not provide any measure for the size of the induced boundary layer. It is conjectured by R. Rannacher in [30] to be of magnitude $O(\sqrt{\nu k})$. This has been verified by W. E & J.G. Liu in a joint paper [7] for a restricted model configuration. Recently, this conjecture was validated for more general flow problems by C. Schwab in [32], where he derived error estimates for the pressure in exponentially weighted Lebesgue-norms. Nevertheless, additional regularity assumptions for the Navier-Stokes solution are necessary in his analysis. Moreover, pointwise statements for the error decay close to the boundary are not yet available under the general assumptions (A1), (A2) and (A3).

Now, we give a brief outline of the proof of Theorem 6.1: In order to get an equality for the auxiliary function $\tilde{u}^{m+1} \in \mathbf{H}_0^1$ which is not divergence-free we shift the index of the iteration in (6.2) by -1 and add the resulting identity to equation (6.1). By using (6.3), we are led to the following equations, with

$\varepsilon = k,$

$$d_t \tilde{u}^{m+1} - \nu \Delta \tilde{u}^{m+1} + (u^m \cdot \nabla) \tilde{u}^{m+1} + \nabla p^m = f^{m+1},$$
$$\operatorname{div} \tilde{u}^{m+1} - \varepsilon \Delta p^{m+1} = 0, \qquad \partial_n p^{m+1}|_{\partial\Omega} = 0. \tag{6.6}$$

Note that the leading term in the convective part is divergence-free. This will be of fundamental relevance for the stability of the system, as well as for its well-posedness. In this context, we refer to the remark below and the first section of chapter 3 for a related discussion. Thanks to (6.6), we get an interpretation of the Chorin scheme as a semi-explicit pressure stabilization method of the Navier-Stokes equations. This approach has first been proposed by R. Rannacher in [30]. Due to the homogeneous boundary conditions for the pressure, this system is a singular perturbation of the incompressible system (1.1). As a result, the main part of the error analysis of Chorin's scheme consists of estimating this effect in Chorin's algorithm. Towards that end, we have to start with the investigation of the following system,

$$d_t u_\varepsilon^{m+1} - \nu \Delta u_\varepsilon^{m+1} + (P_{J_0} u_\varepsilon^m \cdot \nabla) u_\varepsilon^{m+1} + \nabla p_\varepsilon^{m+1} = f^{m+1},$$
$$\operatorname{div} u_\varepsilon^{m+1} - \varepsilon \Delta p_\varepsilon^{m+1} = 0, \qquad \partial_n p_\varepsilon^{m+1}|_{\partial\Omega} = 0, \tag{6.7}$$

together with $u_\varepsilon^0 = u_0$.

Remark 6.2 *The introduction of the projector P_{J_0} is necessary in order to guarantee the well-posedness of system (6.7). This prevents the appearance of a nonlinear mechanism, that will otherwise lead to an unstable system. Analytically speaking, the trilinear form that is related to the nonlinearity in (6.7) has the following form,*

$$\hat{b}(\phi, \cdot, \cdot) = b(P_{J_0}\phi, \cdot, \cdot), \qquad \phi \in H_0^1. \tag{6.8}$$

This nonlinearity is again skew-symmetric with respect to the last two arguments, i.e.,

$$\hat{b}(\phi, u, v) = \hat{b}(\phi, v, u), \qquad \forall u, v \in H_0^1. \tag{6.9}$$

We recall that Temam in [41] introduced another modification of the "original" trilinear form,

$$\tilde{b}(\phi, u, v) = \frac{1}{2}\{b(\phi, u, v) - b(\phi, v, u)\}, \qquad \forall \phi, u, v, \in H_0^1, \tag{6.10}$$

in order to assure that a corresponding skew-symmetry rule holds for this trilinear form on the triple $(\mathbf{H}_0^1)^3$. *Of course, this stabilization technique increases the numerical complexity of the scheme and is not necessary in the case of projection schemes, as the leading solenoidal function can control the influence of the nonlinearity. We further elaborate on this problem subsequently.*

The analysis for (6.7) is split into three steps to handle consistency and perturbation error independently of the error effects that are caused by the nonlinear convection term of the Navier-Stokes operator. Therefore, we will start with an analysis of the perturbation contributions that stem from the relaxed incompressibility constraint, restricted to the Stokes case, i.e., we omit the convection term. In this modified form, the scheme reads as follows,

$$
\begin{aligned}
d_t \bar{u}_\varepsilon^{m+1} - \nu \Delta \bar{u}_\varepsilon^{m+1} + \nabla \bar{p}_\varepsilon^{m+1} &= f^{m+1}, \\
\mathrm{div} \bar{u}_\varepsilon^{m+1} - \varepsilon \Delta \bar{p}_\varepsilon^{m+1} &= 0, \qquad \partial_n \bar{p}_\varepsilon^{m+1}|_{\partial\Omega} = 0,
\end{aligned}
\tag{6.11}
$$

with the solution $\{\bar{u}_\varepsilon^{m+1}, \bar{p}_\varepsilon^{m+1}\}$. Error estimates as well as a-priori estimates for this tuple will be presented in the subsequent Section 6.2. Secondly, the objective of Section 6.3 is to transfer the statements of 6.2 for the Stokes equations (6.11) to corresponding ones for the Navier-Stokes system (6.7). Section 6.4 is then concerned with the presentation of error estimates for the solutions of system (6.6) as the equivalent reformulation of Chorin's scheme. We finish this chapter with computational studies in Section 6.5.

6.2   Analysis for a Pressure Stabilized Stokes Problem

6.2.1   Part 1. Consistency Error

We start our analysis of the system (6.11) by first quantifying the influence that is induced by the time-discretization via the implicit Euler method. Ignoring the pressure stabilization in (6.6) for a moment we arrive at the following linear equations,

$$
\begin{aligned}
d_t \bar{u}^{m+1} - \nu \Delta \bar{u}^{m+1} + \nabla \bar{p}^{m+1} &= f^{m+1}, \\
\mathrm{div} \bar{u}^{m+1} &= 0,
\end{aligned}
\tag{6.12}
$$

together with the given initial function $u^0 = u_0$. — In order to make this discrete system well-defined for all tuples $\{\bar{u}^{m+1}, \bar{p}^{m+1}\}_{m=-1}^{M}$, let us define $\bar{p}^0$ in the canonical way

$$(\nabla \bar{p}^0, \nabla \phi) = \nu(\Delta u_0, \nabla \phi) + (f^0, \nabla \phi), \qquad \forall \phi \in H^1/\mathbb{R}. \tag{6.13}$$

which determines a unique function $\bar{p}^0 \in H^1/\mathbb{R}$. The discretization scheme (6.12) has already been analyzed in the full Navier-Stokes context, see e.g. [16], and the following error estimates are valid,

$$\max_{0 \leq m \leq M} \left\{ \|u(t_{m+1}) - \bar{u}^{m+1}\| + \sqrt{\tau_{m+1}} \|u(t_{m+1}) - \bar{u}^{m+1}\|_1 \right\} \leq Ck, \tag{6.14}$$

with $\{u, p\}$ the solution (2.3), and for the pressure function,

$$\max_{0 \leq m \leq M} \sqrt{\tau_{m+1}} \left\{ \sqrt{\tau_{m+1}} \|p(t_{m+1}) - \bar{p}^{m+1}\| + \|p(t_{m+1}) - \bar{p}^{m+1}\|_{-1} \right\} \leq Ck. \tag{6.15}$$

Further, the following a-priori estimates regarding the solution of (6.12) can be shown, cf. [16], for $i \in \{0, 1, 2\}$, $r \in \{1, 2, 3\}$ and $i + 2r \leq 7$,

$$\max_{r-1 \leq m \leq M} \left\{ \tau_m^{r-1+i/2} \|d_t^r \bar{u}^{m+1}\|_i \right\} + k \sum_{m=r}^{M} \tau_m^{2r-1+i} \|d_t^{r+1} \bar{u}^{m+1}\|_i^2 \leq C, \tag{6.16}$$

for $r - 1 + \frac{i}{2} \geq 0$, and

$$\max_{r \leq m \leq M} \left\{ \tau_m^r \|\nabla d_t^r \bar{p}^{m+1}\| \right\} + k \sum_{m=r}^{M} \tau_{m+1}^{2r+1} \|\nabla d_t^{r+1} \bar{p}^{m+1}\|^2 \leq C. \tag{6.17}$$

Although not necessary for the following, let us mention here that the choice $i = -1$ in formula (6.16) is also correct in the sense that this negative norm is induced by $\mathbf{J}_1$-functions.

Because of the splitting of the error $\{u(t_{m+1}) - u_\varepsilon^{m+1}\}$ (and correspondingly for the pressure) in the parts $\{u(t_{m+1}) - u^{m+1}\}$ and $\{u^{m+1} - u_\varepsilon^{m+1}\}$, the first error contribution is now controlled by the estimates (6.14) and (6.15). The goal of the next subsection is the presentation of optimal estimates for the errors $\{u^{m+1} - u_\varepsilon^{m+1}\}$ and $\{p^{m+1} - p_\varepsilon^{m+1}\}$.

6.2.2   Part 2. Perturbation Error

The difficulty in analyzing system (6.11) consists in the coupling of velocity and pressure. The equations are to be considered on the pair of spaces $l^\infty(0, t_{M+1}; \mathbf{L}^2) \cap l^2(0, t_{M+1}; \mathbf{H}_0^1) \times l^2(0, t_{M+1}; H^1/\mathbb{R})$ and no Helmholtz-like splitting is available to treat the involved functions independently. Thus, in order to get striking a-priori estimates for the solutions of this system we are forced to employ corresponding results for the incompressible problems. We collect the first results regarding a-priori bounds in the following lemma.

Lemma 6.1 *Let the tuple $\{\overline{u}_\varepsilon^{m+1}, \overline{p}_\varepsilon^{m+1}\}$ be the solution of (6.11). Then, there exists a constant C that is only dependent on the given data of the problem, and the following a-priori bounds are valid:*

i) $\quad \max_{0 \leq m \leq M} \|d_t \overline{u}_\varepsilon^{m+1}\| + \varepsilon k \sum_{m=0}^{M} \|\nabla d_t \overline{p}_\varepsilon^{m+1}\|^2 \leq C,$

ii) $\quad k \sum_{m=0}^{M} \left\{ \|\Delta \overline{u}_\varepsilon^{m+1}\|^2 + \|\nabla \overline{p}_\varepsilon^{m+1}\|^2 \right\} \leq C t_{M+1}.$

Remark 6.3 *Note, that the last term on the left hand side of the inequality sign in i) carries a weight ε. Parameter-independent bounds for this term — and even more pointwise in time statements — can only be proven after having reached optimal error bounds for the functions $\overline{u}^{m+1} - \overline{u}_\varepsilon^{m+1}$ (and $\overline{p}^{m+1} - \overline{p}_\varepsilon^{m+1}$), see Lemma 6.3 below. The same facts are valid for the averaged quantities in ii). Again, pointwise in time a-priori estimates will be presented in Lemma 6.3. — Nevertheless, it is important to note that just these results suffice for the subsequent error analysis.*

Proof:
i): This result follows easily from (6.11) by employing the operator d_t onto the first equation and afterwards testing with $d_t \overline{u}_\varepsilon^{m+1}$. Using the second identity in (6.11) and an elementary study of the initial behavior of the involved functions complete the proof.

ii): By introduction of the error functions $\overline{e}^{m+1} := \overline{u}^{m+1} - \overline{u}_\varepsilon^{m+1}$ resp. $\overline{\eta}^{m+1} := \overline{p}^{m+1} - \overline{p}_\varepsilon^{m+1}$, we are led to the following system,

$$d_t \overline{e}^{m+1} - \nu \Delta \overline{e}^{m+1} + \nabla \overline{\eta}^{m+1} = 0,$$
$$\text{div} \overline{e}^{m+1} - \varepsilon \Delta \overline{\eta}^{m+1} = -\varepsilon \Delta \overline{p}^{m+1}, \quad \partial_n \overline{\eta}^{m+1}|_{\partial\Omega} = \partial_n \overline{p}^{m+1}|_{\partial\Omega}. \tag{6.18}$$

We test the first identity in (6.18) with $\bar{e}^{m+1}$, and the second one with $\bar{\eta}^{m+1}$. After integration by parts, adding of both equalities and summing over all time-steps, we arrive at

$$\|\bar{e}^{M+1}\|^2 + 2\nu \sum_{m=0}^{M} \|\nabla\bar{e}^{m+1}\|^2 + \varepsilon k \sum_{m=0}^{M} \|\nabla\bar{\eta}^{m+1}\|^2 \leq C\varepsilon. \qquad (6.19)$$

Here, we used the fact that the initial velocity error vanishes. Inequality (6.17) now gives a bound for the right hand side of the last inequality. This proves the a-priori bound for the pressure in ii). — The second one for the velocity in the equivalent $l^2(0, t_{M+1}; \mathbf{H}^2)$-norm can now easily be obtained by testing the first identity in (6.11) with $-\Delta\bar{u}_\varepsilon^{m+1}$ and using the last result and part i). $\qquad\square$

The main ingredient of this section is formulated in the next lemma, which is of independent relevance.

Lemma 6.2 *Let $\{\bar{u}(t_{m+1}), \bar{p}(t_{m+1})\}$ be the solution of the nonstationary incompressible Stokes equations (2.3). Further, the tuple $\{\bar{u}_\varepsilon^{m+1}, \bar{p}_\varepsilon^{m+1}\}$ is the solution of (6.11). Then, the following error estimates are satisfied, with a constant C that depends only on the given data of the problem,*

$i)$ $\max_{0 \leq m \leq M} \{\|\bar{u}(t_{m+1}) - \bar{u}_\varepsilon^{m+1}\|$
$\qquad + \tau_{m+1}\|\bar{p}(t_{m+1}) - \bar{p}_\varepsilon^{m+1}\|_{-1}\} \leq C\{\varepsilon + k\},$

$ii)$ $\max_{0 \leq m \leq M} \{\|\bar{u}(t_{m+1}) - \bar{u}_\varepsilon^{m+1}\|_1$
$\qquad + \sqrt{\tau_{m+1}}\|\bar{p}(t_{m+1}) - \bar{p}_\varepsilon^{m+1}\|\} \leq C\{\sqrt{\varepsilon}(1 + \sqrt{\tfrac{\varepsilon}{k}}) + \sqrt{k}\}.$

Further, the following estimate is valid,

$iii)$ $\max_{0 \leq m \leq M} \sqrt{\tau_{m+1}} \{\|\bar{u}(t_{m+1}) - \bar{u}_\varepsilon^{m+1}\|_1$
$\qquad + \sqrt{\tau_{m+1}}\|\bar{p}(t_{m+1}) - \bar{p}_\varepsilon^{m+1}\|\} \leq C\{\sqrt{\varepsilon} + \sqrt{k}\}.$

Remark 6.4 *Part ii) is already sufficient for our investigations in the context of Chorin's method — where we put $\varepsilon = k$. On the other hand, the bound ii) leads to severe restrictions in the choice of the parameters if we* *are interested in related statements for more general pressure stabilization formulations, setting $\varepsilon = \mathcal{O}(h^2)$ for instance. Here, we take h as a parameterization for the underlying spatial mesh that is generated in a corresponding finite element context. This justifies the significance of part iii) in the above lemma. A direct comparison of ii) and iii) shows that the choice $k \ll \varepsilon$ only dominates the approximation behavior of (6.11) along the initial phase whereas intrinsic damping properties of this dynamical system lead to good guesses $\{\overline{u}_\varepsilon^{m+1}, \overline{p}_\varepsilon^{m+1}\}$ of the actual solution, at times $t_{m+1} \gg 0$. The proof of assertion iii) follows from results that will be achieved in the framework of verifying statement ii).*

Proof:
Due to the results (6.14) and (6.15), we can confine ourselves to the estimation of the gap between the solutions of (6.12) and (6.11). Thus, it remains to show the validity of the following results,

$$i')\quad \max_{0\le m\le M}\Big\{\|\overline{u}^{m+1} - \overline{u}_\varepsilon^{m+1}\|$$
$$+\tau_{m+1}\|\overline{p}^{m+1} - \overline{p}_\varepsilon^{m+1}\|_{-1}\Big\} \le C\{\varepsilon + k\},$$

$$ii')\quad \max_{0\le m\le M}\Big\{\|u^{m+1} - \overline{u}_\varepsilon^{m+1}\|_1$$
$$+\sqrt{\tau_{m+1}}\|\overline{p}^{m+1} - \overline{p}_\varepsilon^{m+1}\|\Big\} \le C\{\sqrt{\varepsilon}(1 + \sqrt{\tfrac{\varepsilon}{k}}) + \sqrt{k}\}.$$

$$iii')\quad \max_{0\le m\le M}\sqrt{\tau_{m+1}}\Big\{\|\overline{u}^{m+1} - \overline{u}_\varepsilon^{m+1}\|_1$$
$$+\sqrt{\tau_{m+1}}\|\overline{p}^{m+1} - \overline{p}_\varepsilon^{m+1}\|\Big\} \le C\{\sqrt{\varepsilon} + \sqrt{k}\}.$$

In order to verify these estimates, we distinguish between the diverse acting error sources in system (6.11). Firstly, there is the perturbation of the incompressibility constraint which will be investigated in an auxiliary problem introduced in the following. Secondly, the evolution process of this perturbation that is given by the first identity in (6.11) has to be analyzed. Therefore, the succeeding proof decouples in two steps.

1st step: Let us start with the introduction of an auxiliary problem:

$$-\nu\Delta U_\varepsilon^{m+1} + \nabla P_\varepsilon^{m+1} = f^{m+1} - d_t\overline{u}^{m+1},$$
$$\mathrm{div}U_\varepsilon^{m+1} - \varepsilon\Delta P_\varepsilon^{m+1} = 0,\qquad \partial_n P_\varepsilon^{m+1}|_{\partial\Omega} = 0. \tag{6.20}$$

If we subtract the corresponding equations in (6.12) and (6.20), using the abbreviations $E^{m+1} := \bar{u}^{m+1} - U_\varepsilon^{m+1}$ and $\Pi^{m+1} := \bar{p}^{m+1} - P_\varepsilon^{m+1}$, we get

$$- \nu \Delta E^{m+1} + \nabla \Pi^{m+1} = 0,$$
$$\operatorname{div} E^{m+1} - \varepsilon \Delta \Pi^{m+1} = -\varepsilon \Delta \bar{p}^{m+1}, \quad \partial_n \Pi^{m+1}|_{\partial\Omega} = \partial_n \bar{p}^{m+1}|_{\partial\Omega}. \qquad (6.21)$$

The following error estimates for this stationary problem are proved e.g. in [30],

$$\|E^{m+1}\| + \|\Pi^{m+1}\|_{-1} \leq C\varepsilon \|\nabla p^{m+1}\|, \qquad (6.22)$$

$$\|E^{m+1}\|_1 + \|\Pi^{m+1}\| \leq C\sqrt{\varepsilon} \|\nabla p^{m+1}\|. \qquad (6.23)$$

Note, that system (6.21) is well-defined on the whole range $-1 \leq m \leq M$. Together with the results (6.22) and (6.23), this enables a compatible extension of system (6.20), with functions $U_\varepsilon^0 := \bar{u}_\varepsilon^0 - E^0$ and $P_\varepsilon^0 := \bar{p}_\varepsilon^0 - \Pi^0$.

2nd step: In order to complete the error analysis, we have to bound the second error part in the decompositions $\bar{e}^{m+1} := E^{m+1} + \bar{e}_\varepsilon^{m+1}$ resp. $\bar{\eta}^{m+1} := \Pi^{m+1} + \bar{\eta}_\varepsilon^{m+1}$. It is determined as the solution of the system obtained by subtracting (6.20) from (6.11),

$$d_t \bar{e}_\varepsilon^{m+1} - \nu \Delta \bar{e}_\varepsilon^{m+1} + \nabla \bar{\eta}_\varepsilon^{m+1} = d_t E^{m+1},$$
$$\operatorname{div} \bar{e}_\varepsilon^{m+1} - \varepsilon \Delta \bar{\eta}_\varepsilon^{m+1} = 0, \qquad \partial_n \bar{\eta}_\varepsilon^{m+1}|_{\partial\Omega} = 0, \qquad (6.24)$$

with initial error $\bar{e}_\varepsilon^0 = E^0$, and therefore $\|\bar{e}_\varepsilon^0\| \leq C\varepsilon$. — Notice, that the singular perturbation character of system (6.11) that has been analyzed in the first step of the proof is now absent, and the object is now to qualify its influence onto the accuracy of the solution of (6.7) through the evolution process. — Therefore, we test the first identity in (6.24) with $\bar{e}_\varepsilon^{m+1}$ and arrive at the following inequality, using the second equation in (6.24),

$$\frac{1}{2} d_t \|\bar{e}_\varepsilon^{m+1}\|^2 + \nu \|\nabla \bar{e}_\varepsilon^{m+1}\|^2 + \varepsilon \|\nabla \bar{\eta}_\varepsilon^{m+1}\|^2 = (d_t E^{m+1}, \bar{e}_\varepsilon^{m+1}). \qquad (6.25)$$

Now, in order to get an optimal error estimation for the right hand side by making use of the "pointwise" estimate (6.22) that leads us to $\|d_t E^{m+1}\| \leq$

$C\varepsilon\|\nabla d_t\overline{p}^{m+1}\|$, we have to multiply (6.25) with a time weight τ_m to benefit from (6.17). This gives the estimate

$$\frac{1}{2}\tau_{M+1}\|\overline{e}_\varepsilon^{m+1}\|^2 + \nu k \sum_{m=0}^{M}\tau_{m+1}\|\nabla\overline{e}_\varepsilon^{m+1}\|^2 + k\sum_{m=0}^{M}\tau_{m+1}\|\nabla\overline{\eta}_\varepsilon^{m+1}\|^2$$

$$\leq C\Big\{k\sum_{m=0}^{M}\tau_{m+1}^2\|d_tE^{m+1}\|^2 + k\sum_{m=0}^{M}\|\overline{e}_\varepsilon^{m+1}\|^2\Big\} \tag{6.26}$$

$$\leq C\Big\{\varepsilon^2\tau_{M+1} + k\sum_{m=0}^{M}\|\overline{e}_\varepsilon^{m+1}\|^2\Big\}.$$

Here, we employed the stronger $\mathbf{L}^2$-norm for the term on the right hand side instead of a negative one, without loosing essential information for the further error analysis. This is connected with the singular error portion E^{m+1}, allowing no improvement with respect to the order of convergence in negative norms. - Once more, we mention the fact that we can sum over the whole range $0 \leq m \leq M$, because of the compatible extensions to $\overline{p}^0$ and $\overline{p}_\varepsilon^0$. — The remaining task is to estimate the last term on the right hand side of (6.26). Therefore, we introduce the following (evolutionary) backward auxiliary problem: Given a time $t_{M+1} > 0$, find the solution $\{\overline{w}^m, \overline{q}^m\} \in \mathbf{H}_0^1 \cap \mathbf{H}^2 \times H^1/\mathbb{R}$ of

$$d_t\overline{w}^{m+1} + \nu\Delta\overline{w}^m - \nabla\overline{q}^m = \overline{e}^{m+1},$$

$$\text{div}\,\overline{w}^m = 0, \qquad \overline{w}^{M+1} = 0. \tag{6.27}$$

Notice, that we have used the error function $\overline{e}^{m+1}$ instead of e_ε^{m+1} in the formulation of (6.27), but an estimation for this error function in $l^2(0, t_{M+1}; \mathbf{L}^2)$ is sufficient — thanks to (6.22). — Before we continue the error analysis, let us recapitulate an a-priori estimate for the nonstationary Stokes equations given in (6.12): By testing the first identity with $-\Delta\overline{u}^{m+1}$, we get

$$\|\nabla\overline{u}^{M+1}\|^2 + \nu k\sum_{m=0}^{M}\|\Delta\overline{u}^{m+1}\|^2 \leq \|\nabla\overline{u}^0\|^2 + \frac{1}{\nu}k\sum_{m=0}^{M}\|f^{m+1}\|^2,$$

and

$$k\sum_{m=0}^{M}\|\nabla\overline{p}^{m+1}\|^2 \leq C\Big\{\nu\|\nabla\overline{u}^0\|^2 + k\sum_{m=0}^{M}\|f^{m+1}\|^2\Big\}. \tag{6.28}$$

Let us come back to problem (6.27): We test the first equation with $\bar{e}^{m+1}$ and get

$$\|\bar{e}^{m+1}\|^2 = (d_t\bar{w}^{m+1}, \bar{e}^{m+1}) - \nu(\nabla\bar{w}^m, \nabla\bar{e}^{m+1}) - (\nabla\bar{q}^m, \bar{e}^{m+1}).$$
(6.29)

Further, testing the first equation in (6.18) with $\bar{w}^m$ gives:

$$(d_t\bar{e}^{m+1}, \bar{w}^m) + \nu(\nabla\bar{w}^m, \nabla\bar{e}^{m+1}) = 0.$$
(6.30)

Now, by adding the last two equalities and noticing the following identity that can be verified by a simple calculation,

$$(d_t\bar{w}^{m+1}, \bar{e}^{m+1}) + (\bar{w}^m, d_t\bar{e}^{m+1}) = d_t(\bar{w}^{m+1}, \bar{e}^{m+1}),$$

we find the relation

$$\|\bar{e}^{m+1}\|^2 = d_t(\bar{w}^{m+1}, \bar{e}^{m+1}) - (\nabla\bar{q}^m, \bar{e}^{m+1}).$$
(6.31)

Further, we can use the second equality in (6.18) — to be tested with $\bar{q}^m$ — to arrive at the following estimate, after summing over all time-steps and using Cauchy's inequality:

$$k\sum_{m=0}^{M}\|\bar{e}^{m+1}\|^2 \leq C\varepsilon^2\delta k\sum_{m=0}^{M}\|\nabla\bar{p}_\varepsilon^{m+1}\|^2 + \frac{1}{4\delta}k\sum_{m=0}^{M}\|\nabla\bar{q}^m\|^2,$$

with $\delta > 0$. In order to control the last term of this inequality we employ the a-priori estimate (6.28). If we choose δ sufficiently large, it can be absorbed by the term on the left hand side, thanks to stability result (6.28). By that, we arrive at the estimation

$$k\sum_{m=0}^{M}\|\bar{e}^{m+1}\|^2 \leq Ct_{M+1}\varepsilon^2.$$
(6.32)

If we insert this result in (6.26), combining the resulting estimation with (6.22), the first part of statement $i')$ in the above Lemma 6.2 is proved.

In order to verify the first part of $ii')$, we test the first equation in (6.24) with $d_t\bar{e}^{m+1}$ and finally sum over all time-steps,

$$k\sum_{m=0}^{M}\|d_t\bar{e}_\varepsilon^{m+1}\|^2 + \nu\|\nabla\bar{e}_\varepsilon^{M+1}\|^2 + 2\varepsilon\|\nabla\bar{\eta}_\varepsilon^{M+1}\|^2$$

$$\leq C\Big\{k\sum_{m=0}^{M}\|d_t E^{m+1}\|^2 + \varepsilon\Big\}.$$
(6.33)

The first term on the right hand side can be estimated by $\frac{C}{k}\varepsilon^2$, whereas the second one represents the magnitude of the initial errors. Recalling the inequalities (6.22), (6.23), the proof for the velocity components is complete.

We proceed with the derivation of error estimates for the pressure $\bar{p}_\varepsilon^{m+1}$: By using a stability result of the div-operator we arrive at

$$\|\bar{\eta}^{m+1}\| \leq C\{\nu\|\nabla\bar{e}^{m+1}\| + \|d_t\bar{e}^{m+1}\|_{-1}\}, \tag{6.34}$$

where the negative norm $\|\cdot\|_{-1}$ is generated by functions in $\mathbf{H}_0^1$. Whereas the first term on the right hand side of (6.34) can be bounded by $C\{\frac{\varepsilon}{\sqrt{k}} + \varepsilon\}$, the remaining second term has now to be controlled. Again, we emphasize the fact that we cannot take benefit from negative norms for the further analysis, because of the quasi-compressibility constraint that leads us to the coupled investigation on the pair of spaces $H_0^1 \times L^2/\mathbb{R}$. Moreover, this assertion can also be motivated by a related perception that owing to the second equation in (6.11) the pressure is forced to be the zero function at time $t = 0$, owing to the solenoidal initial function. This causes an initial phase to be overcome in order to get good approximations for the pressure. — After this insertion, let us proceed with the estimation of the norm $\|d_t\bar{e}^{m+1}\|$: After employing the discrete operator d_t onto the first equation of (6.18), testing with $\tau_m d_t\bar{e}^{m+1}$ and finally summing over all time-steps, we easily obtain

$$\tau_{M+1}\|d_t\bar{e}^{M+1}\|^2 + \nu k \sum_{m=1}^{M} \tau_{m+1}\|\nabla d_t\bar{e}^{m+1}\|^2 + \varepsilon k \sum_{m=1}^{M} \tau_{m+1}\|\nabla d_t\bar{\eta}^{m+1}\|^2$$
$$\leq C\frac{\varepsilon^2}{k} + C\Big\{\varepsilon k \sum_{m=1}^{M} \|\nabla d_t\bar{p}^{m+1}\|^2 + k \sum_{m=1}^{M} \|d_t\bar{e}^m\|^2\Big\}. \tag{6.35}$$

We can use (6.17) in order to bound the second term on the right hand side by $C\varepsilon$. Therefore, it remains to estimate the last term in (6.35). To do this, we can use the decomposition $\bar{e}^{m+1} := \bar{e}_\varepsilon^{m+1} + E^{m+1}$ and result (6.33). Thus, we finally arrive at

$$\sqrt{\tau_{M+1}}\|\bar{\eta}^{M+1}\| \leq C\Big\{\frac{\varepsilon}{\sqrt{k}} + \sqrt{\varepsilon}\Big\}. \tag{6.36}$$

To complete the error analysis for the solution of (6.11), it remains to derive an error estimate for the pressure in a negative norm. An error control for the pressure in a negative norm is efficient for problems like (6.11) that are singularly perturbed. — We proceed as above and use the splitting $\overline{\eta}^{m+1} := \overline{\eta}_\varepsilon^{m+1} + \Pi^{m+1}$. Owing to (6.22), it suffices to estimate the part $\overline{\eta}_\varepsilon^{m+1}$, which can be controlled as follows,

$$\|\overline{\eta}_\varepsilon^{m+1}\|_{-1} \leq C\{\|d_t\overline{e}_\varepsilon^{m+1}\|_{-2} + \nu\|\overline{e}_\varepsilon^{m+1}\| + \|d_t E^{m+1}\|_{-2}\}, \tag{6.37}$$

with $\mathbf{H}^{-2} \equiv \left(\mathbf{H}_0^1 \cap \mathbf{H}^2\right)'$. Again, it is not possible to take advantage from the negative norm of the velocity error, differentiated in time. Therefore, we start with employing the difference operator d_t onto the first identity in (6.24) and testing with $\tau_m^3 d_t\overline{e}_\varepsilon^{m+1}$. Summing over all time-steps and dividing by τ_{M+1}, this leads to

$$\tau_{M+1}^2\|d_t\overline{e}_\varepsilon^{M+1}\|^2 + \frac{1}{\tau_{M+1}} k \sum_{m=1}^M \tau_{m+1}^3\{\nu\|\nabla d_t\overline{e}_\varepsilon^{m+1}\|^2 + \varepsilon\|\nabla d_t\overline{\eta}_\varepsilon^{m+1}\|^2\}$$

$$\leq \frac{C}{\tau_{M+1}}\{\varepsilon^2 k + k \sum_{m=1}^M \tau_{m+1}^4\|d_t^2 E^{m+1}\|^2$$

$$+ k \sum_{m+1}^M \tau_{m+1}^2\|d_t\overline{e}_\varepsilon^{m+1}\|^2\}. \tag{6.38}$$

The first term in the brackets stems from the initial velocity errors. If we use error estimate (6.22) in combination with statement (6.17), we can easily bound the second term in the brackets by $C\varepsilon^2$. Thus, it remains to control the last sum on the right hand side of (6.38). To do so, we point at (6.33) and use an analog procedure to obtain the following time weighted relation,

$$\frac{1}{\tau_{M+1}}\sum_{m=1}^M \tau_{m+1}^2\|d_t\overline{e}_\varepsilon^{M+1}\|^2 + \nu\tau_{M+1}\|\nabla\overline{e}_\varepsilon^{M+1}\|^2 + \varepsilon\tau_{M+1}\|\nabla\overline{\eta}_\varepsilon^{M+1}\|^2$$

$$\leq \frac{C}{\tau_{M+1}}\{k \sum_{m=0}^M \tau_{m+1}^2\|d_t E^{m+1}\|^2 + k \sum_{m=0}^M \tau_{m+1}\|\nabla\overline{e}_\varepsilon^{m+1}\|^2$$

$$+ \varepsilon k \sum_{m=0}^M \tau_{m+1}\|\nabla\overline{\eta}_\varepsilon^{m+1}\|^2\}. \tag{6.39}$$

Owing to (6.26), the last two sums can be estimated by $C\varepsilon^2$. We get the same bound for the remaining term on the right hand side by applying the inequality (6.22), together with (6.17). It is now easy to see that the results (6.38) and (6.39) give an optimal bound for the right hand side of (6.37), which proves the second assertion in i').

In order to prove the estimates iii'), we employ (6.14), (6.23) and (6.39) that provide the verification of the velocity error, measured in the Dirichlet norm. The demonstration of the upper bound for the pressure error employs the inequality

$$\tau_{m+1}\|\bar{\eta}_\varepsilon^{m+1}\| \le C\tau_{m+1}\big\{\|d_t\bar{e}_\varepsilon^{m+1}\|_{-1} + \nu\|\nabla\bar{e}_\varepsilon^{m+1}\| + \|d_t E^{m+1}\|_{-1}\big\}. \tag{6.40}$$

To control the first two terms on the right hand side, we benefit from (6.38) and (6.39). Thanks to an estimate corresponding to (6.22) and the a-priori bound (6.17) this completes the verification of statement iii'). $\qquad\square$

Based on these error bounds for the solution of (6.11) it is now possible to derive sharpened a-priori estimates, compared to the ones presented in Lemma 6.1. Firstly, if we test the second equation in (6.18) with $\bar{\eta}^{m+1}$, we get the following estimate,

$$\|\nabla\bar{\eta}^{m+1}\| \le \|\nabla\bar{p}^{m+1}\| + \frac{1}{\varepsilon}\|e^{m+1}\| \le C,$$

which gives an a-priori bound for the pressure gradient $\nabla\bar{p}_\varepsilon^{m+1}$ which is independent of ε and k. This result, together with the one in i) of Lemma 6.1 complete the proof of the statements, that are collected in the subsequent lemma. The verification of the second result in the next lemma uses an analogous argument.

Lemma 6.3 *Additionally to the results presented in Lemma 6.1, the following a-priori estimates are valid for the solution of problem (6.11), with a constant C that only depends on the given data of the problem,*

i) $\max_{0\le m\le M}\big\{\|\bar{u}_\varepsilon^{m+1}\|_2 + \|\nabla\bar{p}_\varepsilon^{m+1}\|\big\} \le C,$

ii) $\max_{0\le m\le M}\big\{\tau_{m+1}\|\nabla d_t\bar{p}_\varepsilon^{m+1}\|\big\} \le C.$

6.3 Analysis of the Pressure Stabilized Navier-Stokes Problem

The objective of this section is to establish error estimates for the nonlinear system (6.7), corresponding to the ones that have been formulated for problem (6.11) in Lemma 6.2. — We start with the time-discretization

$$d_t u^{m+1} - \nu \Delta u^{m+1} + (u^m \cdot \nabla) u^{m+1} + \nabla p^{m+1} = f^{m+1},$$
$$\mathrm{div}\, u^{m+1} = 0, \qquad u^{m+1}|_{\partial \Omega} = 0, \tag{6.41}$$

supplemented with the initial function $u^0 = u_0$. We mention an analogous determination of $p^0 \in H^1/\mathbb{R}$ as in the linear case (cf. Subsection 6.2.1) by solving

$$(\nabla p^0, \nabla \phi) = \nu(\Delta u_0, \nabla \phi) - ((u_0 \cdot \nabla) u_0, \nabla \phi) + (f^0, \nabla \phi), \quad \forall \phi \in H^1/\mathbb{R}.$$

Corresponding to (6.14), (6.15), the following error estimates hold true, but now with a constant C that is in general exponentially growing in time, cf. [16],

$$\max_{0 \le m \le M} \left\{ \| u(t_{m+1}) - u^{m+1} \| + \sqrt{\tau_{m+1}} \| u(t_{m+1}) - u^{m+1} \|_1 \right\} \le Ck, \tag{6.42}$$

with $\{u, p\}$ the solution of (1.1), and

$$\max_{0 \le m \le M} \sqrt{\tau_{m+1}} \left\{ \sqrt{\tau_{m+1}} \| p(t_{m+1}) - p^{m+1} \| \right.$$
$$\left. + \| p(t_{m+1}) - p^{m+1} \|_{-1} \right\} \le Ck, \tag{6.43}$$

together with the a-priori estimates for $i \in \{0, 1, 2\}$, $r \in \{1, 2, 3\}$, with $i + 2r \le 7$,

$$\max_{r-1 \le m \le M} \left\{ \tau_{m+1}^{r-1+i/2} \| d_t^r u^{m+1} \|_i \right\} + k \sum_{m=r}^{M} \tau_{m+1}^{2r-1+i} \| d_t^{r+1} u^{m+1} \|_i^2 \le C, \tag{6.44}$$

for values $r - 1 + \frac{i}{2} \ge 0$, and

$$\max_{r \le m \le M} \left\{ \tau_{m+1}^r \| \nabla d_t^r p^{m+1} \| \right\} + k \sum_{m=r}^{M} \tau_{m+1}^{2r+1} \| \nabla d_t^{r+1} p^{m+1} \|^2 \le C. \tag{6.45}$$

By that, in order to provide an error estimate for the gap between the solutions of (1.1) and (6.7), it is sufficient to compare (6.41) and (6.7). Therefore, let us introduce the following auxiliary problem: For $m \ge 0$, find the solution $\{\bar{v}_\varepsilon^{m+1}, \bar{q}_\varepsilon^{m+1}\} \in \mathbf{H}_0^1 \times H^1/\mathbb{R}$ of the equations

$$d_t \bar{v}_\varepsilon^{m+1} - \nu \Delta \bar{v}_\varepsilon^{m+1} + \nabla \bar{q}_\varepsilon^{m+1} = f^{m+1} - (u^m \cdot \nabla) u^{m+1},$$
$$\mathrm{div}\, \bar{v}_\varepsilon^{m+1} - \varepsilon \Delta \bar{q}_\varepsilon^{m+1} = 0, \qquad \partial_n \bar{q}_\varepsilon^{m+1}|_{\partial \Omega} = 0, \tag{6.46}$$

together with $\bar{v}_\varepsilon^0 = u_0$. Again, we can extend this by setting $\bar{q}_\varepsilon^0 = 0$. Because of the analyses of the linear problem (6.11) in Section 6.2, we get the following error bounds for $\{\bar{v}_\varepsilon^{m+1}, \bar{q}_\varepsilon^{m+1}\}$,

$$\max_{0 \le m \le M} \{\|u^{m+1} - \bar{v}_\varepsilon^{m+1}\| + \tau_{m+1}\|p^{m+1} - \bar{q}_\varepsilon^{m+1}\|_{-1}\} \le C\varepsilon, \qquad (6.47)$$

$$\max_{0 \le m \le M} \{\|u^{m+1} - \bar{v}_\varepsilon^{m+1}\|_1 + \sqrt{\tau_{m+1}}\|p^{m+1} - \bar{q}_\varepsilon^{m+1}\|\} \le C\{\frac{\varepsilon}{\sqrt{k}} + \sqrt{\varepsilon}\}, \qquad (6.48)$$

$$\max_{0 \le m \le M} \{\sqrt{\tau_{m+1}}\|u^{m+1} - \bar{v}_\varepsilon^{m+1}\|_1 + \tau_{m+1}\|p^{m+1} - \bar{q}_\varepsilon^{m+1}\|\} \le C\sqrt{\varepsilon}. \qquad (6.49)$$

Thus, in the following proof we can restrict on bounding the error $u_\varepsilon^{m+1} - \bar{v}_\varepsilon^{m+1}$ and $p_\varepsilon^{m+1} - \bar{q}_\varepsilon^{m+1}$, which measure the effect of the nonlinear "machinery" onto the pressure stabilization of the Navier-Stokes equations. We start with a-priori results for the solution of (6.7).

Lemma 6.4 *We assume (A1), (A2) and (A3) to be valid. Then, for sufficiently small parameters $k \le k_0(T)$, the following a-priori estimates for the solution tuple $\{u_\varepsilon^{m+1}, p_\varepsilon^{m+1}\}$ of (6.7) hold true, with a constant C that depends on the given data of the problem only,*

i) $\max_{0 \le m \le M} \|d_t u_\varepsilon^{m+1}\| + k \sum_{m=0}^{M} \{\|\nabla d_t u_\varepsilon^{m+1}\|^2 + \varepsilon\|\nabla d_t p_\varepsilon^{m+1}\|^2\} \le C,$

ii) $\max_{-1 \le m \le M} \{\|u_\varepsilon^{m+1}\|_2 + \|p_\varepsilon^{m+1}\|_1\} \le C,$

iii) $k \sum_{m=1}^{M} \tau_{m+1} \|d_t^2 u_\varepsilon^{m+1}\|^2$

$\qquad + \max_{0 \le m \le M} \{\sqrt{\tau_{m+1}}\{\|\nabla d_t u_\varepsilon^{m+1}\| + \sqrt{\varepsilon}\|\nabla d_t p_\varepsilon^{m+1}\|\}\} \le C,$

iv) $\max_{1 \le m \le M} \{\tau_{m+1} \|d_t^2 u_\varepsilon^{m+1}\|\}$

$\qquad + k \sum_{m=1}^{M} \tau_{m+1}^2 \{\|\nabla d_t^2 u_\varepsilon^{m+1}\|^2 + \varepsilon\|\nabla d_t^2 p_\varepsilon^{m+1}\|^2\} \le C,$

v) $\max_{0 \le m \le M} \{\tau_{m+1}\|\nabla d_t p_\varepsilon^{m+1}\|\} \le C.$

The main objective of this section is to prove error estimates for the tuple $\{u_\varepsilon^{m+1}, p_\varepsilon^{m+1}\}$ on the basis of the preceding studies.

Lemma 6.5 *We assume (A1), (A2) and (A3) to be valid. Let the tuple $\{u(t_{m+1}), p(t_{m+1})\}$ be the solution of (1.1). Then, for sufficiently small parameters $k \leq k_0(T)$, there exists a constant C that is only dependent on the given data of the problem, such that the following error bounds are valid for the solution $\{u_\varepsilon^{m+1}, p_\varepsilon^{m+1}\}$ of (6.7),*

i) $\max_{0 \leq m \leq M} \big\{ \|u(t_{m+1}) - u_\varepsilon^{m+1}\|$

$\quad + \tau_{m+1} \|p(t_{m+1}) - p_\varepsilon^{m+1}\|_{-1} \big\} \leq C\{\varepsilon + k\}$

ii) $\max_{0 \leq m \leq M} \big\{ \|u(t_{m+1}) - u_\varepsilon^{m+1}\|_1$

$\quad + \sqrt{\tau_{m+1}} \|p(t_{m+1}) - p_\varepsilon^{m+1}\| \big\} \leq C\{\sqrt{\varepsilon}(1 + \sqrt{\tfrac{\varepsilon}{k}}) + \sqrt{k}\}.$

Further, the following estimate is valid,

iii) $\max_{0 \leq m \leq M} \big\{ \sqrt{\tau_{m+1}} \|u(t_{m+1}) - u_\varepsilon^{m+1}\|_1$

$\quad + \tau_{m+1} \|p(t_{m+1}) - p_\varepsilon^{m+1}\| \big\} \leq C\{\sqrt{\varepsilon} + \sqrt{k}\}.$

The proofs of the above lemmata cannot be given independently: Error estimates stemming from the second lemma are needed to show the a-priori estimates of Lemma 6.4.

Proof:
This proof incorporates the verification of the statements presented in the Lemmata 6.4 and 6.5.

Because of the remarks in the beginning of this section, it is sufficient to estimate the errors $\{u_\varepsilon^{m+1} - \bar{v}_\varepsilon^{m+1}\}$ and $\{p_\varepsilon^{m+1} - \bar{q}_\varepsilon^{m+1}\}$. Therefore, we introduce the notations $\xi^{m+1} := u_\varepsilon^{m+1} - \bar{v}_\varepsilon^{m+1}$ and $\chi^{m+1} := p_\varepsilon^{m+1} - \bar{q}_\varepsilon^{m+1}$. Subtracting the equations in (6.46) from those in (6.7), this leads us to the following system of equations:

$$d_t \xi^{m+1} - \nu \Delta \xi^{m+1} + \nabla \chi^{m+1} = -(P_{J_0} u_\varepsilon^m \cdot \nabla) u_\varepsilon^{m+1} + (u^m \cdot \nabla) u^{m+1},$$

$$\operatorname{div} \xi^{m+1} - \varepsilon \Delta \chi^{m+1} = 0, \qquad \partial_n \chi^{m+1}|_{\partial\Omega} = 0, \tag{6.50}$$

together with homogeneous initial and boundary data for the velocity field. The right hand side of the first equation in (6.50) can equivalently be written in the following way,

$$
\begin{aligned}
- (P_{\mathbf{J}_0} u_\varepsilon^m \cdot \nabla) u_\varepsilon^{m+1} + (u^m \cdot \nabla) u^{m+1} &= \\
= -(P_{\mathbf{J}_0} \xi^m \cdot \nabla) u^{m+1} &- (P_{\mathbf{J}_0} u_\varepsilon^m \cdot \nabla) \xi^{m+1} \\
+ (P_{\mathbf{J}_0} u_\varepsilon^m \cdot \nabla) \{ u^{m+1} - \bar{v}_\varepsilon^{m+1} \} &+ (P_{\mathbf{J}_0} \{ u^m - \bar{v}_\varepsilon^m \} \cdot \nabla) u^{m+1}.
\end{aligned}
\tag{6.51}
$$

1st step: We start with the derivation of an a-priori bound for the velocity field u_ε^{m+1} that is independent of ε, measured in the $l^\infty(0, t_{M+1}; \mathbf{H}_0^1)$ norm. This cannot be done right from system (6.7) but only with the help of the inserted auxiliary problems given above. Secondly, we intend to verify the error bound for the velocity, given in part *i*) of Lemma 6.5. Therefore, we look at the equations (6.50)/(6.51) and test the resulting first one with ξ^{m+1}. After taking the sum over the numbers $0 \leq m \leq M$, we obtain:

$$
\frac{1}{2} \| \xi^{M+1} \|^2 + \nu k \sum_{m=0}^{M} \| \nabla \xi^{m+1} \|^2 + \varepsilon k \sum_{m=0}^{M} \| \nabla \chi^{m+1} \|^2 \leq \hat{I} + \widehat{II}.
\tag{6.52}
$$

The resulting nonlinear terms on the right hand side are given by:

$$
\begin{aligned}
\hat{I} := k \sum_{m=0}^{M} \Big\{ & |\hat{b}(\xi^m, u^{m+1}, \xi^{m+1})| + |\hat{b}(u_\varepsilon^m, \xi^{m+1}, \xi^{m+1})| \\
& + |\hat{b}(u^m - \bar{v}_\varepsilon^m, u^{m+1}, \xi^{m+1})| \Big\}
\end{aligned}
\tag{6.53}
$$

and

$$
\widehat{II} := k \sum_{m=0}^{M} |\hat{b}(u_\varepsilon^m, u^{m+1} - \bar{v}_\varepsilon^{m+1}, \xi^{m+1})|.
$$

The terms in $\hat{I}$ can be treated in a quite standard way by using continuity results for the nonlinear term, further a-priori results for the incompressible problem (6.41), see (6.44), and the error estimates (6.47) to (6.49). Let us emphasize that the second term in $\hat{I}$ is now identically zero, owing to (6.9). (Of course, corresponding arguments work for the modification $\tilde{b}(\cdot, \cdot, \cdot)$ of Temam.) By that, we can restrict on the analysis of the second term $\widehat{II}$. For an optimal estimation of this sum, we split it into two parts,

$$
k \sum_{m=0}^{M} |\hat{b}(u_\varepsilon^m, u^{m+1} - \bar{v}_\varepsilon^{m+1}, \xi^{m+1})| \leq \widehat{II}_1 + \widehat{II}_2,
$$

with

$$\widehat{II}_1 := k \sum_{m=0}^{M} |\widehat{b}(u_\varepsilon^m - u^m, u^{m+1} - \bar{v}_\varepsilon^{m+1}, \xi^{m+1})|$$

$$\leq k \sum_{m=0}^{M} \|\xi^m\|_1 \|u^{m+1} - \bar{v}_\varepsilon^{m+1}\|_1 \|\xi^{m+1}\|^{1/2} \|\xi^{m+1}\|_1^{1/2}$$

$$+ k \sum_{m=0}^{M} \|u^m - \widehat{v}_\varepsilon^m\|_1 \|u^{m+1} - \bar{v}_\varepsilon^{m+1}\|_1 \|\xi^{m+1}\|^{1/2} \|\xi^{m+1}\|_1^{1/2} \tag{6.54}$$

and

$$\widehat{II}_2 := k \sum_{m=0}^{M} |\widehat{b}(u^m, u^{m+1} - \bar{v}_\varepsilon^{m+1}, \xi^{m+1})|$$

$$= k \sum_{m=0}^{M} |\widehat{b}(u^m, \xi^{m+1}, u^{m+1} - \bar{v}_\varepsilon^{m+1})|$$

$$\leq Bk \sum_{m=0}^{M} \|\nabla u^m\|^{1/2} \|\Delta u^m\|^{1/2} \|\nabla \xi^{m+1}\| \|u^{m+1} - \bar{v}_\varepsilon^{m+1}\|. \tag{6.55}$$

Here, we have used the fact that the trilinear form $\widehat{b}(\phi, \cdot, \cdot)$, $\forall \phi \in \mathbf{J}_1$ is skew-symmetric with respect to the last two arguments, further continuity results for the Leray operator $P_{\mathbf{J}_0}$, $\|P_{\mathbf{J}_0}\phi\|_i \leq C\|\phi\|_i$, $i \in \{0,1\}$, $\forall \phi \in \mathbf{H}_0^1$, see [5]. Again, the appearing "stability terms" in $\widehat{II}_1$ and $\widehat{II}_2$ stem from the well-analyzed problem (6.41). Now we can proceed in a standard way, using Young's Inequality and the results (6.47), (6.48). By recombination of these estimates in (6.52) and using a discrete version of the Gronwall lemma we arrive at the following inequality, for $k \leq k_0(T)$ sufficiently small,

$$\|\xi^{M+1}\|^2 + \nu k \sum_{m=0}^{M} \|\nabla \xi^{m+1}\|^2 + \varepsilon k \sum_{m=0}^{M} \|\nabla \chi^{m+1}\|^2 \leq C\varepsilon^2. \tag{6.56}$$

This verifies the error estimate in Lemma 6.5, $i)$ for the velocity field. Now, this establishes the second statement in Lemma 6.4, $ii)$, which is an a-priori estimate for the pressure gradient $\nabla p_\varepsilon^{m+1}$. Therefore, we can employ the actual error statement and the second identity in system (6.49), using Lemma 6.3. — We can now benefit from the latter result to verify a corresponding bound for $u_\varepsilon^{m+1} \in l^\infty(0, T; \mathbf{H}_0^1)$. In order to do so, we can proceed in a standard way, testing the first equation in (6.50) with $-\Delta \xi^{m+1}$. By means of a local argument in time, we obtain the desired bound for the velocity field. We leave the details to the reader.

2nd step: In order to verify the remaining statements in Lemma 6.4, i), ii) we proceed in the following way: We start from (6.7) and differentiate the first equation in time, i.e., we employ d_t. Then we test the resulting identity with $d_t u_\varepsilon^{m+1}$ and end up, summing over the range $0 \leq m \leq M$. Now, standard calculations lead us to the following inequality,

$$
\|d_t u_\varepsilon^{M+1}\|^2 + \nu k \sum_{m=0}^{M} \|\nabla d_t u_\varepsilon^{m+1}\|^2 + \varepsilon k \sum_{m=0}^{M} \|\nabla d_t p_\varepsilon^{m+1}\|^2
$$
$$
\leq C \Big\{ 1 + \max_{0 \leq m \leq M} \|\nabla u_\varepsilon^m\|^4 k \sum_{m=0}^{M} \|d_t u_\varepsilon^{m+1}\|^2 \Big\}. \tag{6.57}
$$

We can now use Gronwall's lemma in order to verify part i) of Lemma 6.4, for time-steps $k \leq k_0(T)$ sufficiently small. — This result together with the previous ones is sufficient to bound the right hand side of the following inequality,

$$
\|\Delta u_\varepsilon^{m+1}\| \leq C \big\{ \|d_t u_\varepsilon^{m+1}\| + \|\nabla p_\varepsilon^{m+1}\| + \|f^{m+1}\| + \|\nabla u_\varepsilon^m\|^2 \|\nabla u_\varepsilon^{m+1}\| \big\}.
$$

3rd step: In order to show part iii) of Lemma 6.4 we again employ d_t onto the first equation in (6.7). Afterwards testing this identity with $d_t^2 u_\varepsilon^{m+1}$ and using the second equation in (6.7) gives

$$
\|d_t^2 u_\varepsilon^{m+1}\|^2 + \nu d_t \|\nabla d_t u_\varepsilon^{m+1}\|^2 + \varepsilon d_t \|\nabla d_t p_\varepsilon^{m+1}\|^2
$$
$$
\leq C \big\{ \|d_t f^{m+1}\|^2 + \|\nabla d_t u_\varepsilon^{m+1}\|^2 + \|\nabla d_t u_\varepsilon^m\|^2 \big\}. \tag{6.58}
$$

Here, we have used the a-priori bound for the velocity, given in Lemma 6.4, ii). Because of the limited regularity of time-derivatives of the velocity and pressure functions at times $t_{m+1} \to 0$, we have to multiply the last inequality with a time weight τ_{m+1}. Then we sum up over all numbers, $1 \leq m \leq M$. The application of the results of Lemma 6.4, i) completes the proof. — The proof of statement iv) uses essentially analogous arguments, so we will skip it here.

4th step: This step of the proof is devoted to the derivation of error estimates for the pressure function, as they are given in Lemma 6.5, i) and *ii)*. Again, we will benefit from the results (6.47), (6.48). By employing certain stability properties of the divergence operator,

$$\|\cdot\|_{-p} \le C \sup_{\phi \in H_0^1 \cap H^{p+1}} \frac{(\operatorname{div}\phi, \cdot)}{\|\phi\|_{p+1}}, \qquad p \in \{0,1\},$$

and the results for the velocity errors presented above, we just have to control the error of the velocity differentiated in time. Therefore, we start from (6.50), (6.51), by applying the difference operator d_t on it and finally testing the resulting equation with $d_t \xi^{m+1}$. If we multiply this with τ_m^j, $j \in \{1,2\}$, and sum over all time-steps, $1 \le m \le M$, we arrive at

$$\tau_{M+1}^j \| d_t \xi^{M+1} \|^2 + \nu k \sum_{m=1}^{M} \tau_{m+1}^j \| \nabla d_t \xi^{m+1} \|^2 + \varepsilon k \sum_{m=1}^{M} \tau_{m+1}^j \| \nabla d_t \chi^{m+1} \|^2$$

$$\le C \Big\{ k^j \| d_t \xi^1 \|^2 + k \sum_{m=1}^{M} \tau_{m+1}^{j-1} \| d_t \xi^{m+1} \|^2 + k \sum_{m=1}^{M} \tau_{m+1}^j (NLT)_1 \Big\} \tag{6.59}$$

Here, we have used the abbreviation $(NLT)_1$ for the arising trilinear forms. In order to analyze $(NLT)_1$, we employ the results (6.22), (6.35), (6.38) and the a-priori results for the function u_ε^{m+1} together with its time derivative that are collected in Lemma 6.4. Again, we omit the details that are quite standard, arriving at

$$k \sum_{m=1}^{M} \tau_{m+1}^j (NLT)_1 \le C \Big\{ \varepsilon^2 k^{j-2} + \varepsilon^j \Big\} + Ck \sum_{m=1}^{M} \tau_{m+1}^j \| d_t \xi^{m+1} \|^2. \tag{6.60}$$

The second term on the right hand side of (6.59) can be bounded in the following way: we test the first equation in (6.50)/(6.51) with $\tau_{m+1}^{j-1} d_t \xi^{m+1}$ and finally integrate over the time interval $[0, t_{M+1}]$. This leads us to

$$k \sum_{m=0}^{M} \tau_{m+1}^{j-1} \| d_t \xi^{m+1} \|^2 + \nu \tau_{M+1}^{j-1} \| \nabla \xi^{M+1} \|^2 + \varepsilon \tau_{M+1}^{j-1} \| \nabla \chi^{M+1} \|^2$$

$$\le C \Big\{ \varepsilon^2 + k \sum_{m=0}^{M} \tau_{m+1}^{j-1} (NLT)_2 \Big\}, \tag{6.61}$$

again gathering the trilinear forms in the abbreviation $(NLT)_2$. It has to be remarked that the first term on the right hand side stems from the bounds for ξ^{m+1} in $l^2(0, t_{M+1}; H_0^1)$ and $\sqrt{\varepsilon} \chi^{m+1}$ in $l^2(0, t_{M+1}; H^1/\mathbb{R})$, see (6.56). The sum over the trilinear forms that are collected in $(NLT)_2$ can again be controlled by a standard procedure — apart from the arising two following terms, see (6.51),

$$k \sum_{m=0}^{M} \tau_{m+1}^{j-1} \left\{ |\widehat{b}(u_\varepsilon^m, \xi^{m+1}, d_t\xi^{m+1})| + |\widehat{b}(u_\varepsilon^m, u^{m+1} - \bar{v}_\varepsilon^{m+1}, d_t\xi^{m+1})| \right\}.$$

$$(6.62)$$

We proceed as follows, using result $ii)$ of Lemma 6.4,

$$\leq \frac{1}{4\delta} k \sum_{m=0}^{M} \tau_{m+1}^{j} \|\nabla d_t\xi^{m+1}\|^2$$

$$+ C\delta k \sum_{m=0}^{M} \tau_{m+1}^{j-2} \left\{ \|\xi^{m+1}\|^2 + \|u^{m+1} - \bar{v}_\varepsilon^{m+1}\|^2 \right\},$$

$$(6.63)$$

with an arbitrary positive constant δ. Here, again we used the skew-symmetry rule and an estimation of the second nonlinearity that results from the "Hoelder-triple" $\mathbf{L}^\infty \times (\mathbf{L}^2)^2$. Now, we can combine these considerations to arrive at the following result. If we insert the last result in (6.61) and in (6.59), respectively — choosing $\delta > 0$ sufficiently large -, and recall (6.60), Gronwall's lemma gives

$$\tau_{M+1}^{j} \|d_t\xi^{M+1}\|^2 + \nu k \sum_{m=0}^{M} \tau_{m+1}^{j} \|\nabla d_t\xi^{m+1}\|^2$$

$$+ \varepsilon k \sum_{m=0}^{M} \tau_{m+1}^{j} \|\nabla d_t\chi^{m+1}\|^2 \leq C\{\varepsilon^j + \varepsilon^2 k^{j-2}\}, \qquad (6.64)$$

for $j \in \{1, 2\}$. Summing up, the estimates (6.56), (6.59) and (6.61), together with the corresponding ones in (6.47), (6.48) are the essential ingredients to verify the statements for the pressure in the items $i)$ and $ii)$ of Lemma 6.5, $ii)$. Further, owing to the second identity in (6.50) and corresponding a-priori bounds that are collected in Lemma 6.3, $ii)$, the results (6.61) and (6.64) with $j = 2$ justify the result $v)$ in Lemma 6.4.

5th step: The verification of the remaining part in Lemma 6.5, $iii)$ follows from the results (6.61), (6.64) (setting $j = 2$), in combination with corresponding previous results. The inequality (6.64) provides an upper bound for the first sum in (6.63) that is the "critical" one on the right hand side of

(6.61). This gives the desired result in *iii*) for the velocity error. Together with (6.64) that provides an upper bound for the term $\tau_{m+1}\|d_t\xi^{m+1}\|$, we obtain the corresponding result for the pressure error.

With that, the proofs of the Lemmata 6.4 and 6.5 are complete. $\square$

6.4 Completion of the Analysis for the Chorin Scheme

The main objective in this paragraph is to transfer the error estimates derived in the preceding section for the pressure stabilized Navier-Stokes equations onto the semi-explicit version (6.6), which is Chorin's scheme. In doing so, we have to estimate the difference in the solutions of (6.7) — in the following denoted by $\{u_k^{m+1}, p_k^{m+1}\}$ because of the identification $\varepsilon = k$ in this system - and the solution $\{u_{Cho}^{m+1}, p_{Cho}^{m+1}\}$ of system (6.6). Using the error functions $e^{m+1} := u_k^{m+1} - u_{Cho}^{m+1}$ and $q^{m+1} := p_k^{m+1} - p_{Cho}^{m+1}$, we arrive at the following equations,

$$d_t e^{m+1} - \nu\Delta e^{m+1} + \nabla q^m = -k\nabla d_t p_k^{m+1} - \mathcal{Q}(e^{m+1}),$$
$$\text{div } e^{m+1} - k\Delta q^{m+1} = 0, \qquad \partial_n q^{m+1}|_{\partial\Omega} = 0. \tag{6.65}$$

with $e^0 = 0$ and $q^0 = 0$. Here, we have used the abbreviative notation

$$\mathcal{Q}(e^{m+1}) := (P_{J_0} e^m \cdot \nabla) u_k^{m+1} + (P_{J_0} u_{Cho}^m \cdot \nabla) e^{m+1}.$$

The main results of this section are collected in the following lemma, that completes the error analysis for Chorin's method: If we combine the outcomes (6.42), (6.43), Lemma 6.5 and the following one, the proof of Theorem 6.1 is complete.

Lemma 6.6 *We assume (A1), (A2) and (A3) to be valid. Then, for sufficiently small parameters $k \le k_0(T)$, there exists a global constant C, such that the following error bounds are valid,*

i) $\max_{0\le m\le M}\left\{\|u_{Cho}^{m+1} - u_k^{m+1}\| + \tau_{m+1}\|p_{Cho}^{m+1} - p_k^{m+1}\|_{-1}\right\} \le Ck,$

ii) $\max_{0\le m\le M}\left\{\|u_{Cho}^{m+1} - u_k^{m+1}\|_1 + \sqrt{\tau_{m+1}}\|p_{Cho}^{m+1} - p_k^{m+1}\|\right\} \le C\sqrt{k}.$

Proof:

In order to verify these estimates, we have to be more careful with respect to the treatment of the evolutionary velocity term. Subsequently, we will frequently use the algebraic relation

$$2(a - b, a) = |a|^2 - |b|^2 + |b - a|^2.$$

Then, by testing the first equation in (6.6) with e^{m+1}, we get

$$
\begin{aligned}
\frac{1}{2} d_t \|e^{m+1}\|^2 + \frac{1}{2} k \|d_t e^{m+1}\|^2 &+ \nu \|\nabla e^{m+1}\|^2 + k(\nabla q^m, \nabla q^{m+1}) \\
&= -k(\nabla d_t p_k^{m+1}, e^{m+1}) - \widehat{b}(e^m, u_k^{m+1}, e^{m+1}) \\
&\quad - \widehat{b}(u_{Cho}^m, e^{m+1}, e^{m+1}).
\end{aligned}
\tag{6.66}
$$

Again, the second trilinear form is zero, thanks to (6.9). Next, we replace the last term on the left side of the last equation,

$$
\begin{aligned}
k(\nabla q^m, \nabla q^{m+1}) &= k \|\nabla q^{m+1}\|^2 - k^2 (\nabla q^{m+1}, \nabla d_t q^{m+1}) \\
&= k \|\nabla q^{m+1}\|^2 - k(d_t e^{m+1}, \nabla q^{m+1}) \\
&\geq \frac{1}{2} k \left\{ \|\nabla q^{m+1}\|^2 - \|d_t e^{m+1}\|^2 \right\},
\end{aligned}
\tag{6.67}
$$

owing to the second equation in (6.65) and Young's Inequality. We insert this in (6.66) and sum over all time-steps. After absorbing terms and again using the second identity in (6.65) for the first term on the right hand side of (6.66), we are led to

$$
\begin{aligned}
\|e^{M+1}\|^2 + \nu k \sum_{m=0}^{M} \|\nabla e^{m+1}\|^2 &+ k^2 \sum_{m=0}^{M} \|\nabla q^{m+1}\|^2 \\
&\leq C \left\{ k^4 \sum_{m=0}^{M} \|\nabla d_t p_k^{m+1}\|^2 + Ck \sum_{m=0}^{M} \|e^{m+1}\|^2 \right\} \leq Ck^2.
\end{aligned}
\tag{6.68}
$$

The last inequality takes benefit from Lemma 6.4, *i*) and *ii*). By that, the error behavior for the velocity field is quantified. Besides, (6.68) gives us a striking pointwise a-priori result for the pressure gradient ∇p_{Cho}^{m+1}. - For the verification of error estimates for the pressure we confine ourselves to a sketch of the proof that uses arguments similar to those that have already been employed in earlier corresponding investigations. Thus, in order to prove the error estimates for the pressure we can restrict ourselves to deriving optimal order of convergence for the quantities $\tau_{m+1}^{j/2} d_t e^{m+1}$, $j \in \{1, 2\}$ in the norm $l^\infty(0, t_{M+1}; \mathbf{L}^2)$. Combining them with (6.68), together with stability arguments for the div-operator as presented in the fourth step of the proof of the Lemmata 6.4 and 6.5, we are done then. Let us start with the first equation in system (6.65). We employ the operator d_t to it and continue by multiplying the resulting identity with $\tau_m^{j+1} d_t e^{m+1}$, $j \in \{1, 2\}$. Summing over the range $1 \le m \le M$ and finally dividing by τ_{M+1} leads us to the following inequality,

$$\tau_{M+1}^j \|d_t e^{M+1}\|^2 + \frac{1}{\tau_{M+1}} k \sum_{m=0}^M \tau_{m+1}^{j+1} \left\{ \nu \|\nabla d_t e^{m+1}\|^2 + k \|\nabla d_t q^{m+1}\|^2 \right\}$$

$$\le C \frac{1}{\tau_{M+1}} \left\{ k^4 \sum_{m=1}^M \tau_m^{j+1} \|\nabla d_t^2 p_k^{m+1}\|^2 + k \sum_{m=1}^M \tau_{m+1}^j \|d_t e^{m+1}\|^2 \right.$$

$$\left. + k^{j+1} \|d_t e^1\|^2 + k \sum_{m=1}^M \tau_{m+1}^{j+1} \widehat{(NLT)}_1 \right\}. \tag{6.69}$$

Because of Lemma 6.4, *iv*), the first term on the right hand side is bounded by Ck^j. The same holds true for the third one. Further, the last sum of the trilinear forms denoted as $\widehat{(NLT)}_1$ can be dealt with in the same fashion as presented in the last section, using Gronwall's lemma for sufficiently small time-steps $k \le k_0(T)$. We skip the details of this step. — In order to find a corresponding estimate for the remaining sum on the right side of (6.69), we test the first equation in (6.65) with $\tau_{m+1}^j d_t e^{m+1}$ and sum over the range $0 \le m \le M$:

$$\frac{1}{\tau_{M+1}} k \sum_{m=0}^M \tau_{m+1}^j \|d_t e^{m+1}\|^2 + \tau_{M+1}^{j-1} \|\nabla e^{M+1}\|^2 + k \tau_{M+1}^{j-1} \|\nabla q^{M+1}\|^2$$

$$\le C \frac{1}{\tau_{M+1}} \left\{ k^2 \sum_{m=0}^M \tau_{m+1}^j \|d_t e^{m+1}\|^2 + k^3 \sum_{m=0}^M \tau_{m+1}^j \|\nabla d_t p_k^{m+1}\|^2 \right.$$

$$+ \tau_{M+1} \left\{ k \sum_{m=0}^M \|\nabla e^{m+1}\|^2 + k^2 \sum_{m=0}^M \|\nabla q^{m+1}\|^2 \right\} + Ck^2 \tag{6.70}$$

$$\left. + k \sum_{m=0}^M \tau_{m+1}^j \widehat{(NLT)}_2 \right\} \le Ck^j.$$

Again, we abbreviated the arising nonlinear terms in the last sum inside the brackets with $\widehat{(NLT)}_2$. We will skip the straightforward but technical analysis of it. Further, (6.68) and Lemma 6.4, i), v) justify the last error bound, for $k \leq k_0(T)$ sufficiently small. By that, we have an upper bound in (6.69) of size Ck^j. As described above, (6.68) to (6.70) provide the ingredients to verify the error estimates for the pressure given in i) and ii).   □

By that, the results of this section complete the proof of Theorem 6.1 that establishes first order of convergence for the Chorin Scheme.

6.4.1   Interior Super-Convergence Results for the Pressure

Before we proceed further, let us repeat the idea of the proof given above for Theorem 6.1: we separated the errors caused by the time-discretization of the momentum equation from those that are initiated by the quasi-compressibility constraint. The evolution of the last error source was then studied. In a further step the amplification mechanism of these errors caused by the convection term was analyzed. Finally, we investigated the influence on the solution that is caused by the explicit treatment of the pressure in the Chorin scheme. — If we look again at the several steps of the proof, we recognize that only in the second one there is a "fundamental" difference in the error estimates for the pressure error in the norms $l^\infty(0, t_{M+1}; L^2(\Omega)/\mathbb{R})$ and $l^\infty(0, t_{M+1}; H^{-1}(\Omega))$ that cannot be overcome in a way of damping the pressure error with time weights — which can be done in the other steps. The reason for this is of course the singularly perturbed character of the stationary error equations (6.21) that prevents us from getting a better error estimate for the pressure in the norm $l^\infty(0, t_{M+1}; L^2(\Omega)/\mathbb{R})$ than those already presented. Therefore, in order to improve it we have to eliminate the contribution to the global error that stems from a local boundary layer and is caused by the "wrong" Neumann data for the pressure in Chorin's scheme.

Owing to these considerations we can focus on a local analysis of the auxiliary problem (6.20) that results in super-convergence estimates for the error of the pressure as well as the gradient of the velocity field on interior subdomains. This will be sufficient for the sketch of a proof for Theorem 6.2. — The succeeding estimate has been proved by R. Rannacher, viz. [30]: For interior subdomains $\Omega' \subset\subset \Omega$ there exists a constant $\tilde{C}$ that depends on $\mathrm{dist}(\partial\Omega, \Omega')$ such that the following error estimate is valid,

$$\|E^{m+1}\|_{1,\Omega'} + \|\Pi^{m+1}\|_{\Omega'} + \sqrt{\varepsilon}\|\nabla\Pi^{m+1}\|_{\Omega'} \leq \tilde{C}\varepsilon. \tag{6.71}$$

$k =$	u_2	divu bef. proj.	divu after proj.	∇u_2	p
0.1	1.731	$2.28 - 3$	$1.15 - 4$	6.445	0.721
$5. - 2$	1.336	$1.82 - 3$	$9.02 - 5$	4.995	0.571
$2.5 - 2$	0.921	$1.32 - 3$	$6.4 - 5$	3.468	0.411
$1.25 - 2$	0.571	$8.88 - 4$	$4.23 - 5$	2.178	0.273
$6.25 - 3$	0.327	$5.65 - 4$	$2.75 - 5$	1.268	0.172
$3.13 - 3$	0.180	$3.48 - 4$	$1.87 - 5$	0.789	0.101
$1.56 - 3$	$9.51 - 2$	$2.09 - 4$	$1.37 - 5$	0.483	$5.71 - 2$
$7.82 - 4$	$4.92 - 2$	$1.27 - 4$	$9.21 - 6$	0.314	$3.23 - 2$
order	0.92	0.78	0.73	0.81	0.82

Table 6.1: Orders for errors for the Chorin scheme

Here, we have used the notation already introduced in (6.21). — Now, we can combine these local statements with the remaining global in space ones, by imposing additional time weights. We list here the further error contributions that determine the error committed in Chorin's scheme: (6.39)/(6.40); (6.61)/(6.64), for $j = 2$; (6.68) and (6.69)/(6.70), for $j = 2$. This proves the statements in Theorem 6.2.

6.5 Computational Results

Finally, we intend to support the theoretical results collected in the previous Theorems 6.1 and 6.2 for the Chorin method with numerical test calculations that have been carried out for the problem configurations outlined in Section 2.4. We refer to the notations introduced there. Let us recapitulate the fact that the finite element discretization in space is a Q1/Q1-discretization in a conforming approach. The stability of this discretization is secured, provided that the step size condition $k \geq Ch^2$ holds, because of the re-interpretation as a semi-explicit pressure stabilization scheme and the related results in the work of Hughes, Franca and Balestra, see [20]. For further comments, we refer to Remark 6.5 on page 139. The subsequent figures in Figure 6.1 through 6.3 show the anisotropic error structures for the pressure p^{M+1}, while Figure 6.4 through 6.6 show the divergence of the velocity field, $\text{div} \bar{u}^{M+1}$ for decreasing time-steps k. The relative errors that have been achieved for the parameter constellation $\{T, \gamma, \nu\} = \{2, 1, 1\}$ are given in Table 6.1.

First, we notice that optimal convergence behavior starts only for relatively small time-steps k. Further, the convergence behavior for the pressure is much better than predicted in Theorem 6.1. This is due to the local error contribution of the boundary layers arising in the approximation, as a consequence of the homogeneous boundary conditions of Neumann type. In the following, we will give a *heuristic* argument, based on strong practical evidence, that underlines this observation in more specific detail. We have already discussed the fact that the boundary layer Ω_{BD} is assumed to be of measure $\mu(\Omega_{BD}) = \mathcal{O}(\sqrt{k}\log\frac{1}{k})$. Now, if we further suppose that the point-wise errors possess the same convergence behavior (in an asymptotic sense) as the averaged L^2 errors for the pressure, we can proceed as follows (for $\nu = 1$), with a partition $\Omega = \Omega_{BD} \cup \Omega'$,

$$\|p_{Cho}^{m+1} - p(t_{m+1})\|_{\Omega} \leq \|p_{Cho}^{m+1} - p(t_{m+1})\|_{\Omega'} + \|p_{Cho}^{m+1} - p(t_{m+1})\|_{\Omega_{BD}}$$
$$\leq Ck + \mu^{1/2}(\Omega_{BD})\|p_{Cho}^{m+1} - p(t_{m+1})\|_{\infty;\Omega_{BD}}$$
$$\leq C\{k + k^{3/4}\log\frac{1}{k}\}, \tag{6.72}$$

and $p_{Cho}^{m+1}, p(t_{m+1}) \in L_0^2(\Omega)$. Note that we have used the super-convergence result of Theorem 6.2 in order to treat the error contribution from the interior. Let us again note the fact that this relies on an assumption regarding the thickness of the boundary layer, which is not yet completely verified. — These considerations can now be immediately transfered to the results for the approximation $\mathrm{div}\tilde{u}^{m+1}$ or, correspondingly, to the gradient of the approximating velocity field $\tilde{u}^{m+1}$, serving as a justification for the better computational results presented in Table 6.1. Figures 6.1 through 6.3 are to show the impact of the singular perturbation character on the pressure approximation p_{Cho}^{m+1}, depending on the discretization parameter k. The viscosity parameter is kept fix at $\nu = 1$. For this case, the figures show isolines of the error distribution across the domain Ω: They are significant in the vicinity of the boundary and decay exponentially to the interior of the domain, $\Omega' \subset\subset \Omega$. This observation supports the conjecture concerning the error behavior of the pressure approximation p_{Cho}^{m+1},

$$|p(t_{m+1})(x) - p_{Cho}^{m+1}(x)| \leq C\exp\left(-\frac{d(x)}{\sqrt{k}}\right)\sqrt{k} + \mathcal{O}(k), \tag{6.73}$$

with $d(x) = \mathrm{dist}(x, \partial\Omega)$. Quantitatively, we recognize that the boundary layer is significant, even for time-steps of size $k = \mathcal{O}(10^{-3})$. Thus, this gives the error a global structure rather than a local one.

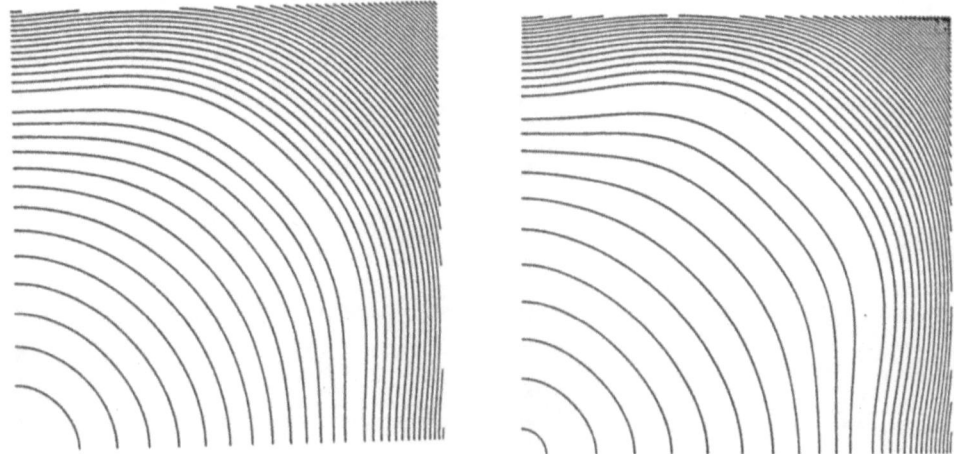

Figure 6.1: Isolines for the error of the pressure ($k = 0.025$ and $k = 0.0125$)

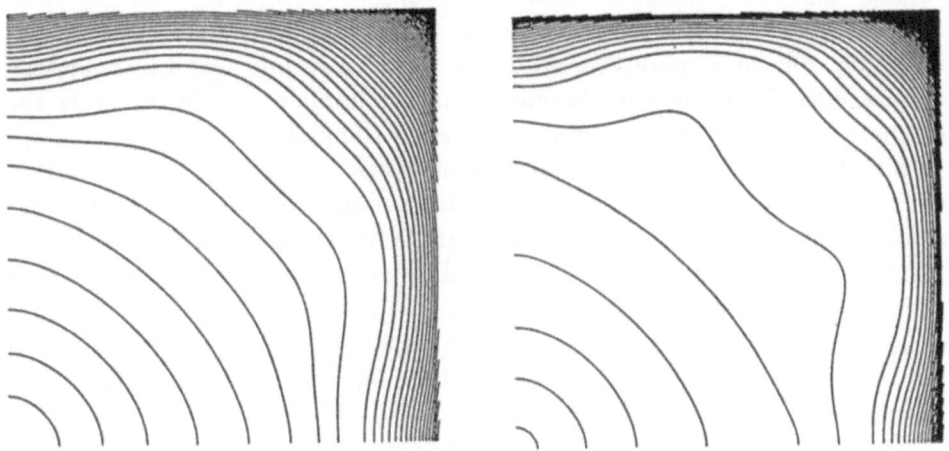

Figure 6.2: Isolines for the error of the pressure ($k = 6.25{-}3$ and $k = 3.13{-}3$)

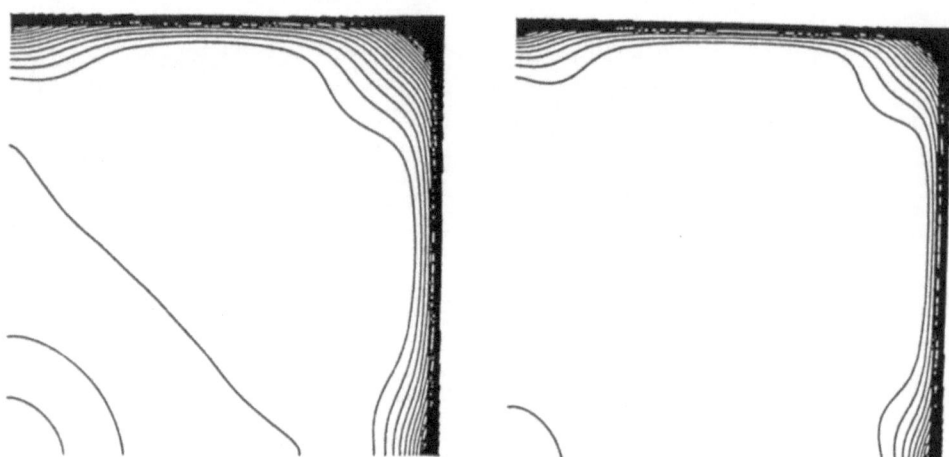

Figure 6.3: Isolines for the error of the pressure ($k = 1.56-3$ and $k = 7.82-4$)

The conjecture that bad approximations of the error in the neighborhood of the boundary are related to the non-vanishing divergence of the velocity field will be illustrated with another sequence of figures, see Figure 6.4 through 6.6. They show the function $\mathrm{div}\tilde{u}_{Cho}^{M+1} = \mathrm{div}\{\tilde{u}_{Cho}^{M+1} - u(t_{M+1})\}$ at time $t_{M+1} = 2$ for $\nu = \gamma = 1$. We see the same pattern of the error distribution, but in a more dramatic way: concentration near the boundary and exponential decay into the interior of the domain. Again, we state the dramatic impact of the no-slip conditions on the global structure of the approximation, even for rather small time-steps ($k = \mathcal{O}(10^{-2})$). Extrapolation strategies to the boundary, as they have been used by Blum in [1] for pressure stabilization formulations of the (stationary) Stokes problem, cannot be applied successfully here. This is due to stability restrictions for the parameters $F(k, h) \geq 0$, which would have to be satisfied otherwise. The last sequence of figures in Figure 6.7 and 6.8 shows the impact of the Reynolds number on the thickness of the numerically caused boundary layer. They motivate the conjecture of its measure being $\mu(\Omega_{BD}) = O(\sqrt{\nu k})$.

We will now give a summary of our investigations of the Chorin method at the end of this chapter. We established the method as a first order accurate scheme through reinterpretation as a semi-explicit pressure stabilization

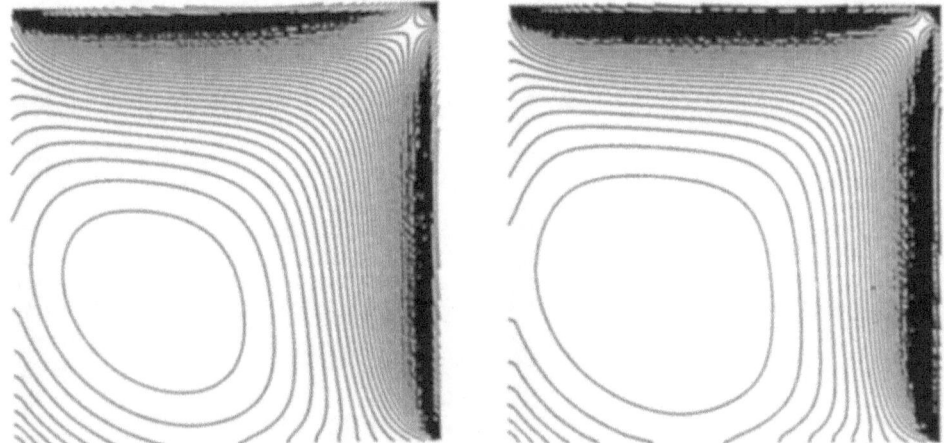

Figure 6.4: Isolines for $\mathrm{div}\tilde{u}_{Cho}^{M+1}$ ($k = 0.025$ and $k = 0.0125$)

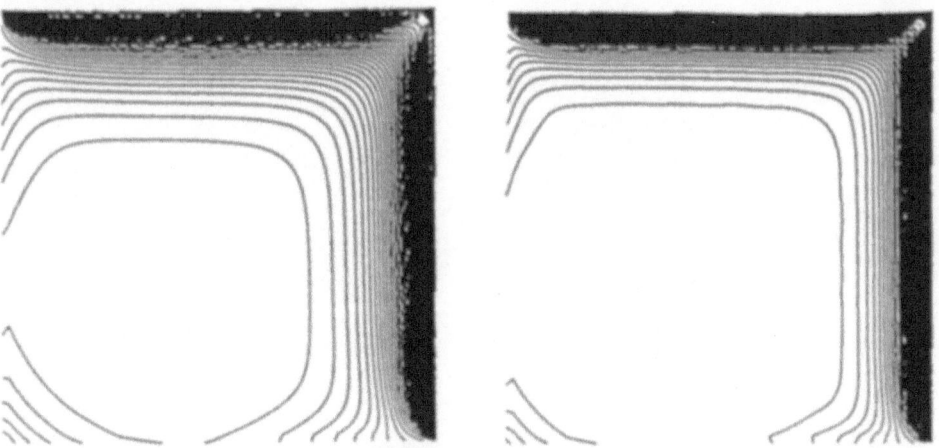

Figure 6.5: Isolines for $\mathrm{div}\tilde{u}_{Cho}^{M+1}$ ($k = 6.25 - 3$ and $k = 3.13 - 3$)

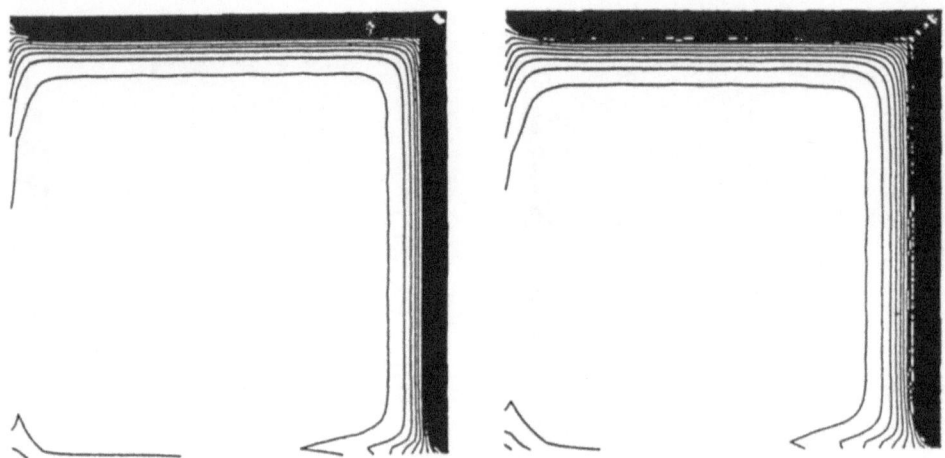

Figure 6.6: Isolines for $\mathrm{div}\tilde{u}_{Cho}^{M+1}$ ($k = 1.56 - 3$ and $k = 7.82 - 4$)

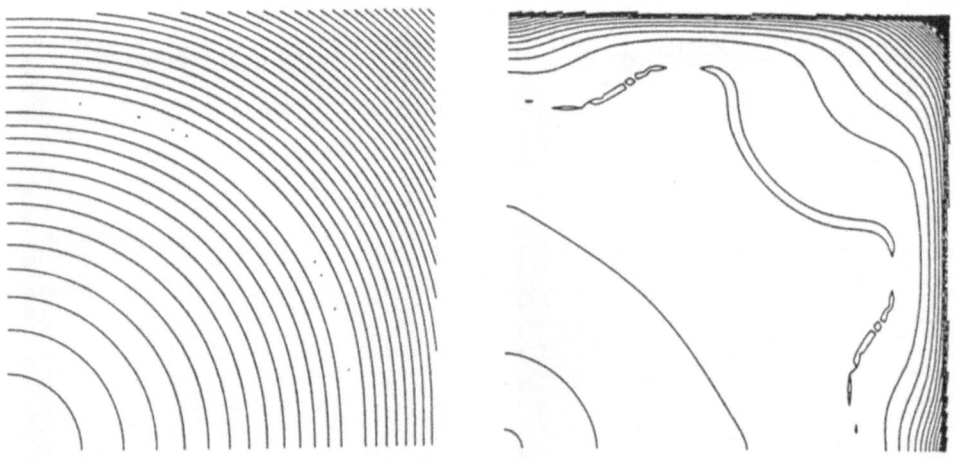

Figure 6.7: Isolines for the error in the pressure ($k = 6. - 3$) for $\nu = 10$ and $\nu = 10^{-1}$

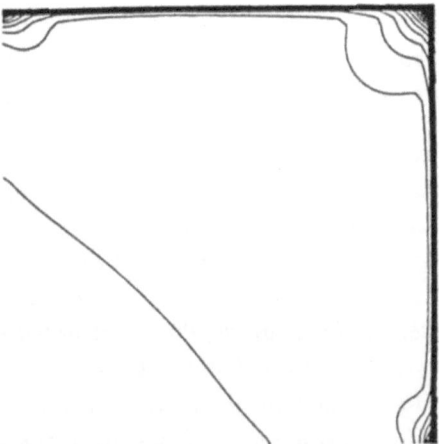

Figure 6.8: Isolines for the error in the pressure $(k = 6. - 3)$ for $\nu = 10^{-2}$

method. This was the basis for the local analysis giving stronger error estimates for the pressure in the interior of the underlying domain Ω. These results have been supported by computational results. They indicate that the Chorin method is a robust scheme and accurate in the sense that we can use the finite element pair Q1/Q1 for spatial discretization, provided we make sure that we have $k \geq Ch^2$ (see remark below). This scheme gives good approximations of the velocity field, whereas the pressure approximation suffers from a marked boundary layer that gives the error a global structure, even for very small time-steps.

Remark 6.5   *1. In our calculations, we have accepted a stability restriction for the choices of k, which is $k \geq Ch^2$, thinking of it as not being restrictive. On the other hand, this stability condition is indeed restrictive for practical applications, lowering the degree of flexibility of our numerical scheme. Therefore, it is useful to construct a projection algorithm that incorporates local spatial features in the quasi-compressibility constraint,*

$$\mathrm{div}u_\varepsilon^{m+1} - \alpha_1 k \Delta p^{m+1} - \alpha_2 \mathrm{div}(h^2(x)\nabla p^{m+1}) = 0, \qquad (6.74)$$

with α_i, $i \in \{1,2\}$ being positive constants. We will not present the detailed formulation of this stabilized Chorin scheme here and refer to a corresponding elaborated modification of the original Van Kan scheme in Section 7.4. — Now, the related modified Chorin scheme enhances the stability properties of the selected Finite Element pairs that do not satisfy the LBB-constraint. In that context, we refer to Chapter 5 for corresponding investigations of the mixed methods.

2. The above test calculations have been done for the Stokes case. Let us mention the fact that the scheme can also be applied to the Navier-Stokes equations in a robust way. But let us emphasize that — due to the splitting character of the scheme — nonlinear error amplification effects are present that are of increasing importance for a decreasing viscosity parameter ν. This can easily be extracted from the above analysis, where we used stability constants $C = C(\nu)$. This dependence is even worse than in the case of algorithms that work with velocity fields which are in $\mathbf{J}_1$ at each time-step. Thus, we think that additional (stationary) $\mathbf{J}_1$-projections at certain time-steps can improve the accuracy of the computed approximation of highly complex flows.

Chapter 7

The Projection Scheme of Van Kan

7.1 Overview and Results

The previous chapter has been concerned with the analysis of the Chorin method. As the main result, we obtained optimal convergence statements valid for a general class of flow problems.

The objective of this chapter is the investigation of a higher order method of projection-type, known as Van Kan's scheme, compare [45]. As a motivation, higher order (semi-)discretization schemes for the incompressible Navier-Stokes problems are of great importance for computational simulations, allowing for more flexibility in comparison to the spatial discretization. In that aspect, they guarantee an equilibration of the portions of errors in the total discretization. Last but not least, higher order projection schemes are expected to reduce the numerical boundary layer that we have observed in the case of Chorin's method. In this context, we expect higher order schemes to reduce its size drastically in the sense of orders of magnitudes of k.

We will start with the presentation and analysis of the Van Kan method, formulated as a semi-discretization ansatz for solving the incompressible Navier-Stokes equations. For the later investigations, let us introduce a parameter β. Then, this projection method reads as follows:

1. Start with initial data $u^0 = u_0 \equiv u(0)$, and an approximation $p^0 \approx p(0)$.

2. For $\beta \geq \frac{1}{2}$ and $m \geq 0$, let $\tilde{u}^{m+1}$ be the solution of the Helmholtz-type

equation

$$\frac{1}{k}\{\tilde{u}^{m+1} - u^m\} - \Delta\overline{\tilde{u}}^{m+1/2} + \mathcal{N}(u^m, \tilde{u}^m, \tilde{u}^{m-1})$$

$$+ (\frac{3}{2} - \beta)\nabla p^m + (\beta - \frac{1}{2})\nabla p^{m-1} = \overline{f}^{m+1/2}, \qquad (7.1)$$

$$\tilde{u}^{m+1}|_{\partial\Omega} = 0.$$

3. Compute $\{\tilde{u}^{m+1}, p^{m+1}\}$ as the solution of

$$\frac{1}{k}\{u^{m+1} - \tilde{u}^{m+1}\} + \beta\nabla\{p^{m+1} - p^m\} = 0$$

$$\text{div}\, u^{m+1} = 0, \qquad u^{m+1}|_{\partial\Omega} \cdot n = 0. \qquad (7.2)$$

Here, $\mathcal{N}(\cdot, \cdot, \cdot)$ stands for the explicit discretization of the nonlinearity,

$$\mathcal{N}(u^m, \tilde{u}^m, \tilde{u}^{m-1}) = \frac{3}{2}(u^m \cdot \nabla)\tilde{u}^m - \frac{1}{2}(u^m \cdot \nabla)\tilde{u}^{m-1}, \qquad (7.3)$$

which is substituted by a trapezoidal discretization for the case $(m = 0)$. — The solution of (7.1) lies in $\mathbf{H}_0^1$ and the second part (7.2) of the projection step admits a unique solution in $\mathbf{J}_0 \times L^2/\mathbb{R}$. Of course, in order to stabilize the scheme, an implicit treatment of the nonlinearity in (1.1) can be employed, cf. [36] and [37] in this respect.

If we set $\beta \neq 1/2$, the prescription of the pressure function p^{-1} is also necessary. This does not cause any additional theoretical or practical problems. We can proceed as follows: In the first iteration step we set $\beta = 1/2$. For $m > 0$ we choose $\beta > 1/2$.

Necessary for an optimal convergence behavior of the Van Kan method — in contrast to Chorin's scheme — is the prescription of sufficiently accurate initial data for velocity field *and* pressure function,

$$\|u(t_0) - u^0\| + k\|p(t_0) - p^0\|_1 \leq Ck^2. \qquad (7.4)$$

The importance of this point will become apparent through the subsequent theoretical and computational results. — Note, that the prescription of the initial pressure function is unusual in the case of the incompressible (Navier-) Stokes equations.

The following investigations will be done for the Stokes equations, i.e., we omit the nonlinearity $\mathcal{N}(\cdot, \cdot, \cdot)$ in the above Van Kan scheme. This is for simplification of the following analysis; let us stress the fact that the following investigations can be carried over to the nonlinear case. Essentially, the missing steps can be copied from the previous chapter, dealing with corresponding arguments like those that have been given for the Chorin method, applied to the full Navier-Stokes equations. In other words, we will focus on the inherent new error mechanisms that are present in the Van Kan method, owing to the different kind of projection, and omit the nonlinearity $\mathcal{N}(\cdot, \cdot, \cdot)$ in the rest of the chapter.

Let us turn back to the choices for β. Setting $\beta = 1/2$ gives the original Van Kan scheme, as it has been proposed in [45]. The choice $\beta > 1/2$ corresponds to a stabilization of this projection scheme of higher order, as it has been introduced by Shen in [36], and leads to additional computational effort in each time-step. This approach is the basis for his error analysis giving averaged error estimates in time, with a stability constant involved that is dependent on the value of $\beta - 1/2$ in a reciprocal way, see below. — The objective of this chapter is to verify optimal error results, *pointwise in time* for the whole parameter range $\beta \geq 1/2$. Compared to the results for the case $\beta > 1/2$ that give striking stability results, we will give corresponding results for algebraic averages of the iterates for the case $\beta = 1/2$. The reason for the necessary separation of these cases is due to the splitting character of the projection method, giving rise to an explicit treatment of the pressure function in the first step (of each iteration), and, therefore, a loss of stability of this scheme. This matter will be discussed in the following.

The algorithm (7.1), (7.2) has the same structure as the one of Chorin: The first part represents a Burgers equation, discretized in time (in the nonlinear case), with a given velocity field $\tilde{u}^{m+1}$ that satisfies the zero Dirichlet boundary conditions. However, this velocity field is not divergence-free. The incompressibility of the velocity field will be considered in the projection step: $u^{m+1} = P_{\mathbf{J}_0}^{VK} \tilde{u}^{m+1}$. Again, in order to ensure the well-posedness of this L^2-type projection, the exact no-slip boundary conditions cannot be imposed any more. Before we start with the theoretical investigation of the Van Kan scheme, we show a reformulation of it, substituting the projection step as a Poisson equation for the pressure and linked up with an update for the velocity field. This can be obtained through application of the divergence operator on the first equation in (7.2),

$$-\Delta q^{m+1} = -\frac{1}{\beta k}\mathrm{div}\tilde{u}^{m+1}, \qquad p^{m+1} = q^{m+1} + p^m,$$

$$\partial_n q^{m+1}|_{\partial\Omega} = 0,$$

$$(7.5)$$

to be followed by an update for u^{m+1} that will be computed by means of

$$u^{m+1} = \tilde{u}^{m+1} - \beta k \nabla q^{m+1}. \qquad (7.6)$$

Again, we expect great errors in the pressure approximation close to the boundary, owing to the homogeneous boundary condition of Neumann type,

$$\partial_n d_t p^{m+1}|_{\partial\Omega} = 0. \qquad (7.7)$$

We will return to this later.

In recent papers, projection schemes of higher order have been treated analytically as well as computationally. We mention the papers of E and Liu, see [7] and [8], which are concerned with the analysis of the related scheme of Kim and Moin, [21], which is formally of second order. Indeed, the requirements with respect to the smoothness of the solution in time and space are too restrictive. E and Liu use an asymptotic analysis based on a classical perturbation ansatz, thus requiring $u \in C(0, T; \mathbf{H}^5)$. Despite of this drawback the major error sources of the algorithm become visible, especially the appearance of numerical boundary layers. — J. Shen has investigated the approximation properties of the Chorin and the Van Kan scheme in a series of papers, cf. [33] through [36]. In [36], he verifies error bounds for the Van Kan method *for the case* $\beta > 1/2$,

$$\left(k\sum_{m=0}^{M}\|u(t_{m+1}) - u^{m+1}\|^2\right)^{1/2}$$
$$+ k\max_{0\le m\le M}\left\{\|u(t_{m+1}) - u^{m+1}\|_1 + \|p(t_{m+1}) - p^{m+1}\|_1\right\} \le Ck^2. \quad (7.8)$$

In order to prove these estimates, Shen needs the assumptions (7.4) for the initial function. Further, he assumes the actual solution to be sufficiently regular, i.e., $p_{tt} \in L^2(0, T; H^1/\mathbb{R})$. — The present chapter is concerned with optimal order error estimates for the Van Kan scheme, pointwise in time that are valid for all parameters $\beta \ge 1/2$. The error analysis is based on the weaker compatibility assumption $p_t \in L^\infty(0, T; H^1/\mathbb{R})$. The necessity of such a requirement for the solution to be approximated will be verified by means of numerical test calculations. These results show that only compatibly posed (flow-) problems give rise to an optimal behavior of convergence of the Van Kan scheme, whereas otherwise the accuracy of this discretization method is limited to an order which is strictly less than two, *globally* in time.

The key for the error analysis of scheme (7.1) – (7.2) is the re-interpretation as a semi-explicit pressure correction method: If we shift the index in step three and add the resulting identity to item two, we end up with the system of equations

$$\frac{1}{k}\{\tilde{u}^{m+1} - \tilde{u}^m\} - \Delta\tilde{\bar{u}}^{m+1/2} + \frac{1}{2}\{3\nabla p^{m+1} - \nabla p^m\} = \bar{f}^{m+1/2},$$

$$\operatorname{div}\tilde{u}^{m+1} - \varepsilon\Delta d_t p^{m+1} = 0, \qquad \partial_n d_t p^{m+1}|_{\partial\Omega} = 0,$$

(7.9)

with the notation $\varepsilon = \beta k^2$.

The error analysis for this system is split into two parts, starting with the consistency error analysis. The second part is devoted to the determination of the perturbation error. Compared to the structure of the proof for the Chorin scheme, we use the same pattern here, but the analysis of the diverse auxiliary problems is much more complicated, owing to the evolutionary character of the quasi-compressibility constraint that causes another error-propagation mechanism. In that respect, we refer to the investigations of the continuous analogon and the improvement of its stability properties with the aid of a damping time weight in the quasi-compressibility constraint *(modified pressure correction)*. As has been pointed out there, this widens the range of problem constellations that can be treated successfully. For our further analysis, we need another assumption regarding the solution behavior.

Postulate B_2: The solution of the incompressible Stokes equations (2.1) satisfies the following a-priori bound,

$$\sup_{0 \leq s \leq T} \{\|\Delta u_t(s)\| + \|\nabla p_t(s)\|\} \leq C. \tag{7.10}$$

Of course, this holds true for times $s > 0$. For solutions of the equation (2.1) that satisfy this assumption, the following theorem gives a theoretical justification for the superiority of Van Kan's method over the Chorin scheme.

Theorem 7.1 *Let $\{\tilde{u}^{m+1}, p^{m+1}\}$ be the solution of the Van Kan scheme (7.1), (7.2). We assume (A1), (A2) to be satisfied and the exact solution $\{u(t_{m+1}), p(t_{m+1})\}$ to satisfy Postulate B_2. Further, let the starting conditions (7.4) be satisfied, and $k < 1$. Then, the following statements hold true, with a constant C that only depends on the given problem data,*

i)   *provided that $\beta > \frac{1}{2}$:*

$$\max_{1 \le m \le M}\Big\{\sqrt{\tau_{m-1/2}}\{\|u(t_{m-1/2}) - \overline{\overline{u}}^{m-1/2}\|$$
$$+ k\|u(t_{m+1/2}) - \overline{\overline{u}}^m\|_1\}\Big\} \le Ck^2 \log\tfrac{1}{k},$$

$$\max_{1 \le m \le M}\Big\{\sqrt{\tau_{m+1/2}}\|p(t_{m+1/2}) - \overline{p}^m\|\Big\} \le Ck\log\tfrac{1}{k}.$$

ii) *provided that $\beta = \frac{1}{2}$: Instead of the error estimate for the velocity, given in i), there holds*

$$\max_{1 \le m \le M}\Big\{\sqrt{\tau_{m+1/2}}\,\|u(t_m) - \overline{\overline{u}}^m\|\Big\} \le Ck^2\log\tfrac{1}{k}.$$

The other statements given in i) remain valid. Further, these bounds hold true for the projected velocities.

Remark 7.1   *1. The validation of the estimate*

$$\max_{0 \le m \le M}\Big\{\sqrt{\tau_{m+1}}\|u(t_{m+1/2}) - \overline{\overline{u}}^{m+1/2}\|\Big\} \le Ck^2\log\frac{1}{k}$$

cannot be given on the basis of our error analysis for the original Van Kan scheme (i.e., $\beta = \frac{1}{2}$). This is due to the explicit treatment of the pressure functions in the first equation of (7.9). We will come back to this in the subsequent analysis.

2. *The transfer of these results to the Navier-Stokes equations, semi-discretized with the Van Kan method (7.1), (7.2), is possible and uses techniques analogous to those proposed in the treatment of the Chorin scheme (Chapter 6).*

3. *Overall, in the above convergence statements, algebraic averages of succeeding iterates serve as good approximations of the velocity field at a certain time. A "pointwise" error estimate cannot be given, owing to the structure of the proof.*

4. *The logarithms appearing in the estimates are caused by the decreasing stability features of the pressure correction. This can be skipped in the case of slightly sharpened assumptions regarding the amount of regularity in the beginning (see Postulate B_2), for example: $\exists\, r << \frac{1}{2} : \tau^r p_{tt} \in L^2(0, T; H^1/\mathbb{R})$.*

5. *The scheme of Van Kan is formally of second order, because it is constructed by means of an Adams-Bashforth discretization for the pressure function. This leads to a more restrictive assumption with respect to the smoothness of the solution, even in the context of consistency of the given set of data. Therefore, in order to make use of the improved regularity results which hold with respect to time derivatives for times $t = \mathcal{O}(1)$, we have to employ time-weights fading out the poor approximating properties in the beginning of the calculation process. A robust higher order discretization scheme should master this singular behavior of the actual solution in the beginning, giving rise to smoothing-type error estimates. This is not the case for the Van Kan scheme. These facts can be observed in the theoretical analysis as well as in numerical test calculations. We refer to the next sections that treat this phenomenon in detail. The reason for this is not only the explicit treatment of the pressure functions in equation (7.9), but relies heavily on the evolutionary character of the quasi-compressibility constraint. Therefore, the Van Kan scheme can be either successfully applied for compatibly posed initial data or starting from times $t = \mathcal{O}(1)$. This leads us to the following statement: Due to the evolutionary perturbation character of this projection method, it does not possess the classical smoothing property and we have to assure compatible initial guesses for pressure and velocity.*

The remainder of this chapter is devoted to the verification of Theorem 7.1. It splits into three main parts: The first is devoted to the study of the influence of those error portions that are due to the second order discretization of the momentum equation in (7.9), while the second part of the proof investigates the perturbation character of the quasi-compressibility constraint in (7.9).

Finally, in the third part we will control the error contribution that is caused by the semi-explicit sort of pressure-approximation. The results of each of these steps are collected in Theorems 7.2 and 7.3 below. Together with the estimates (7.60) and (7.61), they give the proof of Theorem 7.1.

7.1.1   A-priori Bounds for the Crank-Nicolson Scheme

Let us start with an investigation of the Crank-Nicolson scheme, applied to the incompressible Stokes equations (2.1). As has already been pointed out, this will be the first part in order to verify Theorem 7.1. It is:
Given $m \geq 0$, find the tuple of approximative solutions $\{u^{m+1}, p^{m+1}\} \approx \{u(t_{m+1}), p(t_{m+1})\}$ satisfying

$$d_t u^{m+1} - \Delta \overline{u}^{m+1/2} + \nabla \overline{p}^{m+1/2} = \overline{f}^{m+1/2},$$
$$\operatorname{div} u^{m+1} = 0, \qquad u^{m+1}|_{\partial\Omega} = 0. \tag{7.11}$$

Remark 7.2 *Note, that this formulation of the Crank-Nicolson scheme is slightly deviating from the corresponding one in [18] — where it is applied to the incompressible equations. Right here, the treatment of the pressure function is also in the trapezoidal sense. Let us emphasize that this treatment is crucial for the further analysis, whereas the studies in the cited paper relies on the fact that the involved functions are divergence free in the strong sense. The latter scheme causes the approximation features of the pressure to be totally dependent on the related ones for the velocity.*

Owing to the actual discretization in (7.9) and (7.11), we do need initial values for the pressure. For the following analysis it is necessary to get good approximations for p(0). Especially, the choice $p = 0$ is not sufficient for the further studies. The trick is now to solve the "usual" trapezoidal scheme, with $q^1 \equiv \overline{p}^{1/2}$. Subsequently, we set $p^0 \equiv p^1 \equiv q^1$.

In order to start our error analysis, we need certain a-priori results for the semi-discrete solution that are collected in the following lemma. We omit the easy proof that is based on elementary energy estimates.

Lemma 7.1 *Let (A1), (A2) be satisfied. Then, the following a-priori results are valid for $\{u^{m+1}, p^{m+1}\}$ as the solutions of the system (7.11), with a constant C that is only depending on the given data of the continuous problem (2.1),*

$i)$ $\max_{0 \le m \le M} \left\{ \tau_{m+1}^{i/2+j-1} \| d_t^j u^{m+1} \|_i \right\} + k \sum_{m=j-1}^{M} \tau_{m+1}^{2(j-1)} \| d_t^j u^{m+1} \|_1^2 \le C,$

$ii)$ $k \sum_{m=j-1}^{M} \tau_{m+1}^{2j+i-3} \| d_t^j \overline{u}^{m+1/2} \|_i^2 \le C, \qquad 2j + i - 3 \ge 0,$

$iii)$ $\max_{0 \le m \le M} \left\{ \| \nabla \overline{p}^{m+1/2} \| + \tau_{m+1} \| \nabla d_t \overline{p}^{m+1/2} \| \right\}$

$\qquad + k \sum_{m=1}^{M} \tau_{m+1} \| \nabla d_t \overline{p}^{m+1/2} \|^2 + k \sum_{m=2}^{M} \tau_{m+1}^3 \| \nabla d_t^2 \overline{p}^{m+1/2} \|^2 \le C,$

with $i, j \in \{0, 1, 2\}$.

Now it is possible to quantify the consistency error that is arising through the application of the Crank-Nicolson scheme. This is the contents of the next section.

7.1.2 Error Estimates for the Crank-Nicolson Scheme

The objective of this section is the proposal of convergence results for the "standard" Crank-Nicolson scheme. The verification of the statements in the subsequent Theorem 7.2 can e.g. be found in [31], [18]. — If we use the error functions $e^{m+1} := u(t_{m+1}) - u^{m+1}$ and $\eta^{m+1} := p(t_{m+1}) - p^{m+1}$, they are governed by the equalities

$$d_t e^{m+1} + A \overline{e}^{m+1/2} = P_{J_0} R^{m+1}(u),$$
$$\text{div}\, e^{m+1} = 0, \qquad e^{m+1}|_{\partial\Omega} = 0, \qquad e^0 = 0, \tag{7.12}$$

together with the residual

$$R^{m+1}(u) = \frac{1}{k} \left\{ \int_{t_m}^{t_{m+1}} \beta_m(s) u_{ttt}(s)\, ds \right. $$
$$\left. + \int_{t_m}^{t_{m+1}} \{\alpha_m(s) - \beta_m(s)\} \Delta u_{tt}(s)\, ds \right\} \tag{7.13}$$

and the employed weights

$$s \mapsto \alpha_{m+1}(s) \equiv \frac{1}{12}(t_{m+1} - s)(s - t_m)$$

and

$$s \mapsto \beta_{m+1}(s) \equiv \frac{1}{2} \min \left\{ (t_{m+1} - s)^2, (s - t_m)^2 \right\}.$$

If we consider the second function on the interval $[t_m, t_{m+1}]$, we obtain

$$\beta_{m+1}(s)|_{s \in [t_m, t_{m+1}]} \leq C(t_m - s)^2. \tag{7.14}$$

The main results of this section are collected in the next theorem. They verify second order of convergence for the trapezoidal method in the given framework of problems which are considered.

Theorem 7.2 *Let the tuple $\{u^{m+1}, p^{m+1}\}$ be the solution of the Crank-Nicolson scheme (7.11), whereas $\{u(t_{m+1}), p(t_{m+1})\}$ is determined by (2.1). Provided, the regularity assumptions (A1), (A2) hold true, the following error statements for velocity field and pressure function are satisfied,*

i) $\max_{0 \leq m \leq M} \left\{ \tau_{m+1} \| u(t_{m+1}) - u^{m+1} \| \right\} \leq Ck^2$,

ii) $\max_{0 \leq m \leq M} \left\{ \| u(t_{m+1}) - u^{m+1} \|_1 + \| p(t_{m+1}) - p^{m+1} \| \right\} \leq Ck$.

The applied constant C is depending on the given data of the continuous problem (2.1) only.

The kernel of the consistency analysis for the Crank-Nicolson scheme (7.12) is based on sharp estimates for the residual $R^{m+1}(u)$ in diverse operator norms that are given in Lemma 7.2.

Lemma 7.2 *Let $R^{m+1}(u)$ be the above residual term. Then the following estimates are valid, for $r = 0$ and $j \in \{-3, -2, -1, 0, 1\}$,*

i) $k \sum_{m=\max\{0, j+1\}}^{M} \tau^{3+j-r} \| A^{j/2} R^{m+1}(u) \|^2 \leq Ck^4$.

Furthermore, we have the inequality

ii) $\| R^1(u) \|_{-1} \leq C\sqrt{k}$.

Again, the employed constant C is only depending on the given data of the problem. Provided the pressure function in (2.1) is furthermore sufficiently regular with respect to time derivatives, i.e., $p_t \in L^{\infty}(0, T; H^1/\mathbb{R})$, then the corresponding results in i), for $j \in \{-1, 0, 1\}$ are valid for $r = 2$.

Remark 7.3   1. *The time-weights in i) are increasing for increasing values of j (i.e., depending on the chosen norm). We stress the fact that* *the estimates can only be verified for correspondingly shifted summation indices. This is owing to the construction of the parameter function $s \mapsto \alpha_{m+1}(s)$. The time-differences that are involved in this function represent consistency or stability factors — corresponding to the actual situation. The linearity of these factors leads to shifts for stronger norms with respect to the summation index in order to guarantee that the stability character of $s \mapsto \alpha_{m+1}(s)$ is switched on.*

2. *Statement ii) is crucial for the Crank-Nicolson scheme and its $\mathbf{H^1}$ analysis. It is remarkable that $R^1(u)$ cannot be bounded in a stronger norm.*

3. *In the following proof, we will restrict on the cases $j = -3$ and $j = -2$. The verification of the remaining statements is analogous. In this case the second integral factor that takes care of the stability behavior of the continuous solution needs less time-damping factors.*

Proof:
i) *(case $j = -3$)* The proof of this statement uses another a-priori statement for the Stokes equations (2.1) that are formulated on $C(0, T; \mathbf{J_1} \cap \mathbf{H^2})$,

$$u_{ttt} + A u_{tt} = P_{\mathbf{J_0}} f_{tt}. \tag{7.15}$$

If we test this with $A^{-3} u_{ttt}$, we obtain after integration,

$$\int_0^T \|A^{-1} u_{ttt}(s)\|_{-1}^2 \, ds + \|A^{-1} u_{tt}(T)\|^2 \leq$$
$$\leq \|A^{-1} u_{tt}(0)\|^2 + C \int_0^T \|A^{-1} f_{tt}(s)\|_{-1}^2 \, ds. \tag{7.16}$$

Thus, it remains to bound the first term in the last inequality. In order to do so, let us apply the differential operator ∂_t onto the equation (2.1), test this relation with $A^{-2} u_{tt}$ and evaluate it at time $t = 0$,

$$\|A^{-1} u_{tt}(0)\| \leq C\|A^{-1} f_t(0)\| + \|u_t(0)\| \leq C.$$

Therefore, the right hand side of (7.16) can be bounded without using time-weights. Now, the verification of the corresponding error result can be given in the same matter as it is done for the case $j = -2$ in the following.

i) *(case $j = -2$)* If we apply the Young Inequality as well as Cauchy-Schwarz' Inequality, we obtain

$$k \sum_{m=0}^{M} t_{m+1} \| A^{-1} R^{m+1} \|^2$$

$$\leq \frac{1}{k} \sum_{m=0}^{M} \left\{ t_{m+1} \left\{ \int_{t_m}^{t_{m+1}} \beta_{m+1}^{2(1-p)}(s) \, ds \int_{t_m}^{t_{m+1}} \beta_{m+1}^{2p}(s) \| A^{-1} u_{ttt}(s) \|^2 \, ds \right. \right.$$

$$+ \int_{t_m}^{t_{m+1}} (\alpha_{m+1} - \beta_{m+1})^{2(1-p)}(s) \, ds \times \qquad\qquad (7.17)$$

$$\left. \left. \times \int_{t_m}^{t_{m+1}} (\alpha_{m+1} - \beta_{m+1})^{2p}(s) \| u_{tt}(s) \|^2 \, ds \right\} \right\}.$$

For $p = 0$, we can bound the two terms on the right hand side of (7.17). The further treatment of those integral terms having β_{m+1} as a kernel is quite easy, whereas the ones with α_{m+1} need a more careful analysis. Owing to the fact, that the norm $\| A^{-1} \cdot \|$ is sufficiently weak, a small time-weight is enough to get a stability statement. Therefore, we can set $p = 0$, and the second factor of the second sum is bounded, the first giving optimal order of convergence.   □

We will end this section with the verification of another estimate for the error in the pressure in the beginning of the subsequent item 1). This statement is necessary for the proof of Theorem 7.2 in order to succeed in transferring the approximation result (7.4) to the Crank-Nicolson system (7.11). This permits us to think of the equations (7.11) as the reference equations. Finally, item 2) renders the transfer of the results in *Postulate B_2* to the solution of the semi-discretized equations (7.11). This enables us to refer to the regularity statements in *Postulate $\widetilde{B}_2$* for the solution of system (7.11).

1) *Verification of another convergence result for the pressure of system (7.11) at the initial time-point, using the assumptions of Postulate B_2:*

We employ the regularity results of (7.10) for the solution of the Stokes system (2.1). — The error identities are as follows, in the non-projected version,

$$d_t e^{m+1} - \Delta \bar{e}^{m+1/2} + \nabla \bar{\eta}^{m+1/2} = R_1^{m+1}(u) + R_2^{m+1}(\nabla p),$$
$$\text{div} e^{m+1} = 0, \tag{7.18}$$
$$e^{m+1}|_{\partial \Omega} = 0, \qquad e^0 = 0, \qquad \eta^0 = p(t_0) - p^0.$$

We have $R_1^{m+1}(u) \equiv R^{m+1}(u)$, as it is given in equation (7.13), and

$$R_2^{m+1}(\nabla p) = \frac{1}{2}k\left\{\nabla p_t(\xi')|_{\xi' \in [t_m, t_{m+1/2}]} - \nabla p_t(\xi'')|_{\xi'' \in [t_{m+1/2}, t_{m+1}]}\right\}.$$
$$\tag{7.19}$$

The objective is now the verification of the estimate

$$\|\bar{p}^{1/2} - p(t_{1/2})\|_1 \leq Ck. \tag{7.20}$$

This result can be used in order to obtain a pointwise statement regarding the initial approximation properties of the pressure function, using the strategy given in Remark 7.2,

$$\|p^0 - p(t_0)\|_1 \leq Ck. \tag{7.21}$$

This will be necessary for the remainder of the proof for the Van Kan scheme.

Proof:
This is to verify (7.20). We start from the projected version of the error identities, (7.12), for $m = 0$. This will be tested with $kd_t e^1$, and we have

$$k\|d_t e^1\|^2 + \|\nabla e^1\|^2 = (R_1^1(u), e^1) \leq \|R_1^1(u)\|_{-1}\|\nabla e^1\|,$$

or,

$$\|d_t e^1\|^2 \leq C\frac{1}{k}\|R_1^1(u)\|_{-1}^2 \leq Ck^2. \tag{7.22}$$

Note that we have used the ingredients of *Postulate B_2*, which make it possible to use Lemma 7.2, i), with $j = -1$ and $r = 2$. — Furthermore, if we test the first equation (with $m = 0$) of the same system with $A\bar{e}^{1/2}$, we obtain

$$\|Ae^1\|^2 \leq \|R_1^1(u)\|^2 + \|d_t e^1\|^2 \leq Ck^2. \tag{7.23}$$

The upper bound for the first term on the right hand side of the first inequality is again a result from Lemma 7.2, i), for $j = 0$ and $r = 2$. — We can use the results (7.22) and (7.23) to get (7.20). □

2) Postulate $\widetilde{B_2}$: Verification of additional a-priori bounds for the solution of (7.12), based on Postulate B_2:

Employing the difference operator d_t onto the first equation in (7.12), we get the estimate

$$\|\Delta d_t \bar{e}^{m+1/2}\| \leq \|d_t^2 e^{m+1}\| + \|d_t R^{m+1}(u)\|. \tag{7.24}$$

We will sketch the arguments that lead to statements analogously to those in *Postulate B_2*, but now for problem (7.11). Thanks to Lemma 7.2, *i*), for $j = -1$ and $r = 2$, we may skip the time-weight τ_{m+1}^2 as it has been employed in Theorem 7.2, for the present case $r = 2$, and we have

$$\|e^{m+1}\| \leq Ck^2.$$

This can be inserted in (7.24). If we further remember Lemma 7.2, *i*), for $j = 1$ and $r = 2$, we obtain

$$\begin{aligned}
\|\Delta d_t \bar{u}^{m+1/2}\| &\leq \|\Delta d_t \bar{e}^{m+1/2}\| + \|\Delta d_t u(t_{m+1/2})\| \\
&\leq C\left\{1 + \|\Delta u_t(\xi)\|\Big|_{\xi \in [t_{m-1/2}, t_{m+1/2}]}\right\} \leq C,
\end{aligned} \tag{7.25}$$

owing to *Postulate B_2* and the Taylor evolution formula. — An easy consideration now leads to the discrete version of *Postulate B_2*,

$$\|\Delta d_t \bar{u}^{m+1/2}\| + \|\nabla d_t \bar{p}^{m+1/2}\| \leq C,$$

which will be denoted as *Postulate $\widetilde{B_2}$*. — Let us remark, that the estimates given in Lemma 7.1 can be improved to statements with powers of time-weights that are diminished for solutions that satisfy *Postulate $\widetilde{B_2}$*.

7.2 Proposal of a Pressure Correction Version of the Stokes Equations

In the previous section we have presented a consistency error analysis as a first step in order to verify Theorem 7.1. This section is devoted to quantifying the perturbation character of the Van Kan scheme that results from the relaxed quasi-compressibility method, cf. (7.9). In order to investigate the perturbation effects on the approximation in a closer way, we start with the analysis of a fully implicit version:

Given the tuple $\{u_\varepsilon^m, p_\varepsilon^m\}$, find the solution $\{u_\varepsilon^{m+1}, p_\varepsilon^{m+1}\}$ of the equations

$$d_t u_\varepsilon^{m+1} - \Delta \bar{u}_\varepsilon^{m+1/2} + \nabla \bar{p}_\varepsilon^{m+1/2} = \bar{f}^{m+1/2},$$
$$\mathrm{div} u_\varepsilon^{m+1} - \varepsilon \Delta d_t p_\varepsilon^{m+1} = 0,$$
$$u_\varepsilon^0 = u^0 \in \mathbf{J}_1 \cap \mathbf{H}^2, \qquad \|p_\varepsilon^0 - p^0\|_1 \leq C\sqrt{\varepsilon},$$
$$u_\varepsilon^{m+1}|_{\partial\Omega} = 0, \qquad \partial_n d_t p_\varepsilon^{m+1}|_{\partial\Omega} = 0.$$

(7.26)

We use the identification $\varepsilon = \beta k^2$ in order to refer to the pressure stabilization character of the problem. If we set $e^{m+1} := u^{m+1} - u_\varepsilon^{m+1}$ and $\eta^{m+1} := p^{m+1} - p_\varepsilon^{m+1}$, we get

$$d_t e^{m+1} - \Delta \bar{e}^{m+1/2} + \nabla \bar{\eta}^{m+1/2} = 0,$$
$$\mathrm{div} e^{m+1} - \varepsilon \Delta d_t \eta^{m+1} = -\varepsilon \Delta d_t p^{m+1},$$
$$e^0 = 0, \qquad \|\eta^0\|_1 \leq C\sqrt{\varepsilon}.$$
$$e^{m+1}|_{\partial\Omega} = 0, \qquad \partial_n d_t \eta^{m+1}|_{\partial\Omega} = \partial_n d_t p^{m+1}|_{\partial\Omega}.$$

(7.27)

We will give error bounds for the solution of this system in the following. As has already been pointed out, it is necessary to split the appearing mathematical phenomena in (7.27) and to treat them independently. These investigations will be presented in the next two sections. The results are collected in a theorem.

Theorem 7.3 *Let $\{u_\varepsilon^{m+1}, p_\varepsilon^{m+1}\}$ be the solution of the equations (7.26), with $\varepsilon \equiv \mathcal{O}(k^2)$ and $k < 1$. The exact solution $\{u(t_{m+1}, p(t_{m+1})\}$ of (2.1) is assumed to satisfy (A1), (A2) and Postulate B_2. Then, the following error bounds are valid, with a constant C that is solely depending on the given data of problem (2.1),*

i) $\max_{0 \leq m \leq M} \left\{ \sqrt{\tau_{m+1/2}} \| u(t_{m+1/2}) - \bar{u}_\varepsilon^{m+1/2} \| \right\} \leq C\varepsilon \log\frac{1}{k},$

ii) $\max_{0 \leq m \leq M} \left\{ \sqrt{\tau_{m+1/2}} \| u(t_{m+1/2}) - \bar{u}_\varepsilon^m \|_1 \right\} \leq C\sqrt{\varepsilon} \log\frac{1}{k},$

iii) $\max_{0 \leq m \leq M} \left\{ \sqrt{\tau_{m+1/2}} \| p(t_{m+1/2}) - \bar{p}_k^m \| \right\} \leq C\sqrt{\varepsilon} \log\frac{1}{k}.$

Remark 7.4 *The results presented in Theorem 7.2 for the Crank-Nicolson scheme can easily be transfered in a form corresponding to the one in the actual theorem by means of Taylor's formula.*

7.2.1   Proposal and Analysis of an Auxiliary Problem

It is the subject of this subsection to study the impact of the pressure correction in (7.26), without taking care of the evolutionary transport of the
momentum equation. Thus the problem of consideration is as follows:
Find $\{U_\varepsilon^{m+1}, P_\varepsilon^{m+1}\}$ for a given quadruple $\{U_\varepsilon^m, P_\varepsilon^m, d_t u^{m+1}, \overline{f}^{m+1/2}\}$,

$$
\begin{aligned}
&- \Delta \overline{U}_\varepsilon^{m+1/2} + \nabla \overline{P}_\varepsilon^{m+1/2} = -d_t u^{m+1} + \overline{f}^{m+1/2}, \\
&\operatorname{div} U_\varepsilon^{m+1} - \varepsilon \Delta d_t P_\varepsilon^{m+1} = 0, \\
&U_\varepsilon^{m+1}|_{\partial\Omega} = 0, \qquad \partial_n d_t P_\varepsilon^{m+1}|_{\partial\Omega} = 0, \\
&U_\varepsilon^0 = u^0, \qquad P_\varepsilon^0 = p^0.
\end{aligned}
\tag{7.28}
$$

This system is well-defined for every iteration step. The errors $E^{m+1} :=$
$u^{m+1} - U_\varepsilon^{m+1}$ and $\Pi^{m+1} := p^{m+1} - P_\varepsilon^{m+1}$ which are committed if we change
from (7.11) to (7.28), then read as:

$$
\begin{aligned}
&- \Delta \overline{E}^{m+1/2} + \nabla \overline{\Pi}^{m+1/2} = 0, \\
&\operatorname{div} E^{m+1} - \varepsilon \Delta d_t \Pi^{m+1} = -\varepsilon \Delta d_t p^{m+1}, \\
&E^{m+1}|_{\partial\Omega} = 0, \qquad \partial_n d_t \Pi^{m+1}|_{\partial\Omega} = \partial_n d_t p^{m+1}|_{\partial\Omega}, \\
&E^0 = 0, \qquad \Pi^0 = 0.
\end{aligned}
\tag{7.29}
$$

The first step for an error analysis of problem (7.28) is now to investigate
the stability properties of this system. We will summarize the results of it in
the next lemma.

Lemma 7.3 *The functions* $\{u^{m+1}, p^{m+1}\}$ *that are to be approximated as the
solutions of (7.11) are assumed to satisfy (A1), (A2) and Postulate* $\widetilde{B}_2$. *Then,
setting* $\varepsilon = \mathcal{O}(k^2)$, *we get the following a-priori statements for the solution*
$\{\overline{U}_\varepsilon^{m+1/2}, \overline{P}_\varepsilon^{m+1/2}\}$ *of system (7.28), employing a constant* C *that is only depending on the given data of the problem,*

i) $\max_{0 \leq m \leq M} \|\nabla \overline{P}_\varepsilon^{m+1/2}\| + k \sum_{m=0}^{M} \|\nabla \overline{U}_\varepsilon^{m+1/2}\|^2$

$\qquad\qquad + k^2 \sum_{m=1}^{M} \|\nabla d_t \overline{P}_\varepsilon^{m+1/2}\|^2 \leq C,$

ii) $\max_{1 \leq m \leq M} \|\nabla d_t \overline{P}_\varepsilon^{m+1/2}\| + k \sum_{m=1}^{M} \|\nabla d_t \overline{U}_\varepsilon^{m+1/2}\|^2$

$\qquad\qquad + k^2 \sum_{m=2}^{M} \|\nabla d_t^2 \overline{P}_\varepsilon^{m+1/2}\|^2 \leq C,$

$iii)$ $\max_{1 \le m \le M} \|\nabla d_t \overline{U}_{\varepsilon}^{m+1/2}\| + k \sum_{m=2}^{M} \tau_{m+1} \|\nabla d_t^2 \overline{U}_{\varepsilon}^{m+1/2}\|^2 \le C.$

$iv)$ $k \sum_{m=2}^{M} \tau_{m+1} \|\nabla d_t^2 \overline{P}_{\varepsilon}^{m+1/2}\|^2 \le C.$

Proof:

We start from the error equalities (7.29). If we test the first equation in (7.29) with $\overline{E}^{m+1/2}$ and sum up over all iteration steps, we obtain

$$
k \sum_{m=1}^{M} \|\nabla \overline{E}^{m+1/2}\|^2 + \varepsilon \|\nabla \overline{\Pi}^{M+1/2}\|^2 + \varepsilon k^2 \sum_{m=1}^{M} \|\nabla d_t \overline{\Pi}^{m+1/2}\|^2
$$

$$
\le C\varepsilon \frac{1}{t_{M+1}} k \sum_{m=1}^{M} \|\nabla \overline{\Pi}^{m+1/2}\|^2 + C\varepsilon t_{M+1} k \sum_{m=1}^{M} \|\nabla d_t \overline{p}^{m+1/2}\|^2 \quad (7.30)
$$

$$
+ \varepsilon \|\nabla \overline{\Pi}^{1/2}\|^2.
$$

In order to bound the last term on the right hand side of the inequality sign, we look at the first equality in (7.29), for $m = 0$, and test with E^1. This immediately gives,

$$
k\|\nabla E^1\|^2 + \varepsilon \|\nabla \Pi^1\|^2 \le k\|\nabla E^0\|^2 + \varepsilon \|\nabla \Pi^0\|^2 + C\varepsilon k^2 \|\nabla d_t p^1\|^2. \quad (7.31)
$$

With that, the first term on the right hand side of (7.30) can be controlled by $C\varepsilon k^2$. The first term on the same side can be bounded by means of the discrete Gronwall estimate, and $i)$ is proved.

The verification of the result $ii)$ is analogous. If we apply the difference operator d_t onto the first equation of (7.29) and test the resulting relation with $d_t \overline{E}^{m+1/2}$, we get after summation over all iteration steps up to the time t_{M+1},

$$
k \sum_{m=2}^{M} \|\nabla d_t \overline{E}^{m+1/2}\|^2 + \varepsilon \|\nabla d_t \overline{\Pi}^{M+1/2}\|^2 + \varepsilon k^2 \sum_{m=2}^{M} \|\nabla d_t^2 \overline{\Pi}^{m+1/2}\|^2
$$

$$
\le C\delta \varepsilon k \sum_{m=2}^{M} \|\nabla d_t^2 \overline{p}^{m+1/2}\|^2 + \frac{1}{4\delta} \varepsilon k \sum_{m=2}^{M} \|\nabla d_t \overline{\Pi}^{m+1/2}\|^2 \quad (7.32)
$$

$$
+ \varepsilon \|\nabla d_t \overline{\Pi}^{3/2}\|^2, \qquad \delta > 0.
$$

We have to bound the last term in this inequality. Therefore, we return back to the first identity in (7.32), employ the difference operator d_t onto it and finally set $m = 1$. After having tested this identity with $\overline{E}^{3/2}$, we get

$$\|\nabla \overline{E}^{3/2}\|^2 + k^2 \|\nabla d_t E^{3/2}\|^2 + \varepsilon k \|\nabla d_t \overline{\Pi}^{3/2}\|^2$$
$$\leq \|\nabla \overline{E}^{1/2}\|^2 + \varepsilon k \|\nabla d_t \overline{p}^{3/2}\|^2. \tag{7.33}$$

The first term on the right hand side is bounded by $C \varepsilon k$, owing to (7.31). This allows to control the last term in (7.32) by $C \varepsilon$. — The second sum on the right hand side of inequality (7.32) can be treated, using the Discrete Gronwall inequality. This gives an upper bound $C \varepsilon$, for $\delta = C t_{M+1}$. This finishes the proof of the second assertion in the above lemma.

In order to verify item iii), let us employ the difference operator d_t^2 onto the first equation in (7.29) and test it with $d_t \overline{E}^{m+1/2}$. Finally weighting the relation with τ_{m+1} and summation give

$$\tau_{M+1} \|\nabla d_t \overline{E}^{M+1/2}\|^2 + k^2 \sum_{m=2}^{M} \tau_{m+1} \|\nabla d_t^2 \overline{E}^{m+1/2}\|^2$$

$$+ \varepsilon k \sum_{m=2}^{M} \tau_{m+1} \|\nabla d_t^2 \overline{\Pi}^{m+1/2}\|^2$$

$$\leq C \varepsilon k \sum_{m=2}^{M} \tau_{m+1} \|\nabla d_t^2 \overline{p}^{m+1/2}\|^2 \tag{7.34}$$

$$+ 2k \|\nabla d_t \overline{E}^{3/2}\|^2 + k \sum_{m=2}^{M} \|\nabla d_t \overline{E}^{m+1/2}\|^2,$$

and we can use (7.32) to conclude the assertion.   □

After these preliminary results we are able to proof the following error statements for the solution of system (7.29):

Lemma 7.4 *Let us assume that the divergence free solution of problem (7.11) satisfies (A1), (A2) and Postulate B_2. Then, the solution of (7.28) satisfies the following approximation properties, for $\varepsilon = \mathcal{O}(k^2)$,*

i) $\max_{0 \leq m \leq M} \|\overline{u}^{m+1/2} - \overline{U}_\varepsilon^{m+1/2}\|$

$$+\left(k\sum_{m=1}^{M}\tau_{m+1}\|d_t\{\overline{u}^{m+1/2}-\overline{U}_{\varepsilon}^{m+1/2}\}\|^2\right)^{1/2}\leq Ck^2,$$

$ii)$ $\max_{0\leq m\leq M}\sqrt{\tau_{m+1/2}}\{\|\overline{u}^{m+1/2}-\overline{U}_{\varepsilon}^{m+1/2}\|_1$
$$+\|\overline{p}^{m+1/2}-\overline{P}_{\varepsilon}^{m+1/2}\|\}\leq Ck.$$

Proof:

The results in $ii)$ can be verified, using simple energy estimates. — In order to verify the statements presented in $i)$, we employ stationary duality arguments of the type: For given values $\{w^m,q^m,\overline{g}^{m+1/2}\}$, determine $\{w^{m+1},q^{m+1}\}$ as the solution of the incompressible problem

$$-\Delta\overline{w}^{m+1/2}+\nabla\overline{q}^{m+1/2}=\overline{g}^{m+1/2},$$

$$\text{div}w^{m+1}=0,\qquad q^0=0,\qquad w^0=0. \tag{7.35}$$

In order to verify the first error statement of the theorem, we set $\overline{g}^{m+1/2}=\overline{E}^{m+1/2}$. Testing of the first equation in (7.35) with $\overline{E}^{m+1/2}$ gives

$$(\nabla\overline{w}^{m+1/2},\nabla\overline{E}^{m+1/2})+(\nabla\overline{q}^{m+1/2},\overline{E}^{m+1/2})=\|\overline{E}^{m+1/2}\|^2. \tag{7.36}$$

Now, we test the first equation in (7.29) with the divergence-free velocity function $\overline{w}^{m+1/2}$,

$$(\nabla\overline{w}^{m+1/2},\nabla\overline{E}^{m+1/2})=0.$$

This leads us to the following conclusion, with $\delta>0$ arbitrary,

$$\|\overline{E}^{m+1/2}\|^2\leq C\delta\varepsilon^2\|\nabla d_t\overline{P}_{\varepsilon}^{m+1/2}\|^2+\frac{1}{4\delta}\|\nabla\overline{q}^{m+1/2}\|^2,$$

$$\leq C\delta\varepsilon^2\|\nabla d_t\overline{P}_{\varepsilon}^{m+1/2}\|^2. \tag{7.37}$$

The last result is a consequence of absorption for sufficiently large $\delta>0$. The right hand side is now bounded by $C\varepsilon^2$, owing to Lemma 7.3.

In order to verify the second assertion in part $i)$, we again set $\overline{g}^{m+1/2}=\overline{E}^{m+1/2}$ and apply the difference operator d_t onto the first equation in (7.35). Then we test this one with $d_t\overline{E}^{m+1/2}$ and proceed as before. Together with the a-priori statement $iv)$ in Lemma 7.3 we can finish the proof of this lemma.

□

7.2.2   Analysis of the Pressure Correction Method (7.26)

Owing to the error estimates in Lemma 7.4 for the velocity function of
the auxiliary problem (7.28), it is sufficient if we restrict our investigations, instead of dealing with (7.27). Employing the error notations $e_\varepsilon^{m+1} :=$
$u_\varepsilon^{m+1} - U_\varepsilon^{m+1}$ and $\eta_\varepsilon^{m+1} := p_\varepsilon^{m+1} - P_\varepsilon^{m+1}$, the "reduced" equations are

$$d_t e_\varepsilon^{m+1} - \Delta \bar{e}_\varepsilon^{m+1/2} + \nabla \bar{\eta}_\varepsilon^{m+1/2} = d_t E^{m+1},$$
$$\operatorname{div} e_\varepsilon^{m+1} - \varepsilon \Delta d_t \eta_\varepsilon^{m+1} = 0,$$
$$e_\varepsilon^0 = 0, \qquad \|\nabla \eta_\varepsilon^0\| \le C\sqrt{\varepsilon}, \tag{7.38}$$
$$e_\varepsilon^{m+1}|_{\partial\Omega} = 0, \qquad \partial_n d_t \eta_\varepsilon^{m+1}|_{\partial\Omega} = 0.$$

In order to be able to treat the term on the right hand side of the first
equation by means of an estimate as it is given in Lemma 7.4, let us average
the first equation, summing the corresponding identity for the actual and
the previous step and dividing by 2. Finally, testing with $\bar{\bar{e}}_\varepsilon^m$ and taking the
time-weight τ_m leads to

$$\tau_{M+1}\|\bar{e}_\varepsilon^{M+1/2}\|^2 + k \sum_{m=2}^M \tau_m \|\nabla \bar{\bar{e}}_\varepsilon^m\|^2$$

$$+ \varepsilon \tau_{M+1}\|\nabla \bar{\bar{\eta}}_\varepsilon^M\|^2 + \varepsilon k^2 \sum_{m=2}^M \tau_{m+1}\|\nabla d_t \bar{\bar{\eta}}_\varepsilon^m\|^2$$

$$\le 2\varepsilon k\|\nabla \bar{\bar{\eta}}_\varepsilon^1\|^2 + 2k\|\bar{e}_\varepsilon^{3/2}\|^2 + Ck \sum_{m=2}^M \tau_{m+1}^2\|d_t \bar{E}^{m+1/2}\|^2 \tag{7.39}$$

$$+ Ck \sum_{m=2}^M \|\bar{e}_\varepsilon^{m+1/2}\|^2 + C\varepsilon k \sum_{m=2}^M \|\nabla \bar{\bar{\eta}}_\varepsilon^m\|^2.$$

It is clear, that a corresponding inequality holds true at time $t_{M-1/2}$. If
we add these both relations, we can estimate the resulting left side by the
(double) of the right hand side of (7.39). We mention these evident facts,
because they are needed in the following. We will refer to this inequality,
without causing misunderstandings. - We will begin with the analysis of
the first two terms on the right hand side of inequality (7.39). Owing to
inequality (7.31), an analysis of problem (7.27) is sufficient therefore. To
reach our aim, we start with investigations of the error accumulation in the
first time-step. Thus, setting $m = 1$ in (7.27) and testing with ke^1 gives,

$$\|e^1\|^2 + k\|\nabla e^1\|^2 + k\|\nabla \bar{e}^{1/2}\|^2 + \varepsilon\|\nabla \eta^1\|^2$$
$$\le \|e^0\|^2 + k\|\nabla e^0\|^2 + \varepsilon\|\nabla \eta^0\|^2 + C\varepsilon k^2\|\nabla d_t p^1\|^2 \le C\varepsilon k^2. \tag{7.40}$$

The last term in front of the last inequality sign is identically zero, owing to the definition of the start pressure in (7.11). A corresponding analysis can be carried out for the case $m = 2$. This permits us to bound the first two terms on the right hand side of (7.39) through $C\varepsilon k^3$. Owing to Lemma 7.4 it remains to analyze the last two integral terms in (7.39).

The term $k\sum_{m=2}^{M}\|\bar{e}_{\varepsilon}^{m+1/2}\|^2$: In order to get an optimal bound for it, it is enough to bound the sum $k\sum_{m=2}^{M}\|\bar{e}^{m+1/2}\|^2$, owing to Lemma 7.4. — In order to succeed in that, we will introduce another backward Stokes problem, for velocities that are divergence-free: Given $t_{M+1} > 0$, find the solution $\{w^m, q^m\}$ of

$$
\begin{aligned}
d_t w^{m+1} + \Delta\bar{w}^{m+1/2} - \nabla\bar{q}^{m+1/2} &= \bar{g}^{m+1/2}, \\
\mathrm{div}w^m = 0, \qquad w^{M+1} = 0, \qquad q^{M+1} &= 0.
\end{aligned}
\tag{7.41}
$$

We set $\bar{g}^{m+1/2} = \bar{e}^{m+1/2}$ and test the first identity with $\bar{e}^{m+1/2}$. Also, let us test the first equation in (7.27) with $\bar{w}^{m+1/2}$. Owing to the incompressibility of the latter function we obtain, after adding both identities and subsequent summation over all iteration steps, thanks to absorption on the left hand side,

$$
k\sum_{m=0}^{M}\|\bar{e}^{m+1/2}\|^2 \le C\varepsilon^2 k\sum_{m=1}^{M}\|\nabla d_t\bar{p}_\varepsilon^{m+1/2}\|^2.
\tag{7.42}
$$

Therefore, the problem of getting optimal error statements is reduced to the derivation of an a-priori estimate for the pressure function of system (7.27). In order to get a parameter independent a-priori bound, we return back to (7.27), applying the difference operator d_t onto the first identity. Then, we test it with $d_t\bar{e}^{m+1/2}$. After summation over all iteration steps, we arrive at

$$
\|d_t e^{M+1}\|^2 + k\sum_{m=1}^{M}\|\nabla d_t\bar{e}^{m+1/2}\|^2 + \varepsilon\|\nabla d_t\bar{\eta}^{M+1/2}\|^2
$$

$$
+ \varepsilon k^2 \sum_{m=1}^{M}\|\nabla d_t^2\bar{\eta}^{m+1/2}\|^2
$$

$$
\le \|d_t e^1\|^2 + \varepsilon\|\nabla d_t\bar{\eta}^{3/2}\|^2 + \varepsilon k\sum_{m=1}^{M}\delta\|\nabla d_t^2\bar{p}^{m+1/2}\|^2
\tag{7.43}
$$

$$
+ \varepsilon k\sum_{m=1}^{M}\frac{1}{4\delta}\|\nabla d_t\bar{\eta}^{m+1/2}\|^2, \qquad \delta > 0.
$$

If we set $\delta \equiv \delta(t_{m+1}) = (1 + t_M + \log\frac{1}{k})$, we can control the first sum on the right hand side of this inequality through $C(1 + t_M + \log\frac{1}{k})\varepsilon$. This is owing to Lemma 7.1, in its weaker form which holds thanks to Postulate $\widetilde{B}_2$. Now, we can apply a standard statement for harmonic series, while applying the Gronwall lemma: Owing to the logarithmic term that appears in the choice for δ, the argument of the exponential function is now of order $\mathcal{O}(1)$, because of the fact that we have

$$\left(1 + \log\frac{t_M}{k}\right)^{-1} \sum_{m=1}^{M} \frac{1}{m} < 1.$$

By means of the parallelogram identity and the inequality (7.40) we obtain the boundedness of the first two terms on the right hand side of (7.43) by $C\varepsilon$. — Owing to the a-priori bound for the divergence-free Stokes-problem (7.11), as it is given in Lemma 7.1, we gain a pointwise bound for the pressure function gradient of problem (7.26), differentiated in time. We can now use corresponding statements for the solution of (7.11) to apply (7.43) for the derivation of a-priori bounds for $\{u_\varepsilon^{m+1}, p_\varepsilon^{m+1}\}$. They are collected in another lemma, which is of importance for the further analysis.

Lemma 7.5 *Suppose that the solution $\{u^{m+1}, p^{m+1}\}$ of problem (7.11) satisfies the assumptions (A1), (A2) and Postulate $\widetilde{B}_2$. Then, we have the following a-priori bounds for the solution of problem (7.26), with a constant C that depends on the given data of the actual problem only, for $\varepsilon = \mathcal{O}(k^2)$ and $0 < k < 1$,*

i) $\max_{1 \leq m \leq M} \left\{ \|d_t u_\varepsilon^{m+1}\| + \|\nabla d_t \overline{p}_\varepsilon^{m+1/2}\| \right\} \leq C(1 + \log\frac{1}{k}),$

ii) $k \sum_{m=1}^{M} \|\nabla d_t \overline{u}_\varepsilon^{m+1/2}\|^2 + k^2 \sum_{m=2}^{M} \|\nabla d_t^2 \overline{p}_\varepsilon^{m+1/2}\|^2 \leq C(1 + \log\frac{1}{k}).$

This allows us to bound the right hand side of inequality (7.42) through $C\varepsilon^2$, and it remains to analyze the last terms in (7.39).

Remarks on the term $\varepsilon k \sum_{m=2}^{M} \|\nabla \overline{\eta}_\varepsilon^m\|^2$ in (7.39): In order to bound this sum in an optimal way, we apply the following duality argument: Given $t_{M+1} > 0$, find the solution $\{\kappa^m, \rho^m\}$ of the system

$$d_t \overline{\kappa}^{m+1/2} + \Delta \overline{\kappa}^m - \nabla \overline{\rho}^m = 0,$$

$$\operatorname{div} \overline{\kappa}^{m-1/2} + \varepsilon \Delta d_t \overline{\rho}^{m+1/2} = -\varepsilon \Delta \overline{\eta}_\varepsilon^{m-1/2},$$

$$\partial_n d_t \overline{\rho}^{m+1/2}|_{\partial\Omega} = -\partial_n \overline{\eta}_\varepsilon^{m-1/2}|_{\partial\Omega}, \qquad \kappa^m|_{\partial\Omega} = 0,$$

$$\overline{\kappa}^{M+1/2} = 0, \qquad \overline{\rho}^{M+1/2} = 0.$$

$$(7.44)$$

Before we continue with our investigations, let us present some remarks.

Remark 7.5 *Note, that this auxiliary problem has been chosen to be dually posed to the equations (7.38) and not dually to problem (7.27), like in the above discussion of the integrally averaged velocity error. This auxiliary problem takes benefit from the formulation of the boundary conditions in system (7.44) in an essential way, see below. This requires a-priori estimates for the system (7.44), that will be derived in the next insertion.*

The subsequent section is dealing with the verification of some a-priori estimates for the solution of system (7.44). They will be employed for the derivation of an optimal error control of the term in question.

   Insertion: This insertion is devoted to the investigation of the stability properties of system (7.44). In order to simplify the study, we turn around the direction of the evolution process in (7.44) and arrive at equalities of type: Find solutions $\{\bar{\bar{\kappa}}^{m+1}, \bar{\bar{\rho}}^{m+1}\}$ of

$$d_t\bar{\bar{\kappa}}^{m+1/2} - \Delta\bar{\bar{\kappa}}^m + \nabla\bar{\bar{\rho}}^m = 0,$$

$$\mathrm{div}\bar{\bar{\kappa}}^{m-1/2} - \varepsilon\Delta d_t\bar{\bar{\rho}}^{m+1/2} = \varepsilon\Delta\bar{\bar{g}}^{m+1/2},$$

$$\partial_n d_t\bar{\bar{\rho}}^{m+1/2}|_{\partial\Omega} = -\partial_n\bar{\bar{g}}^{m+1/2}|_{\partial\Omega}, \qquad \bar{\bar{\kappa}}^{m+1/2}|_{\partial\Omega} = 0,$$

$$\bar{\bar{\kappa}}^{1/2} = 0, \qquad \bar{\bar{\rho}}^{1/2} = 0.$$

$$(7.45)$$

If we test the first equation with $\bar{\bar{\kappa}}^m$, we obtain after summation,

$$\|\bar{\bar{\kappa}}^{M+1/2}\|^2 + k\sum_{m=2}^{M}\|\nabla\bar{\bar{\kappa}}^m\|^2 + \varepsilon\|\nabla\bar{\bar{\rho}}^M\|^2 + \varepsilon k^2\sum_{m=2}^{M}\|\nabla d_t\bar{\bar{\rho}}^m\|^2$$

$$\leq \varepsilon\|\nabla\bar{\bar{\rho}}^1\|^2 + \|\bar{\bar{\kappa}}^{3/2}\|^2 + Ct_{M+1}\varepsilon k\sum_{m=2}^{M}\|\nabla\bar{\bar{g}}^m\|^2. \qquad (7.46)$$

We have used Gronwall's lemma, together with a normalizing time-weight. The first two terms on the right hand side of (7.46) are controllable, owing to the vanishing initial data in (7.45). This gives us

$$\bar{\bar{\kappa}}^{3/2} - k\Delta\bar{\bar{\kappa}}^{3/2} + \nabla\bar{\bar{\rho}}^{3/2} = 0. \qquad (7.47)$$

The second relation in (7.45) gives us, for the case $m = 1$ by means of the same reason,

$$k\mathrm{div}\bar{\bar{\kappa}}^{3/2} - \varepsilon\Delta\bar{\bar{\rho}}^{3/2} = -\varepsilon k\Delta\bar{\bar{g}}^{3/2}.$$

If we test the equation (7.47) with $\overline{\overline{\kappa}}^{3/2}$ and make use of the last identity, we obtain the inequality

$$\|\overline{\overline{\kappa}}^{3/2}\|^2 + k\|\nabla\overline{\overline{\kappa}}^{3/2}\|^2 + \varepsilon\|\nabla\overline{\overline{\rho}}^{3/2}\|^2 \leq C\varepsilon k^2 \|\nabla\overline{g}^{3/2}\|^2.$$

We can proceed equally in the same way for the second time-step, such that inequality (7.46) reads as:

$$\|\overline{\overline{\kappa}}^{M+1/2}\|^2 + k\sum_{m=2}^{M}\|\nabla\overline{\overline{\kappa}}^{m}\|^2 + \varepsilon\|\nabla\overline{\overline{\rho}}^{M}\|^2 + \varepsilon k^2 \sum_{m=2}^{M}\|\nabla d_t\overline{\overline{\rho}}^{m}\|^2$$

$$\leq C\varepsilon k^2 + C\varepsilon t_{M+1} k \sum_{m=2}^{M}\|\nabla\overline{g}^{m}\|^2. \tag{7.48}$$

This completes our analysis for the system (7.45). This a-priori bound will be used for the duality argument (7.44).

We continue our proof after this insertion. Two times applying the average operator and testing the second equation in (7.44) with $\overline{\overline{\eta}}_\varepsilon^{m-1}$ gives

$$\varepsilon\|\nabla\overline{\overline{\eta}}_\varepsilon^{m-1}\|^2 = -\varepsilon(\nabla d_t\overline{\overline{\rho}}^{m}, \nabla\overline{\overline{\eta}}_\varepsilon^{m-1}) - (\overline{\overline{\kappa}}^{m-1}, \nabla\overline{\overline{\eta}}_\varepsilon^{m-1}). \tag{7.49}$$

On the other hand, correspondingly testing the second identity in (7.38) with $\overline{\overline{\rho}}^{m}$ after having applied the average operator twice, there holds

$$(\overline{\overline{e}}_\varepsilon^{m}, \nabla\overline{\overline{\rho}}^{m}) - \varepsilon(\nabla\overline{\overline{\rho}}^{m}, \nabla d_t\overline{\overline{\eta}}_\varepsilon^{m}) = 0. \tag{7.50}$$

If we add both identities (7.49) and (7.50), we get

$$\varepsilon\|\nabla\overline{\overline{\eta}}_\varepsilon^{m-1}\|^2 = \varepsilon d_t(\nabla\overline{\overline{\rho}}^{m}, \nabla\overline{\overline{\eta}}_\varepsilon^{m}) - (\overline{\overline{\kappa}}^{m-1}, \nabla\overline{\overline{\eta}}_\varepsilon^{m-1}) + (\overline{\overline{e}}_\varepsilon^{m}, \nabla\overline{\overline{\rho}}^{m}).$$

Now, we sum over all time-steps and make use of the a-priori estimate (7.48) that gives a bound for the pressure gradient. Thus, we obtain after absorption,

$$\varepsilon k \sum_{m=2}^{M}\|\nabla\overline{\overline{\eta}}_\varepsilon^{m-1}\|^2 \leq C\varepsilon\|\nabla\overline{\overline{\eta}}_\varepsilon^{1}\|^2$$

$$- k \sum_{m=2}^{M}\left\{(\overline{\overline{\kappa}}^{m-1}, \nabla\overline{\overline{\eta}}_\varepsilon^{m-1}) - (\overline{\overline{e}}_\varepsilon^{m}, \nabla\overline{\overline{\rho}}^{m})\right\}. \tag{7.51}$$

Because of the results (7.48), (7.31) and (7.40), we continue with our estimates,

$$
\leq C \varepsilon k^2 - k \sum_{m=1}^{M} \left\{ (\overline{\overline{\kappa}}^m, \nabla \overline{\overline{\eta}}_\varepsilon^m) - (\overline{\overline{e}}_\varepsilon^m, \nabla \overline{\overline{\rho}}^m) \right\} + k(\overline{\overline{e}}_\varepsilon^1, \nabla \overline{\overline{\rho}}^1)
$$

$$
\leq C \varepsilon k^2 - k \sum_{m=1}^{M} \left\{ (\overline{\overline{\kappa}}^m, \nabla \overline{\overline{\eta}}_\varepsilon^m) - (\overline{\overline{e}}_\varepsilon^m, \nabla \overline{\overline{\rho}}^m) \right\}
$$

$$
+ \frac{1}{4\delta} \|\overline{\overline{e}}_\varepsilon^1\|^2 + \delta k^2 \|\nabla \overline{\overline{\rho}}^1\|^2,
$$

(7.52)

and $\delta > 0$. We choose δ sufficiently small and apply the a-priori statement (7.48), owing to the identification $\varepsilon = \mathcal{O}(k^2)$. Together with another argument that is analogous to (7.40) we can bound the last but one term on the right hand side by $C\varepsilon^2$. — In order to treat the second term on the right hand side, we use the first equation in (7.44) and (7.38). At first applying the average operator and subsequently testing with $\overline{\overline{e}}_\varepsilon^m$ and $\overline{\overline{\kappa}}^m$, respectively, summation of the resulting identities gives,

$$
d_t(\overline{e}_\varepsilon^{m+1/2}, \overline{\kappa}^{m+1/2}) + (\nabla \overline{\overline{\eta}}_\varepsilon^m, \overline{\overline{\kappa}}^m) - (\overline{\overline{e}}_\varepsilon^m, \nabla \overline{\overline{\rho}}^m)
$$

$$
= (d_t \overline{E}^{m+1/2}, \overline{\overline{\kappa}}^m).
$$

(7.53)

After summing over all iteration steps we arrive at

$$
k \sum_{m=1}^{M} \left\{ (\nabla \overline{\overline{\eta}}_\varepsilon^m, \overline{\overline{\kappa}}^m) - (\overline{\overline{e}}_\varepsilon^m, \nabla \overline{\overline{\rho}}^m) \right\} \leq \frac{1}{4\delta_1} \|\overline{e}_\varepsilon^{1/2}\|^2 + C \delta_1 \|\overline{\overline{\kappa}}^1\|^2
$$

$$
+ k \sum_{m=1}^{M} \left\{ \delta_2 \|d_t \overline{E}^{m+1/2}\|^2 + \frac{1}{4\delta_2} \|\overline{\overline{\kappa}}^m\|^2 \right\}, \qquad \delta_1, \delta_2 > 0.
$$

(7.54)

The first term on the right hand side can be controlled by $C \frac{1}{4\delta_1} \varepsilon^2$, owing to Lemma 7.2 and formula (7.40). Because of the a-priori statement (7.48) we can control the second term for sufficiently small δ_1 through the left hand side of (7.52). — Finally, the last sum in (7.54) will be dealt with independently. The first expression in the brackets is only bounded for values

$\delta_2 \equiv \delta_2(t_{m+1}) = \tilde{\delta}_2 \tau_{m+1}$. We can then proceed as follows,

$$k \sum_{m=1}^{M} \left\{ \delta_2 \|d_t \overline{E}^{m+1/2}\|^2 + \frac{1}{4\delta_2} \|\overline{\overline{\kappa}}^m\|^2 \right\}$$

$$\leq \tilde{\delta}_2 k \sum_{m=1}^{M} \tau_{m+1} \|d_t \overline{E}^{m+1/2}\|^2 + \frac{1}{4\tilde{\delta}_2} k \sum_{m=1}^{M} \frac{1}{4\tau_{m+1}} \|\overline{\overline{\kappa}}^m\|^2, \qquad \tilde{\delta}_2 > 0,$$

$$\leq C\tilde{\delta}_2 \varepsilon^2 + \frac{1}{4\tilde{\delta}_2} \max_{1 \leq m \leq M} \|\overline{\overline{\kappa}}^m\|^2 \sum_{m=1}^{M} \frac{1}{m+1}, \qquad (7.55)$$

$$\leq C\tilde{\delta}_2 \varepsilon^2 + \frac{1}{4\tilde{\delta}_2} \max_{1 \leq m \leq M} \|\overline{\overline{\kappa}}^m\|^2 (1 + \log \frac{t_M}{k}).$$

Again, we have made benefit from an upper bound for the harmonic series. Now, we choose $\tilde{\delta}_2 = (1 + t_M + \log \frac{1}{k})$. We remember (7.54), (7.55) and (7.52) and the a-priori statement (7.48) to finally arrive at the result

$$\varepsilon k \sum_{m=1}^{M} \|\nabla \overline{\eta}^m\|^2 \leq C\varepsilon^2 (1 + \log \frac{1}{k}). \qquad (7.56)$$

Now, the auxiliary results (7.42) and (7.56) that give optimal upper bounds for the terms on the right hand side of the basic inequality (7.39) and Lemma 7.4 complete the proof of Theorem 7.3.

7.3   Error Estimates for the Van Kan Scheme

This section completes the analysis for the Van Kan scheme, based on the results that have been derived in the previous sections. Therefore, we set $\varepsilon = \beta k^2$ for the perturbation parameter in (7.26) and set $u_k^{m+1} \equiv u_\varepsilon^{m+1}|_{\varepsilon = \beta k^2}$ and $p_k^{m+1} \equiv p_\varepsilon^{m+1}|_{\varepsilon = \beta k^2}$. — It remains to investigate the following error identities, using the abbreviative notation $e_{VK}^{m+1} := u_k^{m+1} - u_{VanKan}^{m+1}$ and $\eta_{VK}^{m+1} := p_k^{m+1} - p_{VanKan}^{m+1}$, respectively:

$$d_t e_{VK}^{m+1} - \Delta \overline{e}_{VK}^{m+1/2} + \nabla \overline{\eta}_{VK}^{m+1/2} = \frac{1}{2} k^2 \nabla d_t^2 \eta_{VK}^{m+1} - \frac{1}{2} k^2 \nabla d_t^2 p_k^{m+1},$$

$$\text{div} e_{VK}^{m+1} - \beta k^2 \Delta d_t \eta_{VK}^{m+1} = 0, \qquad \beta \geq 1/2,$$

$$e_{VK}^0 = 0, \qquad \|\eta_{VK}^0\|_1 \leq Ck, \qquad (7.57)$$

$$e_{VK}^{m+1}|_{\partial\Omega} = 0, \qquad \partial_n d_t \eta_{VK}^{m+1}|_{\partial\Omega} = 0.$$

Remark 7.6 *In this part of the analysis, the parameter β plays a role as a stabilizing parameter. The resulting error effects in the Van Kan scheme once stem from the sole perturbation and consistency error contributions, and on the other hand from the choice of β as a measure for the implicit weighting of the pressure function. In the following, we will start with a discussion for the case $\beta > \frac{1}{2}$.*

Let us start with an error analysis for the system (7.57). After taking the average over two subsequent iteration steps, we test the resulting identities in (7.57) with $\overline{\overline{e}}_{VK}^m$. After summation over all time-steps we get

$$
\begin{aligned}
\|\overline{e}_{VK}^{M+1/2}\|^2 &+ k \sum_{m=2}^{M} \|\nabla \overline{\overline{e}}_{VK}^m\|^2 + \frac{1}{2}\beta k^2 \|\nabla \overline{\overline{\eta}}_{VK}^M\|^2 + \frac{1}{2}\beta k^4 \sum_{m=2}^{M} \|\nabla d_t \overline{\overline{\eta}}_{VK}^m\|^2 \\
&\leq \frac{1}{2}\beta k^4 \|\nabla d_t \overline{\eta}_{VK}^{M+1/2}\|^2 + \frac{1}{2}\beta k^4 \delta k \sum_{m=2}^{M} \|\nabla d_t^2 \overline{p}_k^{m+1/2}\|^2 \\
&+ \frac{1}{2}\beta k^4 \frac{1}{4\delta} k \sum_{m=2}^{M} \|\nabla d_t \overline{\overline{\eta}}_{VK}^m\|^2 + \|\overline{e}_{VK}^{3/2}\|^2 + \beta k^2 \|\nabla \overline{\overline{\eta}}_{VK}^1\|^2,
\end{aligned}
\tag{7.58}
$$

choosing $\delta > 0$. The terms on the right hand side of this inequality can be controlled or absorbed by the left hand side term: If we choose $\delta = k$, the third term on the right hand side can be absorbed by the corresponding one on the left hand side. Owing to Lemma 7.5, the second term on the right hand side of the inequality sign is bounded by Ck^4. Apart from the first norm on the right hand side there remains an estimation of the last two terms in (7.58). The derivation of optimal order error bounds is "canonical". In order to control the first term on the right side of the inequality sign we employ the estimate

$$
\beta^2 k^4 \|\nabla d_t \overline{\eta}^{m+1/2}\|^2 \leq \|\overline{e}^{m+1/2}\|^2,
\tag{7.59}
$$

which is an easy consequence of the second relation in (7.57). Dealing with this inequality, (7.58) looks like

$$
\begin{aligned}
(1 - \frac{1}{2\beta})\|\overline{e}_{VK}^{M+1/2}\|^2 &+ k \sum_{m=2}^{M} \|\nabla \overline{\overline{e}}_{VK}^m\|^2 + \beta k^2 \|\nabla \overline{\overline{\eta}}_{VK}^M\|^2 \\
&+ \beta k^4 \sum_{m=2}^{M} \|\nabla d_t \overline{\overline{\eta}}_{VK}^m\|^2 \leq Ck^4.
\end{aligned}
\tag{7.60}
$$

This completes the proof of assertion i) in Theorem 7.1, i), owing to the
Theorems 7.2 and 7.3, for the case $\beta > \frac{1}{2}$. - For the case of the original
Van Kan scheme, i.e., $\beta = \frac{1}{2}$, we do not get any optimal error statement
from (7.60) in the $l^\infty(0, T; \mathbf{L}^2)$-norm, and we have to take another way of
verification.

In this case we apply the averaging operator twice onto the first equation
in (7.57) and finally test it with $\overline{\overline{e}}_{VK}^{m-1/2} := \frac{1}{2}\{\overline{\overline{e}}_{VK}^m + \overline{\overline{e}}_{VK}^{m-1}\}$. Now we can
continue as in the following of (7.58), such that the following inequality can
be achieved,

$$\|\overline{\overline{e}}_{VK}^M\|^2 + k\sum_{m=3}^M \|\nabla\overline{\overline{e}}_{VK}^{m-1/2}\|^2 + \frac{1}{2}\beta k^2\|\nabla\overline{\overline{\eta}}_{VK}^{M-1/2}\|^2$$

$$+ \frac{1}{2}\beta k^4 \sum_{m=3}^M \|\nabla d_t\overline{\overline{\eta}}_{VK}^{m-1/2}\|^2 \le Ck^4 + \frac{1}{4}k^4\|\nabla d_t\overline{\overline{\eta}}_{VK}^M\|^2. \qquad (7.61)$$

The boundedness of the remaining term on the right hand side of (7.61) now
gives (7.60), and Theorem 7.1 is verified at all.

7.4   Computational Results for the Van Kan Scheme

We recall that the stability of the classical Chorin method is assured, using
a Q1/Q1 finite element discretization, provided the assumption $k \ge Ch^2$
is satisfied. Now, dealing with Q1/Q1 in case of the Van Kan scheme is a
much more delicate thing. On the one hand, the stabilization effect is only
switched on for the special finite element tuple for nonstationary flows, i.e.,
those that have not yet reached their stationary limit (if they have any).
Moreover, provided the flow under consideration exhibits an nonstationary
behavior, we are led to the following condition, $k \ge Ch$. We can conclude
from this that a rapid decay with respect to stability properties occurs and
more restrictive step-size conditions $F(k, h) > 0$ with respect to the time-
step parameter k are necessary the higher the order of a chosen projection
method is. Thus, one way to circumvent these problems is to strengthen the
stability properties of the chosen finite element pair.

Here, we would like to propose another way to enforce the stability behavior of the fully discretized projection algorithm of Van Kan (via Q1/Q1)

by means of modifying the equations. In the following, we will refer to the original Van Kan scheme (7.1)/(7.2) as $VK_k\{u^{m+1}, p^{m+1}; u^0, p^0\}$. The modification to be proposed below will be denoted by $\overline{VK}_{k,h}\{u_h^{m+1}, p_h^{m+1}; u_h^0, p_h^0\}$, with the solution $u_h^{m+1} \in V_h \subset H_0^1$ and $p_h^{m+1} \in Q_h \subset L_0^2$. We will propose it for $\beta = \frac{1}{2}$ and the "Stokes-case".

1. For $m \geq 0$, determine $\tilde{u}_h^{m+1} \in V_h$ by solving

$$\frac{1}{k}\{\tilde{u}_h^{m+1} - u_h^m\} - \Delta_h \overline{\tilde{u}}_h^{m+1/2} + (1 - \frac{h^2}{2k})\nabla_h p_h^m = \overline{f}_h^{m+1/2},$$

$$\tilde{u}_h^{m+1}|_{\partial\Omega} = 0. \tag{7.62}$$

2. Given $\{\tilde{u}_h^{m+1}, p_h^m\}$, find the tuple $\{u_h^{m+1}, p_h^{m+1}\} \in V_h \times Q_h$ by solving

$$\frac{1}{k}\{u_h^{m+1} - \tilde{u}_h^{m+1}\} + \frac{1}{2}\nabla_h\{(1 + \frac{h^2}{k})p_h^{m+1} - p_h^m\} = 0,$$

$$\text{div}_h u_h^{m+1} = 0, \qquad u_h^{m+1}|_{\partial\Omega} \cdot n = 0, \tag{7.63}$$

with $\Omega_h \subset \Omega$. The discrete Laplacian $-\Delta_h$ is defined by its action on V_h, namely by $-(\Delta v_h, \phi_h) = (\nabla v_h, \nabla \phi_h)$, $\forall\ v_h, \phi_h \in V_h$, whereas div_h and ∇_h are defined as the canonical restrictions of the continuous operators on the subspaces V_h and Q_h.

If we replace u_h^m in the first equality (7.62), we arrive at the following identities,

$$d_t\tilde{u}_h^{m+1} - \Delta_h\overline{\tilde{u}}_h^{m+1/2} + \frac{1}{2}\nabla_h\{3p_h^m - p_h^{m-1}\} = \overline{f}_h^{m+1/2},$$

$$\text{div}_h\tilde{u}_h^{m+1} - \frac{1}{2}\{h^2\Delta_h p_h^{m+1} + k^2\Delta_h d_t p_h^{m+1}\} = 0, \tag{7.64}$$

$$\partial_{n,h} d_t p_h^{m+1}|_{\partial\Omega} = 0,$$

with the canonical restriction of the boundary operator to the finite element space Q_h. These equations can be discretized with Q1/Q1 in a stable way, without demanding restrictive conditions for the time-step size k or worrying about stationary behavior of the actual solution. For the rigorous error analysis in the continuous case we refer to Chapter 5, which deals with *mixed quasi-compressibility methods*.

Remark 7.7 *Again, the spatial discretization has been characterized by a parameter h, but it is easy to construct a modification of the presented algorithm $\overline{VK}_{k,h}\{u_h^{m+1}, p_h^{m+1}; u_h^0, p_h^0\}$ for the case when $h : x \mapsto h(x)$ is an appropriate grid function.*

$k =$	u_2	divu bef. proj.	divu after proj.	∇u_2	p
0.4	2.652	1.42 − 2	9.84 − 4	8.044	0.481
0.28	1.843	8.6 − 3	5.6 − 4	4.567	0.307
0.2	1.024	5.2 − 3	3.27 − 4	2.571	0.198
0.14	0.571	2.97 − 3	1.81 − 4	1.367	0.114
0.1	0.286	1.75 − 3	1.08 − 4	0.762	7.2 − 2
7. − 2	0.151	9.72 − 4	6.05 − 5	0.395	4.3 − 2
5. − 2	8.2 − 2	5.6 − 4	4.4 − 5	0.241	2.6 − 2
3.6 − 2	4.4 − 2	3.5 − 4	3.7 − 5	0.142	1.6 − 2
order	1.87	1.69	1.61	1.78	1.62

Table 7.1: Orders for errors for the Van Kan scheme

The remainder of this section is devoted to the presentation of computational results to illustrate the theoretical results above. These have been achieved for the parameter selection $\beta = \frac{1}{2}$. — The above analyses have shown that the scheme of Van Kan is of second order, i.e., the velocity errors $u(t_{m+1}) - \bar{u}_{VK}^{m+1}$ decrease in the norm $l^\infty(0, t_{M+1}; \mathbf{L}^2)$ of second order in k, whereas the errors converge of first order in the stronger norm $l^\infty(0, t_{M+1}; \mathbf{H}^1)$. The latter holds true even for the error in the pressure in the norm $l^\infty(0, t_{M+1}; L^2/\mathbb{R})$. Now, the objective of the present section is to illustrate these properties for our model problem at hand, and to discuss flows resulting from configurations with incompatible initial data, leading to a global loss of order.

Let us start with the summary of the results of convergence obtained for the model configuration $\{T, \gamma, \nu\} = \{2, 1, 1\}$. They are collected in Table 7.1.

A direct comparison with the results for the Chorin method given in Table 6.1 shows that the history of convergence is starting relatively early. Further, we have results like those predicted by the theory. Again, the error in the pressure, measured in the norm $l^\infty(0, t_{M+1}; L^2/\mathbb{R})$ and the velocity error, evaluated in the norm $l^\infty(0, t_{M+1}; \mathbf{H}^1)$ give better results than predicted by the theory. In order to give reasons for that phenomenon, we can come back to the local portion of the singular perturbation error. In this algorithm and the employed example this source is dominant for the error contributions (in

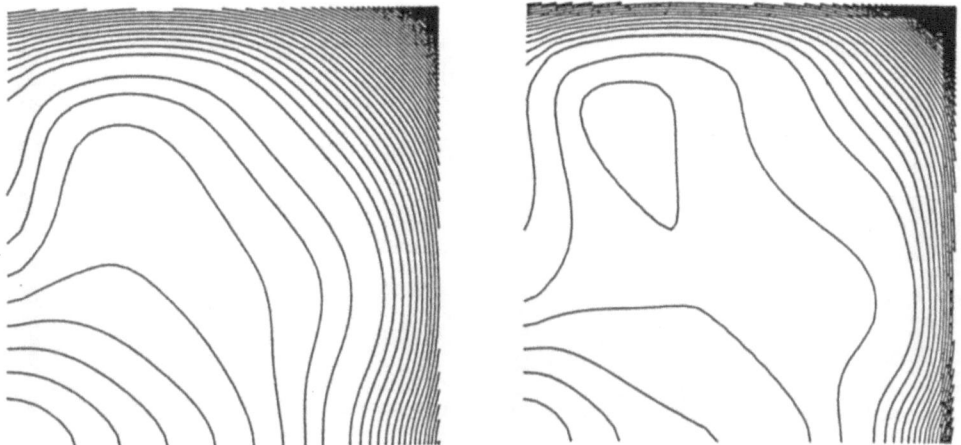

Figure 7.1: Isolines for the pressure error for $k = 0.28$ and $k = 0.20$

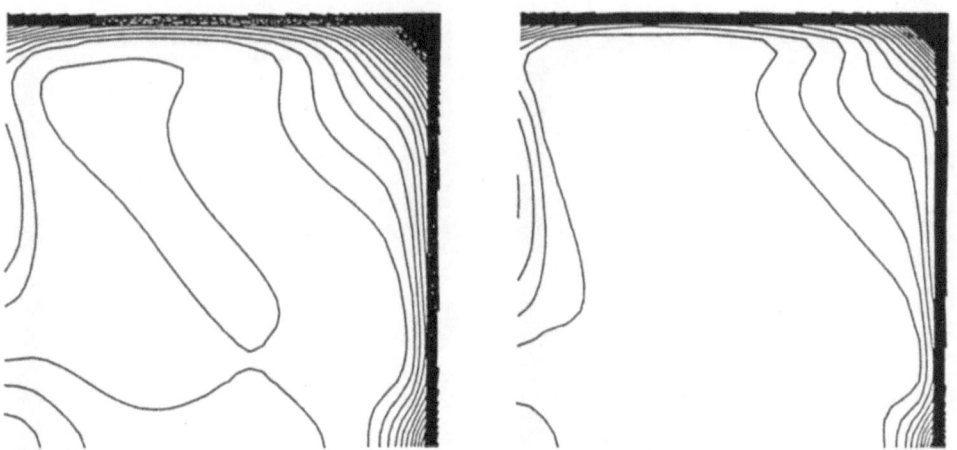

Figure 7.2: Isolines for the pressure error for $k = 0.14$ and $k = 0.10$

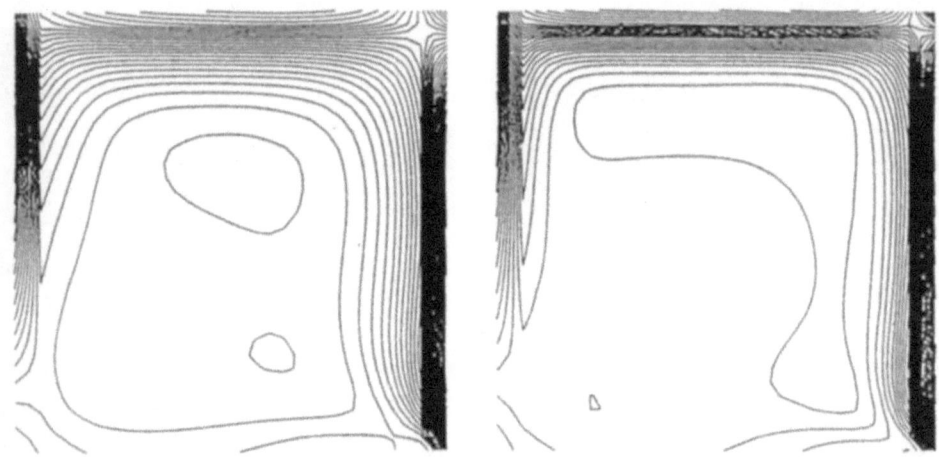

Figure 7.3: Isolines for $\mathrm{div}\tilde{u}_{VK}^{M+1}$ ($k = 0.28$ and $k = 0.20$)

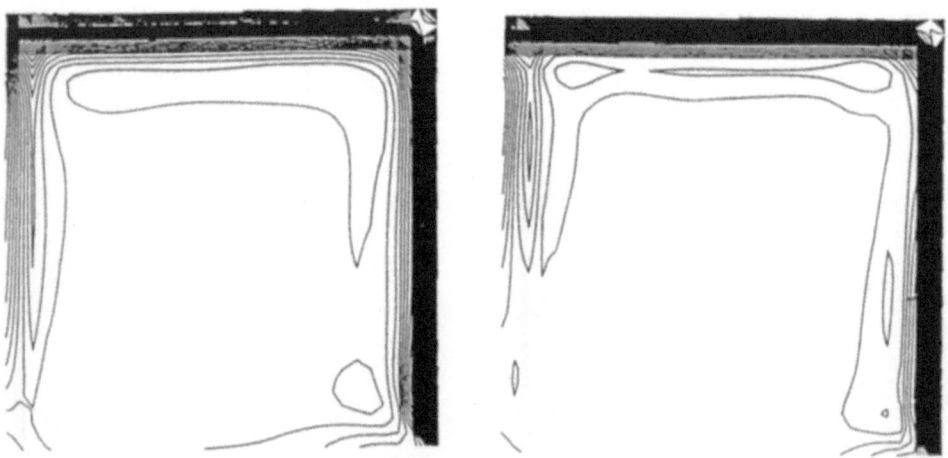

Figure 7.4: Isolines for $\mathrm{div}\tilde{u}_{VK}^{M+1}$ ($k = 0.14$ and $k = 0.10$)

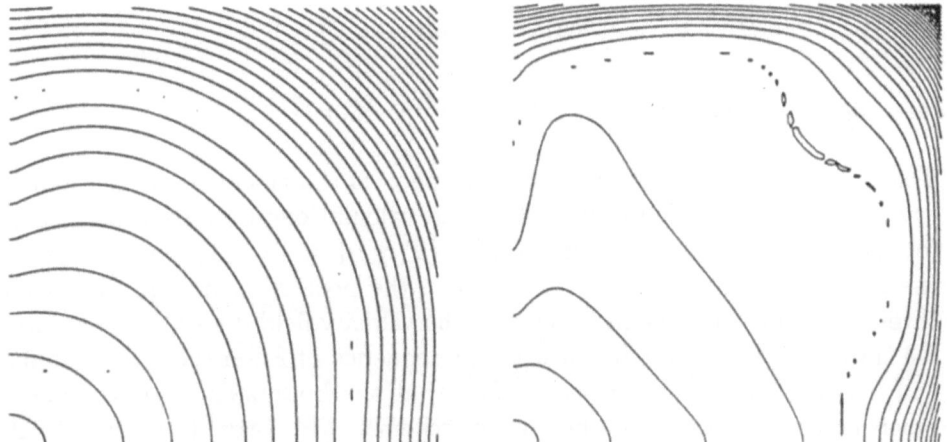

Figure 7.5: Isolines for the pressure error for $\nu = 1$ and $\nu = 10^{-1}(k = 0.4)$

Figure 7.6: Isolines for the pressure error for $\nu = 10^{-2}$ $(k = 0.4)$

this context, we refer to the (time-)damped pressure correction methods as they have been proposed in Chapter 4, and the super-convergence results for them, given in Theorem 4.4.). This causes a localization of the error in the pressure close to the boundary. Due to the pressure correction formulation we expect the error to be described as follows,

$$|p(t_{m+1/2})(x) - \overline{p}_{VK}^m(x)| \le C\exp\left(-\frac{d(x)}{k}\right)k + \mathcal{O}(k^2). \tag{7.65}$$

This would give us a statement with respect to the thickness of the boundary layer to be of magnitude $\mathcal{O}(k)$. The calculations that have been done for the compatible model configuration $\{T, \gamma, \nu\} = \{2, 1, 1\}$ supply us with the following results, once for the error in the pressure, compare Figure 7.1 through 7.2, once for the divergence of the velocity field $\operatorname{div} u_{VK}^{m+1}$, see Figures 7.3 through 7.4. A comparison with the sequence of Figures 6.1 through 6.3 and 6.4 through 6.6 evidences the fact that the boundary layers disappear much faster than those in the Chorin scheme. Moreover, the relative values are much smaller compared to the ones that we have calculated for the Chorin method. As an illustration, we can see, by comparing two figures in each of the sequences, that for a reduction of the time-step size by a factor one half leads to half the thickness of the boundary layer.

In the next part of our discussion we intend to study the impact of the viscosity ν. The calculations have been done for the cases $\{T, \gamma, \nu\} = \{2, 1, \nu_i\}|_{i \in \{1,2,3\}}$, with values $\nu_i \in \{1, 10^{-1}, 10^{-2}\}$. The results are given in Figures 7.5 through 7.6. In comparison to the corresponding results in Chapter 6, compare Figures 6.7 through 6.8, the boundary layer disappears in ν equally fast. These reflections coincide with *heuristic* considerations, that can be made correspondingly to [30], see Lemma 1 there.

Remark 7.8 *A consequence of this reflection might be to select the projection algorithm for the Navier-Stokes equations with high Reynolds numbers, because no significant boundary layers are present in that case. However, this conclusion is not quite correct: Despite the moderate impact of the errors that are concentrated near the boundary, other error mechanisms become dominant that determine the accuracy of the approximation. Especially, we mention the error amplification mechanism due to the nonlinearity. It causes a dramatic growth in the approximation errors for the velocity field in the case of increasing Reynolds numbers. In that context, we recall that also the analysis presented in Chapter 6 for the Chorin method indicates a rather*

$k =$	u_2	div$u$ bef. proj.	div$u$ after proj.	∇u_2	p
0.4	2.439	$1.05 - 2$	$7.29 - 4$	5.983	0.431
0.28	1.355	$6.17 - 3$	$3.95 - 3$	3.202	0.271
0.2	0.732	$3.68 - 3$	$2.29 - 4$	1.791	0.167
0.14	0.414	$2.09 - 3$	$1.29 - 4$	0.953	0.104
0.1	0.267	$1.22 - 3$	$7.88 - 5$	0.556	$6.52 - 2$
$7. - 2$	0.174	$6.83 - 4$	$4.97 - 5$	0.316	$4.11 - 2$
$5. - 2$	$9.7 - 2$	$3.9 - 4$	$3.4 - 5$	0.192	$2.49 - 2$
$3.6 - 2$	$7.4 - 2$	$2.39 - 4$	$2.69 - 5$	0.142	$1.80 - 2$
order	1.65	1.70	1.64	1.71	1.57

Table 7.2: Errors for the Van Kan scheme for $\gamma = 0.75$

"sensitive" behavior of the error for the solution with respect to increasing Reynolds numbers.

The numerical results that have been presented up to now favor the Van Kan algorithm as a fast and accurate method for fluid flow simulation. But in the previous studies we have already stressed the fact that problems arise, if the initial data are not compatibly posed. These stability problems lead to a global (in time) reduction of order from two to one for complex flows. It is our aim to support this evidence through test calculations. In order to proceed in that way, we will look at the following test problem, $T = 2$, $\nu = 1$, together with the pressure functions $p^{\gamma_i}|_{i \in \{1,2,3,4,5\}}$, with $\gamma_i \in \{1, 0.75, 0.5, 0.25, 0.05\}$. Therefore, the pressure becomes increasingly singular, depending on the index γ_i: $\lim_{s \to 0} \|\nabla p_t^{\gamma_i}(s)\| = \infty$. — The errors that result in the case $\gamma_i = 1$ are presented in Table 7.1, whereas the remaining cases are treated in Tables 7.2 through 7.5.

It is significant that the deviated orders of convergence for the velocity approximation decay, from 1.87 to 1.22. This reduction of order can be verified through the theoretical investigations in the first part of this chapter.

$k =$	u_2	divu bef. proj.	divu after proj.	∇u_2	p
0.4	1.422	$7.48 - 3$	$5.13 - 4$	3.221	0.382
0.28	0.933	$4.22 - 3$	$2.57 - 4$	2.077	0.235
0.2	0.509	$2.49 - 3$	$1.55 - 4$	1.217	0.142
0.14	0.317	$1.42 - 3$	$8.96 - 5$	0.669	$9.17 - 2$
0.1	0.224	$8.16 - 4$	$5.63 - 4$	0.403	$6.06 - 2$
$7. - 2$	0.153	$4.57 - 4$	$3.73 - 5$	0.250	$3.99 - 2$
$5. - 2$	$9.1 - 2$	$2.61 - 4$	$2.64 - 5$	0.198	$2.45 - 2$
$3.6 - 2$	$6.3 - 2$	$1.78 - 4$	$2.31 - 5$	0.161	$1.88 - 2$
order	1.52	1.71	1.57	1.48	1.51

Table 7.3: Errors for the Van Kan scheme for $\gamma = 0.5$

$k =$	u_2	divu bef. proj.	divu after proj.	∇u_2	p
0.4	0.851	$5.0 - 3$	$3.29 - 4$	2.726	0.321
0.28	0.442	$2.69 - 3$	$1.46 - 4$	1.188	0.197
0.2	0.336	$1.61 - 3$	$1.02 - 4$	0.782	0.124
0.14	0.273	$9.04 - 4$	$6.08 - 5$	0.473	$8.3 - 2$
0.1	0.221	$5.15 - 4$	$4.04 - 5$	0.315	$5.9 - 2$
$7. - 2$	0.169	$2.9 - 4$	$2.86 - 5$	0.239	$4.1 - 2$
$5. - 2$	0.114	$1.74 - 4$	$2.15 - 5$	0.174	$2.6 - 2$
$3.6 - 2$	$7.3 - 2$	$1.61 - 4$	$1.49 - 5$	0.152	$2.1 - 2$
order	1.31	1.63	1.46	1.33	1.48

Table 7.4: Errors for the Van Kan scheme for $\gamma = 0.25$

$k =$	u_2	divu bef. proj.	divu after proj.	∇u_2	p
0.4	0.572	$3.39 - 3$	$2.06 - 4$	1.714	0.269
0.28	0.340	$1.82 - 3$	$8.99 - 5$	0.731	0.155
0.2	0.299	$1.11 - 3$	$7.19 - 5$	0.578	0.107
0.14	0.258	$6.17 - 4$	$4.52 - 5$	0.384	$7.9 - 2$
0.1	0.216	$3.51 - 4$	$3.23 - 5$	0.296	$6.1 - 2$
$7. - 2$	0.158	$2.07 - 4$	$2.48 - 5$	0.233	$4.4 - 2$
$5. - 2$	0.103	$1.46 - 4$	$2.1 - 5$	0.199	$2.75 - 2$
$3.6 - 2$	$7.7 - 2$	$1.03 - 4$	$1.75 - 5$	0.146	$2.24 - 2$
order	1.22	1.59	1.32	1.12	1.36

Table 7.5: Errors for the Van Kan scheme for $\gamma = 0.05$

Therefore, let us come back to estimate (7.43): The last but one term in this inequality causes a loss of convergence order in the other estimates, if it cannot be bounded in an optimal fashion. A simple application of this estimate to the actual test problem gives the following theoretical orders of convergence for the velocity approximation, in the named four latter cases (compare with the above tables): 1.75, 1.5, 1.25 and ≈ 1. In general, these results underscore the necessity of the requirements that are given in *Postulate B_2* on the approximating solution. To be more specific, the Van Kan scheme does not work in an optimal way, if *Postulate B_2* is not satisfied. This underscores the sharpness of the results presented in Theorem 7.1.

Chapter 8

Two Modified Chorin Schemes

8.1 Overview and Results

In the previous chapters we have analyzed the projection methods of Chorin and of Van Kan, based on their reformulation as semi-explicit quasi-compressibility methods. Due to the resulting Poisson equation for the pressure function in combination with homogeneous boundary conditions, these schemes possess a singular perturbation character. This chapter is devoted to the question of whether it is possible to construct projection type algorithms that do not suffer from numerically induced boundary layers for the pressure approximation. The complete removal of the boundary layers would give rise to a regular perturbation formulation of the incompressible Stokes operator.

Let us start with the analysis of a variant of Chorin's scheme, proposed by Timmermans et al. in [44]. (To be more precise, these authors have dealt with a modified Van Kan scheme in the cited paper. In order to fix the ideas, we will restrict ourselves to a corresponding formulation for the Chorin method). The main idea of this scheme is to start with Chorin's scheme. Now, in order to improve the pressure calculation in each iteration step, an additional step is joined to the original scheme. Therefore, let us denote by $\{\tilde{u}^{m+1}, u^{m+1}, \tilde{p}^{m+1}\}$ the outcome of the original Chorin scheme, then the corrected pressure p^{m+1} will be computed by

$$p^{m+1} = \tilde{p}^{m+1} - \nu \mathrm{div}\tilde{u}^{m+1}. \tag{8.1}$$

It is evident that this pure update will not cause a severe additional numerical effort. Thus, the complexity of the projection scheme is asymptotically the same. The next discussion is related to the question of whether this modification leads to a reduction or removal of the boundary layers. Let us remember the fact that this damage of the pressure approximation was due to the fact that the Lagrange multiplier in the projection step has been *interpreted as a pressure approximation*. The above modification seeks to replace this unphysical numerical modeling by determining the pressure approximation through a correction of the Lagrange multiplier.

Remark 8.1 *Let us stress the fact that the "normalizing" ν in (8.1) is essential for getting optimal results of convergence.*

One objective of the present chapter is to find out, whether this modified scheme gives an improvement of the pressure approximation, especially we will be concerned with the question of whether this leads to the disappearance of the boundary layers. Unfortunately, we can only give a negative answer to the last question. Boundary layers are always present. (In that respect, we want to mention the work of Timmermans et al. [44], that arrives at just the opposite result, based on computational experiments; they conclude that their modification gives pressure approximations that are even better close to the boundary.) On the other hand, this new strategy has another advantage: The pressure error portions of higher frequency disappear "of higher order" in the interior of the domain. We will refer to this as a smoothing effect.

    For the following investigation, we will again restrict ourselves to the Stokes equations. Nevertheless, we stress the fact that corresponding analyses can immediately be done for the Navier-Stokes equations. — The first main result in this chapter is the presentation of error statement for the iterates of Timmermans' modification. It is clear, that it has no influence on the computed velocity fields.

Theorem 8.1 *Let $\{\tilde{u}^{m+1}, p^{m+1}\}$ be the solution of the modified Chorin scheme with the update (8.1), and $\{u, p\}$, with $p \in L^\infty(0, T; L_0^2)$, be the solution of the incompressible Stokes equations (2.1), with given data that satisfy the assumptions (A1), (A2). Then we have,*

i)   *global error statements:*

$$\max_{0 \leq m \leq M} \left\{ \|u(t_{m+1}) - \tilde{u}^{m+1}\| + \tau_{m+1} \|p(t_{m+1}) - p^{m+1}\|_{-1} \right\} \leq Ck,$$

$$\max_{0 \leq m \leq M} \left\{ \|u(t_{m+1}) - \tilde{u}^{m+1}\|_1 + \sqrt{\tau_{m+1}} \|p(t_{m+1}) - p^{m+1}\| \right\} \leq C\sqrt{k},$$

ii)   *local error statements: On compact subdomains $\Omega' \subset\subset \Omega$, we additionally have,*

$$\max_{0 \le m \le M} \sqrt{\tau_{m+1}} \big\{ \|u(t_{m+1}) - \tilde{u}^{m+1}\|_{1;\Omega' \subset\subset \Omega} \\ + \|p(t_{m+1}) - p^{m+1}\|_{\Omega' \subset\subset \Omega} \big\} \le Ck,$$

$$\max_{0 \le m \le M} \big\{ \tau_{m+1} \|p(t_{m+1}) - p^{m+1}\|_{1;\Omega' \subset\subset \Omega} \big\} \le Ck.$$

Note that the employed constant C only depends on the given data of the actual problem.

Remark 8.2   *1. Throughout the present chapter, the pressure function p of the incompressible Stokes problem (2.1) will be regarded as an element of $L^\infty(0, T; L_0^2(\Omega))$.*

2. The global error statements are identical to the ones that have been obtained in the case of the original Chorin scheme, compare Theorem 6.1. The second local statement represents an improvement in comparison with the original scheme; the order of convergence is raised from $\frac{1}{2}$ to 1; compare Theorem 6.2 in that respect. Due to this improvement of the order of convergence for the pressure iterates p^{m+1}, this ansatz gives results that are superior to the ones obtained by the original projection method of Chorin. This is the mathematical justification for the above assertion that error portions of high frequency are damped with increasing time on interior subdomains.

The proof of the statements in Theorem 8.1 uses again the same patterns as presented in previous proofs. Primarily, we will be concerned with the investigation of the pressure stabilization effects, before an analysis completes the proof that is dealing with the explicit pressure treatment in the momentum equation. These analyses are topic of the Section 8.2.

The *second part* of this chapter is related to the proposal of another modified Chorin scheme, which — in contrast to (8.1) — does not cause any boundary layers any more, despite of having quite the same trifling complexity as Chorin's method. Due to the similarity of the now given

algorithm in the third step with the iterative Uzawa method, that is used
for solving stationary saddle-point problems, we propose to call it *Chorin-
Uzawa scheme.* Alternatively, we could have named it Chorin-"Artificial
Compressibility" scheme. It reads as follows: Start with the given triple
$\{u^0, p^0; \tilde{p}^0\}$, which has to satisfy

$$\|u^0 - u_0\| + \sqrt{k}\|p^0 - p_0\| \le Ck, \qquad \tilde{p}^0 \equiv 0, \qquad p^0 \in L_0^2. \qquad (8.2)$$

Given $m \ge 0$, we start with the following sequence of iteration steps:

1. Determine the auxiliary function $\tilde{u}^{m+1} \in \mathbf{H}_0^1$ as the solution of

    $$\frac{1}{k}\{\tilde{u}^{m+1} - u^m\} - \Delta\tilde{u}^{m+1} + \nabla\{p^m - \tilde{p}^m\} = f^{m+1}. \qquad (8.3)$$

2. Calculate $u^{m+1} = P_{\mathbf{J}_0}\tilde{u}^{m+1}$ by means of solving

    $$\frac{1}{k}\{u^{m+1} - \tilde{u}^{m+1}\} + \nabla\tilde{p}^{m+1} = 0,$$
    $$\text{div}u^{m+1} = 0, \qquad u^{m+1}|_{\partial\Omega} \cdot n = 0. \qquad (8.4)$$

3. The pressure function p^{m+1} will be computed as follows,

    $$p^{m+1} = p^m - \alpha\text{div}\tilde{u}^{m+1}, \qquad \alpha < 1. \qquad (8.5)$$

Due to the insignificance of the accuracy requirements for the initial pressure
and velocity functions, this algorithm is suited for an improvement of the
pressure functions that the basic Chorin method gives us, for times $t_{m+1} = \mathcal{O}(1)$. If we compare this scheme to the above modification, we see that an
additional function $\tilde{p}^{m+1}$ has been introduced in the algorithm, a fact, which
will give a uniform accuracy of p^{m+1} over the domain Ω up to the boundary.
We will come back to this point later on. Further more, we will see that the
requirements (8.2) are necessary in order to guarantee an optimal behavior
of convergence. The requirement for accurate initial data follows from (8.5),
which is an nonstationary quasi-compressibility constraint,

$$\text{div}\tilde{u}^{m+1} + \frac{1}{\alpha}kd_t p^{m+1} = 0. \qquad (8.6)$$

Now, if we combine the Chorin-Uzawa algorithm with the basic Chorin
method in the sense that we start with the original Chorin scheme that

is replaced by the Chorin-Uzawa scheme at times $t_{m+1} = \mathcal{O}(1)$, the compatibility assumption (8.2) is satisfied. These requirements are even satisfied by the Chorin method. This approach, which is one representative for *multi-component schemes* that will be further elaborated in the following chapter, combines the stability properties of the original Chorin scheme with the gain of accuracy of the Chorin-Uzawa method. In this chapter, we will limit ourselves to the case of given approximate initial data, as it is demanded in (8.2).

Again, our motivation for the proposal of the Chorin-Uzawa scheme is the removal of the boundary layers of the pressure function. In order to reach this goal, we distinguish between $\tilde{p}^{m+1}$ as a Lagrange multiplier and the pressure approximation p^{m+1}. The coincidence of both tasks in one iterate in the scheme of Chorin or in the modification by Timmermans et al. is the reason for the appearing boundary layers. Especially, the step (8.1) of the modified Chorin scheme cannot serve as a means to reduce the boundary layer, because the physical pressure function is determined via the Lagrange multiplier $\tilde{p}^{m+1}$ — which is performed with a "bad structure" in the vicinity of the boundary.

Another advantage of the Chorin-Uzawa scheme is that it is *consistent* in the stationary limit. This can be extracted from (8.6): Provided the pressure function runs into a stationary limit for $t_{m+1} \to \infty$, this implies the incompressibility of the corresponding velocity function $\tilde{u}^{m+1}$. This property is not shared by the Chorin method.

In our computations, we fixed the parameter α in (8.5) to $\alpha = 0.8$. Corresponding studies will be presented in Section 8.4. The reason for the application of a stabilizing parameter in this case in contrast to the original Chorin method is the explicit update of the actual pressure function. In order to equilibrate the perturbation effects of the incompressibility constraint with the explicit treatment of the pressure gradient in the first step (8.3), the condition $\alpha < 1$ is sufficient. The computational results that are given in Section 8.4 illustrate these observations.

Corresponding to (8.6), we can think of the splitting scheme (8.3) through (8.5) as a semi-explicit quasi-compressibility technique for solving the Stokes equations (2.1). Therefore, let us shift the index of the first equation in (8.4) with -1 and add this identity to the "momentum equation" in (8.3). This gives the first of the subsequent equations,

$$d_t\tilde{u}^{m+1} - \Delta\tilde{u}^{m+1} + \nabla p^m = f^{m+1},$$

$$\mathrm{div}\,\tilde{u}^{m+1} + \frac{k}{\alpha}d_t p^{m+1} = 0, \qquad \alpha < 1, \qquad \tilde{u}^{m+1}|_{\partial\Omega} = 0, \tag{8.7}$$

together with starting conditions for velocity and pressure function, cf. (8.2). The functions that are determined by system (8.7) are $\{\tilde{u}^{m+1}, p^{m+1}\}$. Especially, the step (8.4) in the Chorin-Uzawa scheme giving $\bar{p}^{m+1}$ is not used here. This gives rise to the fact that the formulation of the quasi-compressibility constraint is not related to an $H^1/\mathbb{R}$-elliptic operator demanding the prescription of boundary conditions for the pressure function any more but only with the time derivative of the pressure function. This observation leads us to the conjecture that *no numerical boundary layer* is present any more. This assertion will be verified in Section 8.3, giving the proof of the next theorem.

Theorem 8.2 *Let* $\{\tilde{u}^{m+1}, p^{m+1}\}$ *be computed by the Chorin-Uzawa scheme to approximate the solution* $\{u(t_{m+1}), p(t_{m+1})\}$ *of (2.1), for* $k < 1$*. We will start this projection-type algorithm at time* $t_{m_1} = \mathcal{O}(1)$*. We assume the approximation property (8.2) to be valid at that time and the given data to satisfy the assumptions (A1) and (A2). Then, the following statements of convergence hold true, for* $m \geq m_1$:

i) $\max_{m_1 \leq m \leq M} \Big\{ \|\tilde{u}^{m+1} - u(t_{m+1})\|$
$\qquad\qquad + \sqrt{k}\|p^{m+1} - p(t_{m+1})\| \Big\} \leq C(1 + \log\frac{1}{k})k,$

ii) $\Big(k\sum_{m=m_1}^{M} \|\tilde{u}^{m+1} - u(t_{m+1})\|_1^2 \Big)^{1/2} \leq C(1 + \log\frac{1}{k})k.$

The constant C *in the estimates reflects the stability properties of the actual problem, solely depending on the given data.*

The crucial point is statement *ii*), which gives a gain of the convergence order from $\frac{1}{2}$ to 1 compared with the corresponding result for the Chorin method (compare Theorem 6.1). If we recall the diverse error sources that appear in the Chorin scheme, we recognize that result *ii*) in the above theorem reflects the *regular spatial* perturbation character of the method, which means: there are no boundary layers apparent. The errors that determine the approximation features of this scheme are caused by the evolutionary character of the second equation in (8.7).

These observations are of crucial importance for the classification of the Chorin-Uzawa method and give rise to the question of whether there is a projection method that does not suffer from the named error source in the Chorin-Uzawa scheme. In the next chapter that deals with *multi-component schemes* we will propose another modification of the Chorin-Uzawa scheme that gives approximations $\{\tilde{u}^{m+1}, p^{m+1}\}$ that do *not* suffer from singular perturbation in space and nonstationary perturbation effects as well, and we obtain first order convergence for both, velocity and pressure, in relevant norms (i.e., pointwise in time).

Remark 8.3 1. *The statements of Theorem 8.2 are limited to the case that the Chorin-Uzawa scheme will be started at times $t_{m+1} = \mathcal{O}(1)$. This requirement is necessary in order to guarantee the assumption (8.2) to be satisfied, which will be established by means of a precursory calculation, using the basic Chorin scheme. Let us recall the results that we have derived for the Chorin scheme,*

$$\max_{0 \le m \le M} \left\{ \|u_{Cho}^{m+1} - u(t_{m+1})\| + \sqrt{k}\sqrt{\tau_{m+1}}\|p_{Cho}^{m+1} - p(t_{m+1})\| \right\} \le Ck. \tag{8.8}$$

Here, $\{u_{Cho}^{m+1}, p_{Cho}^{m+1}\}$ denotes the approximation that has been obtained via the original Chorin method. These inequalities ensure the validity of assumption (8.2).

2. *The Chorin-/Chorin-Uzawa scheme combines the advantages of the damping (and more stable) Chorin method and those of the more accurate Chorin-Uzawa algorithm, which results in an improved pressure approximation for iterates $m \ge m_1$. For that respect, we refer to the above discussion.*

3. *The Chorin-Uzawa method can also be applied independently of the Chorin algorithm. However, this needs a compatibility condition to be satisfied for the initial data of problem (2.1), which is given in the next Postulate.*

Postulate B_1: *The solution $\{u, p\}$ of the incompressible Stokes equations (2.1) satisfies the following regularity assumption,*

$$\sup_{0 \le s \le T} \left\{ \|\nabla u_t(s)\| + \|p_t(s)\| \right\} \le C.$$

The remainder of this chapter is organized as follows: We start with the proof of Theorem 8.1 in Section 8.2. A sketch of the proof for Theorem 8.2 is subject of Section 8.3. Finally, we conclude this chapter with computational results that enable a direct comparison of the diverse projection schemes of Chorin, modified Chorin and Chorin-Uzawa.

8.2 An Analysis of the Modified Chorin Scheme of Timmermans et al.

8.2.1 Proposal of a Modified Pressure Stabilization Scheme

The modified Chorin scheme that has been formulated in Section 8.1 can again be analyzed in the context of pressure stabilization ansatzes. In a first step, we leave the equation (8.1) unconsidered and obtain the original Chorin scheme for the iterates $\{\tilde{u}^{m+1}, u^{m+1}, p^{m+1}\}$, which corresponds to an explicit pressure stabilization ansatz. The governing equations are in this notation,

$$
\begin{aligned}
&d_t \tilde{u}^{m+1} - \Delta \tilde{u}^{m+1} + \nabla \tilde{p}^m = f^{m+1} \\
&\mathrm{div}\tilde{u}^{m+1} - k\Delta \tilde{p}^{m+1} = 0, \qquad \partial_n \tilde{p}^{m+1}|_{\partial\Omega} = 0.
\end{aligned}
\tag{8.9}
$$

(In the subsequent studies, we set $\nu = 1$ in (8.1).) Remembering the pressure correction step (8.1), we are led to consider the following system of equations,

$$
\begin{aligned}
&d_t \tilde{u}^{m+1} - \Delta \tilde{u}^{m+1} + \{\nabla p^m + \nabla \mathrm{div}\tilde{u}^m\} = f^{m+1}, \\
&\mathrm{div}\tilde{u}^{m+1} - k\,\mathrm{div}\{\nabla \mathrm{div}\tilde{u}^{m+1} + \nabla p^{m+1}\} = 0, \\
&\{\nabla \mathrm{div}\tilde{u}^{m+1} + \nabla p^{m+1}\}|_{\partial\Omega} \cdot n = 0.
\end{aligned}
\tag{8.10}
$$

As before, we will consider in a first step an implicit form of these equations in order to investigate the effects of singular perturbation. Subsequently, we will take care of the explicit character with respect to the pressure approximation in (8.10). The equations we will focus on at first are:

$$
\begin{aligned}
&d_t u_\varepsilon^{m+1} - \Delta u_\varepsilon^{m+1} + \nabla p_\varepsilon^{m+1} + \nabla \mathrm{div}u_\varepsilon^{m+1} = f^{m+1} \\
&\mathrm{div}u_\varepsilon^{m+1} - \varepsilon\,\mathrm{div}\{\nabla \mathrm{div}u_\varepsilon^{m+1} + \nabla p_\varepsilon^{m+1}\} = 0, \qquad \varepsilon = \mathcal{O}(k), \\
&\{\nabla \mathrm{div}u_\varepsilon^{m+1} + \nabla p_\varepsilon^{m+1}\}|_{\partial\Omega} \cdot n = 0.
\end{aligned}
\tag{8.11}
$$

In the following, we will make use of the abbreviative notation,

$$\phi_\varepsilon^{m+1} := \mathrm{div} u_\varepsilon^{m+1} + p_\varepsilon^{m+1} \in L_0^2. \tag{8.12}$$

Clearly, substitution of this sum in system (8.10) gives the pressure stabilization scheme, that has been considered in the context of the Chorin scheme. Especially, existence and uniqueness of a solution $\{u_\varepsilon^{m+1}, \phi_\varepsilon^{m+1}\}$ are established. In order to derive global error statements for the solution of system (8.10), we will benefit from the estimates that have been proved for the Chorin scheme, compare Theorem 6.1,

$$\max_{0 \le m \le M} \left\{ \|\tilde{u}^{m+1} - u(t_{m+1})\| + \sqrt{k}\|\tilde{u}^{m+1} - u(t_{m+1})\|_1 \right\} \le Ck,$$

$$\max_{0 \le m \le M} \sqrt{\tau_{m+1}} \left\{ \sqrt{\tau_{m+1}} \|\tilde{p}^{m+1} - p(t_{m+1})\|_{-1} \right.$$
$$\left. + \sqrt{k}\|\tilde{p}^{m+1} - p(t_{m+1})\| \right\} \le Ck. \tag{8.13}$$

We can use these error estimates for the pressure function $\tilde{p}^{m+1}$, to arrive at

$$\max_{0 \le m \le M} \sqrt{\tau_{m+1}} \left\{ \sqrt{\tau_{m+1}} \|p^{m+1} - p(t_{m+1})\|_{-1} \right.$$
$$\left. + \sqrt{k}\|p^{m+1} - p(t_{m+1})\| \right\} \le Ck. \tag{8.14}$$

The object of the present subsection is the verification of the *local* error statements given in Theorem 8.1 for system (8.11).

Theorem 8.3 *Let* $\{u_\varepsilon^{m+1}, p_\varepsilon^{m+1}\}$ *be the solution of scheme (8.11), whereas the tuple* $\{u^{m+1}, p^{m+1}\}$ *is the solution of the Stokes equations (2.1) that are discretized via implicit Euler scheme. Let the assumptions (A1) and (A2) be satisfied. Then, the following global and local error statements are valid for* $\varepsilon = \mathcal{O}(k)$, *with a constant* C *that only depends on the given data of the problem,*

i) *"global error statements":*

$$\max_{0 \le m \le M} \left\{ \|u^{m+1} - u_\varepsilon^{m+1}\| + \tau_{m+1}\|p^{m+1} - p_\varepsilon^{m+1}\|_{-1} \right\} \le C\varepsilon,$$

$$\max_{0 \le m \le M} \left\{ \|u^{m+1} - u_\varepsilon^{m+1}\|_1 + \sqrt{\tau_{m+1}}\|p^{m+1} - p_\varepsilon^{m+1}\| \right\} \le C\sqrt{\varepsilon},$$

ii) *"local error statements":*

$$\max_{0 \le m \le M} \sqrt{\tau_{m+1}} \big\{ \|u^{m+1} - u_\varepsilon^{m+1}\|_{1,\Omega' \subset\subset \Omega}$$

$$+ \|p^{m+1} - p_\varepsilon^{m+1}\|_{\Omega' \subset\subset \Omega} \big\} \le C\varepsilon,$$

$$\max_{0 \le m \le M} \big\{ \tau_{m+1} \|p^{m+1} - p_\varepsilon^{m+1}\|_{1,\Omega' \subset\subset \Omega} \big\} \le C\varepsilon.$$

Remark 8.4   *1. The global error statements need no further explanation, due to the previous related discussion. The local error bounds are of superior interest, especially the latter one. In contrast to the original pressure stabilization scheme (8.11), using (8.12), that only allows the following result,*

$$\max_{0 \le m \le M} \big\{ \sqrt{\tau_{m+1}} \|p^{m+1} - \phi_\varepsilon^{m+1}\|_{1,\Omega' \subset\subset \Omega} \big\} \le C\sqrt{\varepsilon},$$

high error frequencies in the interior of the domain, $\Omega' \subset\subset \Omega$, will be damped away in the case of the functions $\{u_\varepsilon^{m+1}, p_\varepsilon^{m+1}\}$. This will justify the notation "smoothing effect", which has been used above.

2. In this theorem, we confine ourselves to quantifying the error between the solutions $\{u^{m+1}, p^{m+1}\}$ and $\{u_\varepsilon^{m+1}, p_\varepsilon^{m+1}\}$, with the first being the solution of the Stokes equations (2.1) that are semi-discretized via implicit Euler-scheme. An additional consistency analysis gives corresponding (in the case ii) global) statements i) and ii) for the Euler discretization — by setting $\varepsilon = \mathcal{O}(k)$. We will skip this part here.

Proof:
Proof of Theorem 8.3, ii): We will restrict on the local error statements. Thanks to the identity

$$\mathrm{curl}(\mathrm{curl}\phi) = -\Delta\phi + \mathrm{grad}(\mathrm{div}\phi), \tag{8.15}$$

that holds true for all functions $\phi \in \mathbf{H}^2(\Omega)$ in $\mathbf{L}^2(\Omega)$, the equations (8.11) can be written in the form

$$d_t u_\varepsilon^{m+1} + \mathrm{curl}(\mathrm{curl}u_\varepsilon^{m+1}) + \nabla p_\varepsilon^{m+1} = f^{m+1},$$
$$\mathrm{div}u_\varepsilon^{m+1} - \varepsilon\Delta\phi_\varepsilon^{m+1} = 0, \qquad \partial_n \phi_\varepsilon^{m+1}|_{\partial\Omega} = 0. \tag{8.16}$$

Note that this is correct, because we have $u_\varepsilon^{m+1} \in \mathbf{H}^2(\Omega)$. For a corresponding uniform a-priori bound, see Lemma 6.3. In order to arrive at local statements for the error functions for velocity and pressure in (8.16), we will split the impact of the (stationary) pressure stabilization from the evolutionary character of the problem. The auxiliary problem to be considered below is then:

Find the solution $\{U_\varepsilon^{m+1}, P_\varepsilon^{m+1}\}$ of

$$\begin{aligned}
\mathrm{curlcurl}U_\varepsilon^{m+1} + \nabla P_\varepsilon^{m+1} &= f^{m+1} - d_t u^{m+1}, \\
\mathrm{div}U_\varepsilon^{m+1} - \varepsilon\Delta\Phi_\varepsilon^{m+1} &= 0, \qquad \partial_n\Phi_\varepsilon^{m+1}|_{\partial\Omega} = 0,
\end{aligned}$$
(8.17)

with

$$\Phi_\varepsilon^{m+1} := \mathrm{div}U_\varepsilon^{m+1} + P_\varepsilon^{m+1}.$$
(8.18)

Before we can start the error analysis for (8.17), we need some preliminary considerations concerning the existence and uniqueness of solutions: At first, through insertion of (8.18) in (8.17), we get the pressure stabilization formulation for $\{U_\varepsilon^{m+1}, \Phi_\varepsilon^{m+1}\}$. Owing to the local analysis in chapter 6, we have

$$\|u^{m+1} - U_\varepsilon^{m+1}\|_{1;\Omega'\subset\subset\Omega} \le C\varepsilon.$$
(8.19)

Thanks to the second equation in (8.17), this gives the a-priori bound

$$\|\Delta(\sigma\Phi_\varepsilon^{m+1})\| \le C,$$
(8.20)

$\sigma \in C_0^\infty(\Omega)$ being a cut-off function, with the following properties: $\sigma|_{\Omega'\subset\subset\Omega} = 1$ and $\max_{x\in\Omega}|\nabla\sigma(x)| \le C$. Owing to the singular perturbation character of the present problem we can not globalize this a-priori bound. Owing to the regularity shift of the inverse Laplacian we further obtain: $\|\Phi_\varepsilon^{m+1}\|_{2;\Omega'\subset\subset\Omega} \le C\|\Delta(\sigma\Phi_\varepsilon^{m+1})\|$. The derivation of a corresponding statement for $\Delta U_\varepsilon^{m+1} \in \mathbf{H}^1(\Omega')$ is now easy to obtain, if we reformulate system (8.17) in the quantities $\{U_\varepsilon^{m+1}, \Phi_\varepsilon^{m+1}\}$. — Let us now come back to the first equation in (8.17). Thanks to the latter considerations, we can now make use of the identity

$$\mathrm{divcurlcurl}U_\varepsilon^{m+1} \equiv 0, \qquad \text{on } L^2(\Omega').$$

The error statements that arise if we compare problem (8.17) with the related incompressible problem, i.e., the via Euler method semi-discretized equations

(6.12), are as follows, if we introduce the error notations $E^{m+1} := u^{m+1} - U_\varepsilon^{m+1}$, $\Psi^{m+1} := p^{m+1} + \operatorname{div} u^{m+1} - \Phi_\varepsilon^{m+1}$ and $\Pi^{m+1} := p^{m+1} - P_\varepsilon^{m+1}$,

$$\operatorname{curlcurl} E^{m+1} + \nabla \Pi^{m+1} = 0,$$
$$\operatorname{div} E^{m+1} - \varepsilon \Delta \Psi^{m+1} = -\varepsilon \Delta p^{m+1}, \qquad (8.21)$$
$$\partial_n \Psi^{m+1}|_{\partial\Omega} = \{\nabla p^{m+1} - \Delta u^{m+1}\}|_{\partial\Omega} \cdot n.$$

Owing to the above discussion, it is allowed to apply the divergence operator on the first equation in (8.21), finally interpreting this as a functional on $L^2(\Omega')$. This assures the local harmonicity of the pressure error function,

$$-\Delta \Pi^{m+1} = 0, \qquad \text{on } \Omega', \qquad (8.22)$$

for $\operatorname{supp} \Pi^{m+1} \subset \Omega' \subset\subset \Omega$. We will now give local error statements for the pressure function of (8.21), by multiplying this equality with the local weighting function $\sigma^4 \in C_0^\infty(\Omega)$. Again, we take $\Omega' \subset\subset \Omega$, such that $\max_{x \in \Omega} |\nabla \sigma(x)| \leq C$. If we test this weighted version of (8.22) with Π^{m+1}, we get

$$\|\nabla \Pi^{m+1}\|_{L^2(\Omega')} \leq \|\sigma^2 \nabla \Pi^{m+1}\| \leq C \|\sigma \Pi^{m+1}\| \leq C\varepsilon. \qquad (8.23)$$

This chain of inequalities needs further comments. The first inequality is trivially true. The transition to the next upper bound is a consequence of the Young Inequality, together with (8.22). Finally, the last estimate uses the following arguments: We made use of the local statement, given in Theorem 6.2 for the function Ψ^{m+1}, $\|\Psi^{m+1}\|_{\Omega' \subset\subset \Omega} \leq C\varepsilon$, on the other hand a super-convergence estimate for the local velocity error, measured in the Dirichlet-norm. It is of the following topic,

$$\|\sigma \nabla E^{m+1}\|^2 \leq C \|E^{m+1}\|^2 + \left(\Psi^{m+1}, \operatorname{div}(\sigma^2 E^{m+1})\right)$$
$$\leq C \|E^{m+1}\|^2 + \|\sigma \Psi^{m+1}\|^2 + \|E^{m+1}\|^2 \qquad (8.24)$$
$$+ \frac{1}{4\delta} \|\sigma \Psi^{m+1}\|^2 + \delta \|\sigma \operatorname{div} E^{m+1}\|^2, \qquad \delta > 0.$$

For sufficiently small δ, we can absorb the last term on the right hand side of this inequality through the left side, and we obtain, owing to the pressure error statements in Theorem 6.2,

$$\|\sigma \nabla E^{m+1}\| \leq C\varepsilon. \qquad (8.25)$$

In an analogous way, we get

$$\tau_{m+1}\left\{\|d_t E^{m+1}\| + \|\sigma \nabla d_t E^{m+1}\|\right\} \leq C\varepsilon. \tag{8.26}$$

This last estimate will be of importance later on. We can now use estimate (8.25) to confirm the validity of the last bound in (8.23).

Apart from the deviation of inequality (8.23) we can achieve analogous results concerning the error behavior for the pressure function in higher norms. Therefore, if we test the relation (8.22) (which is locally valid) with the function $-\sigma^2 \Delta(\sigma \Pi^{m+1})$, we find

$$\|\Pi^{m+1}\|_{H^2(\Omega')} \leq \|\sigma^2 \Pi^{m+1}\|_{H^2(\Omega)} \leq C\|\Delta(\sigma \Pi^{m+1})\|$$
$$\leq C\left\{\|\sigma \nabla(\sigma \Pi^{m+1})\| + \|\sigma \Pi^{m+1}\|\right\} + C\varepsilon \leq C\varepsilon. \tag{8.27}$$

The validity of the last inequality is owing to (8.22) and (8.23). It is now obvious, that the following can be shown to be valid, owing to the harmonicity of the solution of error equality (8.22),

$$\|\Pi^{m+1}\|_{H^r(\Omega')} \leq C\varepsilon, \qquad r \geq 0. \tag{8.28}$$

We stress the fact that these results are much better than in the case of the original Chorin scheme. The inequalities (8.23) and (8.28) give the reason for the interior smoothing effect. Based on these statements regarding the error in the pressure function we can give another estimate for the divergence of U_ε^{m+1}. In order to do so, we employ the local statement in Theorem 6.2 for the error of the pressure gradient,

$$\|\nabla \operatorname{div} E^{m+1}\|_{L^2(\Omega')} \leq C\sqrt{\varepsilon}\{1 + \sqrt{\varepsilon}\} \leq C\sqrt{\varepsilon}. \tag{8.29}$$

For the derivation of this, we have taken benefit from the error identity that is related to (8.18). We can use this inequality to gain another, sharper estimate. Thus, we make use of relation (8.15) and can write

$$\|E^{m+1}\|_{H^2(\Omega')} \leq C\|\sigma E^{m+1}\|_{H^2(\Omega)} \leq C\|\Delta(\sigma E^{m+1})\|$$
$$\leq \|\nabla \operatorname{div}(\sigma E^{m+1})\| + \|\operatorname{curl}\operatorname{curl}(\sigma E^{m+1})\|$$
$$\leq C\left\{\|\sigma \nabla \operatorname{div} E^{m+1}\| + \|\sigma \operatorname{curl}\operatorname{curl}(\sigma E^{m+1})\|\right.$$
$$\left. + \|\nabla E^{m+1}\|\right\} \leq C\sqrt{\varepsilon}. \tag{8.30}$$

A corresponding error statement can be verified for the pressure stabilization method (viz. Chapter 3). Therefore, (8.30) is no improvement in that
respect. The reason for that is the first term on the right hand side of the
last but one inequality in (8.30), thus limiting the order of convergence.

These statements for the error of the pressure gradient remain valid, if
we intend to bring over the recent results of convergence for (8.17) to the
original error equalities. This requires the study of the following system,

$$d_t e_\varepsilon^{m+1} - \Delta e_\varepsilon^{m+1} + \{\nabla \eta_\varepsilon^{m+1} + \nabla \operatorname{dive}_\varepsilon^{m+1}\} = d_t E^{m+1},$$
$$\operatorname{dive}_\varepsilon^{m+1} - \varepsilon \operatorname{div}\{\nabla \operatorname{dive}_\varepsilon^{m+1} + \nabla \eta_\varepsilon^{m+1}\} = 0, \tag{8.31}$$
$$\{\nabla \operatorname{dive}_\varepsilon^{m+1} + \nabla \eta_\varepsilon^{m+1}\}|_{\partial\Omega} \cdot n = 0.$$

The analysis of this problem, formulated for $\{e_\varepsilon^{m+1}, \theta_\varepsilon^{m+1}\}$, with the pressure function $\theta_\varepsilon^{m+1} := \eta_\varepsilon^{m+1} + \operatorname{dive}_\varepsilon^{m+1}$, has been carried out in Chapter 6.
Especially, we obtain with this substitution in (8.31), compare with (6.39),

$$\frac{k}{\tau_{M+1}} \sum_{m=0}^{M} \tau_m^2 \|d_t e_\varepsilon^{m+1}\|^2 + \tau_{M+1}\|\nabla e_\varepsilon^{M+1}\|^2 + \varepsilon\tau_{M+1}\|\nabla\theta_\varepsilon^{M+1}\|^2 \leq C\varepsilon^2. \tag{8.32}$$

Further, we obtain pointwise estimates for the velocity error, measured in
the $\mathbf{H}^2(\Omega)$-norm, (compare with (6.38)),

$$\tau_{m+1}\|e_\varepsilon^{m+1}\|_2 \leq C\tau_{m+1}\{\|\nabla\theta_\varepsilon^{m+1}\| + \|d_t E^{m+1}\| + \|d_t e_\varepsilon^{m+1}\|\}$$
$$\leq C\sqrt{\varepsilon}. \tag{8.33}$$

These global estimates (8.32) and (8.33) — which can be verified owing to the
fact that the error identities (8.31) exhibit no singular perturbation character
any more — will be employed, in order to verify convergence results the
pressure error η_ε^{m+1}. At first, we make use of inequality (8.32), which gives
rise to an error estimate for the velocity error, measured in the Dirichlet-
norm. Thanks to the second identity in (8.31), we obtain another a-priori
statement,

$$\sqrt{\tau_{m+1}}\|\theta_\varepsilon^{m+1}\|_2 \leq C\sqrt{\tau_{m+1}}\|\Delta\theta_\varepsilon^{m+1}\| \leq C. \tag{8.34}$$

This holds true, owing to a standard regularity statement for the Laplace-
Neumann operator. - This statement will be employed in the following. To
do so, we will reformulate the first equation in (8.31) in the form

$$d_t e_\varepsilon^{m+1} + \operatorname{curlcurl} e_\varepsilon^{m+1} + \nabla \eta_\varepsilon^{m+1} = d_t E^{m+1}. \tag{8.35}$$

We can apply an argument analogously to the one given subsequently to (8.20). Owing to (8.34), we can apply the divergence operator *globally* and arrive at the following identity on L_0^2,

$$- \Delta \eta_\varepsilon^{m+1} = d_t \text{dive}_\varepsilon^{m+1} - d_t \text{div} E^{m+1}. \tag{8.36}$$

In order to benefit from the super-convergence results of the auxiliary problem (8.17), formula (8.26), we are also dependent on local error estimates for the pressure gradient. Thus, we test this relation with $\sigma^2 \eta_\varepsilon^{m+1}$ and arrive at the following inequality,

$$\|\sigma \nabla \eta_\varepsilon^{m+1}\| \leq \|\sqrt{\sigma} \eta_\varepsilon^{m+1}\| + \|\sigma d_t \text{dive}_\varepsilon^{m+1}\|$$
$$+ \|\sigma d_t \text{div} E^{m+1}\| \leq C\{1 + \tau_{m+1}^{-1}\}\varepsilon. \tag{8.37}$$

An upper bound for the last term on the right hand side of the first inequality sign gives (8.26). The verification of an estimation of the second term on the right hand side, $\tau_{m+1}\|d_t e_\varepsilon^{m+1}\|_1 \leq C\varepsilon$, follows easily from (8.31), owing to the regular perturbation character. We will skip the easy detail here, arriving at

$$\sqrt{\tau_{m+1}}\|\eta_\varepsilon^{m+1}\| + \sqrt{\varepsilon \tau_{m+1}}\|\nabla \eta_\varepsilon^{m+1}\| + \tau_{m+1}\|\nabla \eta_\varepsilon^{m+1}\|_{\Omega' \subset\subset \Omega} \leq C\varepsilon. \tag{8.38}$$

This gives the proof of Theorem 8.3, *ii*). The combination of the inequalities (8.19), (8.32) and (8.23), (8.38) give the stated result. □

In order to complete the proof of Theorem 8.1, we will investigate the influence of the transition from the implicit pressure stabilization formulation (8.16) to the explicit version (8.10). This is the subject of the succeeding Subsection 8.2.2.

8.2.2 Transfer of the Error Statements of Subsection 8.2.1 to the Modified Chorin Method

The objective of this section is the transfer of the local statements to system (8.10), setting $\varepsilon = k$ (identification of perturbation and discretization parameter). Therefore, we have to analyze the following error equations, with $e^{m+1} := u_k^{m+1} - \tilde{u}^{m+1}$ and $\eta^{m+1} := p_k^{m+1} - \bar{p}^{m+1}$,

$$d_t e^{m+1} - \Delta e^{m+1} + \nabla \eta^m + \nabla \text{dive}^m = -k \nabla d_t\{p_k^{m+1} + \text{div} u_k^{m+1}\},$$

$$\text{dive}^{m+1} - k \Delta \vartheta^{m+1} = 0, \qquad \vartheta^{m+1} := \eta^{m+1} + \text{dive}^{m+1},$$

$$\partial_n \eta^{m+1}|_{\partial\Omega} = 0. \tag{8.39}$$

In order to proceed in a way that is analogous to the method in Subsection 8.2.1 for the verification of super-convergence statements for the pressure error gradient, we will look at the error identities in a first step after substitution in the first equality, together with (8.12),

$$d_t e^{m+1} - \Delta e^{m+1} + \nabla \vartheta^m = -k \nabla d_t \phi_k^{m+1}. \tag{8.40}$$

These relations (8.39), (8.40), formulated in the variables $\{e^{m+1}, \vartheta^{m+1}\}$, have been analyzed in Chapter 6. For the present purpose to reach a uniform a-priori bound for $\Delta e^{m+1} \in \mathbf{H}^1(\Omega'), \forall \varepsilon > 0$, these investigations have to be extended, what will be done in the following. Let us start with an error statement for the scheme (8.39), (8.40), compare also inequality (6.68) in Section 6.4 that guarantees the conservation of the error in the velocities,

$$\|e^{M+1}\|^2 + k \sum_{m=0}^{M} \|\nabla e^{m+1}\|^2 + k^2 \sum_{m=0}^{M} \|\nabla \vartheta^{m+1}\|^2 \leq Ck^2. \tag{8.41}$$

We intend to derive another, pointwise error statement for the velocity error, measured in the Dirichlet-norm. This requires the identity (8.40). After testing with $d_t e^{m+1}$ we obtain

$$\begin{aligned} \|d_t e^{m+1}\|^2 + d_t \|\nabla e^{m+1}\|^2 + k d_t \|\nabla \vartheta^{m+1}\|^2 \\ \leq k^2 \|\nabla d_t \vartheta^{m+1}\|^2 + Ck^2 \|\nabla d_t \phi_k^{m+1}\|^2. \end{aligned} \tag{8.42}$$

Based on the second identity in (8.39), the first term on the right hand side of (8.42) can be absorbed by the first sum on the left hand side. In order to control the last term on the right hand side of (8.42), we need a time-weight. Thus, we multiply (8.42) with τ_m and finally sum over all iteration steps,

$$\begin{aligned} k \sum_{m=0}^{M} \tau_{m+1} \|d_t e^{m+1}\|^2 + \tau_{M+1} \|\nabla e^{M+1}\|^2 + k \tau_{M+1} \|\nabla \vartheta^{M+1}\|^2 \\ \leq C\Big\{ k \sum_{m=0}^{M} \|\nabla e^{m+1}\|^2 + k^2 \sum_{m=0}^{M} \|\nabla \vartheta^{m+1}\|^2 \\ + k^3 \sum_{m=0}^{M} \tau_m \|\nabla d_t \phi_k^{m+1}\|^2 \Big\} \leq Ck^2. \end{aligned} \tag{8.43}$$

In order to derive an optimal, pointwise in time error estimate for the discretely differentiated velocity error, we employ the operator d_t on equation

(8.40) and finally test with $d_t e^{m+1}$. Then we obtain

$$\frac{1}{2} d_t \|d_t e^{m+1}\|^2 + \frac{1}{2} k \|d_t^2 e^{m+1}\|^2 + \|\nabla d_t e^{m+1}\|^2 + k \|\nabla d_t \vartheta^{m+1}\|^2$$
$$= k^2 (\nabla d_t \vartheta^{m+1}, \nabla d_t^2 \vartheta^{m+1}) + k^2 (\nabla d_t \vartheta^{m+1}, \nabla d_t^2 \phi_k^{m+1}). \qquad (8.44)$$

Owing to the second relation in (8.39), we can write the first term on the right hand side of (8.44) as

$$k^2 (\nabla d_t \vartheta^{m+1}, \nabla d_t^2 \vartheta^{m+1}) = k(\nabla d_t \vartheta^{m+1}, d_t^2 e^{m+1}). \qquad (8.45)$$

After application of Young's Inequality, the terms that result from the right hand side of (8.45) can be absorbed through the second and the fourth term on the left hand side of (8.44). The treatment of the last integral on the right hand side of (8.44) is evident. Now, weighting this relation with τ_m^2 and subsequent summation over all iteration steps gives

$$\tau_{M+1}^2 \|d_t e^{M+1}\|^2 + k^2 \sum_{m=1}^{M} \tau_{m+1}^2 \|d_t^2 e^{m+1}\|^2$$

$$+ k \sum_{m=1}^{M} \tau_{m+1}^2 \|\nabla d_t e^{m+1}\|^2 + k^2 \sum_{m=1}^{M} \tau_{m+1}^2 \|\nabla d_t \vartheta^{m+1}\|^2 \qquad (8.46)$$

$$\leq Ck \sum_{m=1}^{M} \tau_{m+1} \|d_t e^{m+1}\|^2 + Ck^2 \leq Ck^2.$$

Here, we have taken benefit from error statement (8.43). Thanks to a stability condition for the divergence operator we can now control the committed error in the pressure, using (8.39), (8.40), (8.43) and (8.46),

$$\tau_{m+1} \|\vartheta^{m+1}\| \leq C\tau_{m+1} \{\|d_t e^{m+1}\| + \|\nabla e^{m+1}\|\} \leq Ck. \qquad (8.47)$$

With this, we complete the error analysis for the problem (8.39), (8.40) (in substituted form) und come back to the treatment of the equations (8.39), with $\{u^{m+1}, \eta^{m+1}\}$ as the solution. Before giving these equations in another form in order to derive super-convergence results for the gradient of the pressure gradient, we start with some other remarks. First, let us note that the estimates (8.43) and (8.47) give the following error result,

$$\tau_{m+1} \|\eta^{m+1}\| \leq Ck. \qquad (8.48)$$

Secondly, the second equation in (8.39) and the inequality (8.43) give the a-priori bound $\sqrt{\tau_{m+1}}\|\Delta\vartheta^{m+1}\| \leq C$. A regularity statement then gives $\sqrt{\tau_{m+1}}\|\vartheta^{m+1}\|_2 \leq C$. This implies a uniform a-priori bound for e^{m+1} in $\mathbf{H}^3(\Omega')$, for $\Omega' \subset\subset \Omega$, owing to the first equality in (8.39).

Based on these considerations, we can again benefit from the property, that functions ϕ which can be written as $\phi = \text{curl}\Phi \in \mathbf{H}(\text{div}; \Omega')$ are in the kernel of the divergence operator. — We will write the first error equation in (8.39) in the following way,

$$
\begin{aligned}
d_t e^{m+1} &+ \text{curlcurl}\,e^{m+1} - k\nabla\text{div}d_t e^{m+1} + \nabla\eta^m \\
&= -k\nabla d_t\{p_k^{m+1} + \text{div}u_k^{m+1}\}.
\end{aligned}
\tag{8.49}
$$

The application of the div-operator now gives, on $\Omega' \subset\subset \Omega$,

$$
\begin{aligned}
-\Delta\eta^m &= d_t\text{div}e^{m+1} - k\text{div}\nabla\text{div}d_t e^{m+1} \\
&\quad + k\Delta d_t\{p_k^{m+1} + \text{div}u_k^{m+1}\}.
\end{aligned}
\tag{8.50}
$$

If we test this equation with the spatially weighted $\sigma^2\eta^m$, we get

$$
\begin{aligned}
\|\sigma\nabla\eta^m\|^2 &\leq C\big\{\|\eta^m\|^2 + \|d_t e^{m+1}\|^2 + k^2\|\sigma\nabla\text{div}d_t e^{m+1}\|^2 \\
&\quad + Ck^2 + k^2\|\sigma\nabla d_t\{p_k^{m+1} + \text{div}u_k^{m+1}\}\|^2\big\}.
\end{aligned}
\tag{8.51}
$$

In order to verify that the right hand side of this inequality is bounded by Ck^2, we have to employ a time-weight τ_{m+1} in this relation. This gives

$$
\max_{0\leq m\leq M}\big\{\sqrt{\tau_m}\|\nabla\eta^m\|_{\Omega'\subset\subset\Omega}\big\} \leq C\varepsilon,
\tag{8.52}
$$

and the proof of Theorem 8.1 is accomplished.

8.3   Analysis for the Chorin-Uzawa Scheme

In order to derive statements concerning the convergence behavior for the Chorin-Uzawa scheme (8.7), we have to separate and analyze the effects that are caused by the quasi-compressibility constraint and the evolutionary character of the momentum equation that act on the solution $\{u_{CU}^{m+1}, p_{CU}^{m+1}\}$. Thus, we will start our analysis with the investigation of an implicitly shifted auxiliary problem (with respect to the pressure in the "momentum" equation). We will denote the corresponding solution as $\{u_{AC}^{m+1}, p_{AC}^{m+1}\}$, for setting $\varepsilon = \frac{1}{\alpha}k$. The continuous version of these equations — which are the Stokes equations based on the compressibility ansatz — has been investigated in Section 4.2. Owing to the fact that discretization in time gives no new insights compared to those ones that we have already gained in the framework of the study of the Chorin scheme, we will restrict ourselves to a sketch of the proof of Theorem 8.2. Thus, we will concentrate here on the estimation of the error source that comes out due to the explicit approximation of the pressure function in (8.7). This impact on the approximation behavior will be quantified in the present section. Moreover, we will reflect on the necessity of the stabilizing parameter α. — The way of proceeding is illustrated in Figure 8.1 below, which is giving the sketch of the proof. Now, we will employ the following error notations, $e^{m+1} := u_{AC}^{m+1} - u_{CU}^{m+1}$ and $q^{m+1} := p_{AC}^{m+1} - p_{CU}^{m+1}$. In order to be comprehensive, we will write down the equations that are determining the tuple $\{u_{AC}^{m+1}, p_{AC}^{m+1}\}$, for $m \geq m_0$ and $\mathcal{O}(t_{m_0}) = 1$,

$$d_t u_{AC}^{m+1} - \Delta u_{AC}^{m+1} + \nabla p_{AC}^{m+1} = f^{m+1},$$

$$\text{div} u_{AC}^{m+1} + \frac{k}{\alpha} d_t p_{AC}^{m+1} = 0, \qquad \alpha < 1,$$

$$u_{AC}^{m+1}|_{\partial\Omega} = 0,$$

$$u_{AC}^m|_{m=m_0} = u_{AC}^{m_0}, \qquad p_{AC}^m|_{m=m_0} = p_{AC}^{m_0} \in L_0^2(\Omega),$$

$$\tag{8.53}$$

together with an approximation property at time $t_{m_0} = \mathcal{O}(1)$,

$$\|u_{AC}^{m_0} - u(t_{m_0})\| + \sqrt{k}\|p_{AC}^{m_0} - p(t_{m_0})\| \leq Ck.$$

The tuple $\{u_{CU}^{m+1}, p_{CU}^{m+1}\}$ is determined through the Chorin-Uzawa ansatz (8.7), for $m_1 > 0$. The error equations are then as follows, with the notations $e^{m+1} := u_{AC}^{m+1} - u_{CU}^{m+1}$ and $q^{m+1} := p_{AC}^{m+1} - p_{CU}^{m+1}$, for times $t_{m+1} \geq t_{m_1} = \mathcal{O}(1)$,

$$d_t e^{m+1} - \Delta e^{m+1} + \nabla q^{m+1} = k\nabla d_t q^{m+1} - k\nabla d_t p_{AC}^{m+1},$$

$$\text{div} e^{m+1} + \frac{k}{\alpha} d_t q^{m+1} = 0, \qquad \alpha < 1,$$

$$e^{m+1}|_{\partial\Omega} = 0.$$

$$\tag{8.54}$$

Now, we will deal with the problematic nature of stability statements for "reference" problems at times $t_{m+1} \geq t_{m_1}$ and valid initial values at time

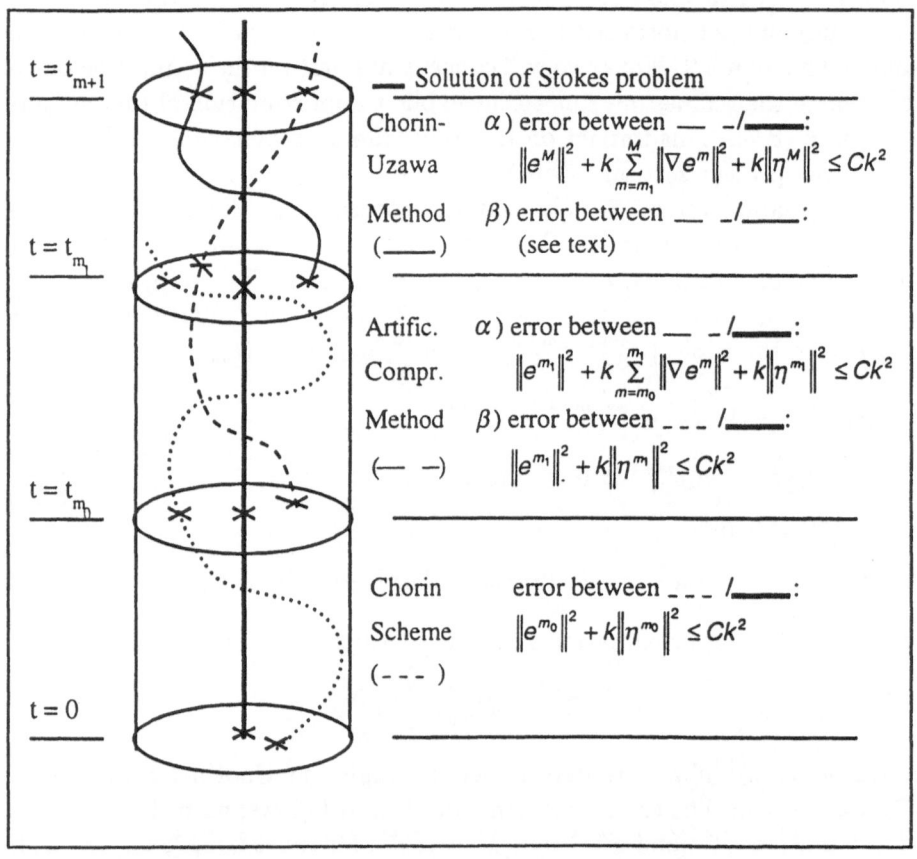

Figure 8.1: Sketch of the proof of Theorem 8.2

$t_{m+1} = t_{m_1}$ for the pressure and the velocity. This requires a series of considerations that will now be performed. Thus, there holds for the scheme (8.53) and its continuous version (4.5), with $t_{m_0} = \mathcal{O}(1)$,

$$\widehat{\tau}_{M+1}^{m_0}\|u_{AC}^{M+1} - u(t_{M+1})\|^2 + k \sum_{m=m_0}^{M} \widehat{\tau}_{m+1}^{m_0}\|u_{AC}^{m+1} - u(t_{m+1})\|_1^2$$

$$+ k\widehat{\tau}_{M+1}^{m_0}\|p_{AC}^{M+1} - p(t_{M+1})\|^2 \leq Ck^2(1 + \log\frac{1}{k}), \tag{8.55}$$

with $\{u(t_{M+1}), p(t_{M+1})\} \in \mathbf{H}_0^1 \times L_0^2$ as the solution of the continuous problem (2.1). Further, we have used the abbreviation $\widehat{\tau}_{M+1}^{m_0} := \tau(t_{M+1} - t_{m_0})$. This estimate has been derived in Section 4.2 for the artificial compressibility method (4.5), based on the requirement (4.8), and with the identification $\varepsilon = k$. Remember the fact that the solution of the incompressible Stokes problem is perturbed in the framework of the accuracies that are given from the Chorin algorithm,

$$\|u(t_{m_0}) - u_{Cho}^{m_0}\| + \sqrt{k}\|p(t_{m_0}) - p_{Cho}^{m_0}\| \leq Ck. \tag{8.56}$$

The estimation (8.55) of the second term on the left hand side implies an a-priori statement for the pressure function, due to the second equation in (8.53),

$$k \sum_{m_0=0}^{M} \widehat{\tau}_{m+1}^{m_0}\|d_t p_{AC}^{m+1}\|^2 \leq C(1 + \log\frac{1}{k}). \tag{8.57}$$

Owing to the time-weight that is appearing in the last result, it is not possible to apply this in the error equalities directly, if there is no further restriction onto the choice of t_{m_0}. The structure of the next way of proof of Theorem 8.2 is illustrated in the above figure.

We can now start with an error analysis for the Chorin-Uzawa scheme for times $t_{m+1} \geq t_{m_1}$, because of the existing approximation statement (8.55), together with the further estimate

$$\|u(t_{m_1}) - u_{Cho}^{m_1}\| + \sqrt{k}\|p(t_{m_1}) - p_{Cho}^{m_1}\| \leq Ck, \tag{8.58}$$

with $p_{Cho}^{m_1} \in L_0^2(\Omega)$, that give the necessary assumptions at the starting time $t_{m+1} = t_{m_1}$ for the succeeding investigations. Additionally, provided the time-point t_{m_1} satisfies $|t_{m_1} - t_{m_0}| = \mathcal{O}(1)$, we obtain another stability estimate for the pressure function of system (8.53), that needs no damping time-weight, owing to (8.56). By means of the triangle inequality we obtain the following estimate, for times $t_{m+1} \geq t_{m_1}$,

$$
\|u(t_{M+1}) - u_{CU}^{M+1}\|^2 + k \sum_{m=m_1}^{M} \|u(t_{m+1}) - u_{CU}^{m+1}\|_1^2 + k\|p(t_{M+1}) - p_{CU}^{M+1}\|^2
$$

$$
\leq \|u(t_{M+1}) - u_{AC}^{M+1}\|^2 + k \sum_{m=m_1}^{M} \|u(t_{m+1}) - u_{AC}^{m+1}\|_1^2
$$

$$
+ k\|p(t_{M+1}) - p_{AC}^{M+1}\|^2 + \|u_{CU}^{M+1} - u_{AC}^{M+1}\|^2 \tag{8.59}
$$

$$
+ k \sum_{m=m_1}^{M} \|u_{CU}^{m+1} - u_{AC}^{m+1}\|_1^2 + k\|p_{CU}^{M+1} - p_{AC}^{M+1}\|^2.
$$

We stress the fact that the functions $\{u_{AC}^{m+1}, p_{AC}^{m+1}\}$ are defined for values $m \geq m_0$, whereas the tuple $\{u_{CU}^{m+1}, p_{CU}^{m+1}\}$ is existent for $m \geq m_1$. Thanks to estimate (8.55), it is sufficient to control the error functions $e^{m+1} := u_{AC}^{m+1} - u_{CU}^{m+1}$ and $q^{m+1} := p_{AC}^{m+1} - p_{CU}^{m+1}$. In order to reach this aim, we test the first equation in (8.54) with e^{m+1} and make use of the uniform bound for $(1 + \log \frac{1}{k})^{1/2} d_t p_k^{m+1} \in L^2(t_{m_1}, t_{M+1}; L_0^2(\Omega))$, together with the statements (8.55), (8.58) for the initial quality of approximation at time t_{m_1},

$$
\|e^{M+1}\|^2 + (1 - \alpha - \delta)k \sum_{m=m_1}^{M} \|\nabla e^{m+1}\|^2 + k\|q^{M+1}\|^2
$$

$$
\leq \|e^{m_1}\|^2 + k\|q^{m_1}\|^2 + C\frac{1}{\delta\alpha^2}k^2 \sum_{m=m_1}^{M} \|d_t p_k^{m+1}\|^2, \tag{8.60}
$$

owing to the second equality in (8.54). Now, we take δ sufficiently small. The right hand side of this inequality can be bounded by Ck^2. In order to guarantee the stability of the Chorin-Uzawa scheme, it is sufficient to choose $\alpha < 1$. This completes the proof of Theorem 8.2.

8.4 Computational Results

8.4.1 The Modified Chorin Scheme

The first part of this section is devoted to the numerical study of the modification of the Chorin scheme, as it is proposed by Timmermans et al. in

$k =$	u_2	div$u$ bef. proj.	div$u$ after proj.	∇u_2	p
0.1	1.731	$2.28 - 3$	$1.15 - 4$	6.445	0.434
$5. - 2$	1.336	$1.82 - 3$	$9.02 - 5$	4.995	0.345
$2.5 - 2$	0.921	$1.32 - 3$	$6.4 - 5$	3.468	0.251
$1.25 - 2$	0.571	$8.88 - 4$	$4.23 - 5$	2.178	0.168
$6.25 - 3$	0.327	$5.65 - 4$	$2.75 - 5$	1.268	0.107
$3.13 - 3$	0.180	$3.48 - 4$	$1.87 - 5$	0.789	$6.21 - 2$
$1.56 - 3$	$9.51 - 2$	$2.09 - 4$	$1.37 - 5$	0.483	$3.47 - 2$
$7.82 - 4$	$4.92 - 2$	$1.27 - 4$	$9.21 - 6$	0.314	$2.06 - 2$
order	0.92	0.78	0.73	0.81	0.82

Table 8.1: Errors for the Modified Chorin Scheme

[44]. Let us recall the theoretical results that give no improvement regarding the suppression of the numerically caused boundary layers but indicate a smoothing of the error in the interior subdomains. Computational results are presented in Table 8.1 and have been obtained for the model problem $\{T, \gamma, \nu\} = \{2, 1, 1\}$. They indicate the validity of the global estimates.

We see that the errors for the pressure function are on average a factor of half less than the corresponding errors in Table 6.1. But again, this ansatz *does not* give any qualitative improvements in the pressure approximation: the boundary layers are still present and of magnitude $\mathcal{O}(\sqrt{k})$ — although being not so marked compared to those that perturb the solution of the Chorin scheme. Nevertheless, they have a significant impact on the approximation quality close to the boundary. These statements will be substantiated through a comparison of Figures 8.2 and 8.3.

8.4.2 The Chorin-Uzawa Scheme

Another modification of the basic Chorin scheme has been proposed in the beginning of this chapter, and was named Chorin-Uzawa method. The investigation of this projection ansatz has highlighted its connection with a semi-explicit artificial-compressibility method. This quasi-compressibility ansatz does not cause any singular perturbation of the pressure function any more, giving rise to accurate pressure values up to the boundary (up to the dis-

$k =$	u_2	divu bef. proj.	divu after proj.	∇u_2	p
0.2	1.498	1.57 – 3	1.02 – 4	5.645	0.262
0.1	0.781	8.08 – 4	5.36 – 5	2.977	0.144
5. – 2	0.393	4.05 – 4	2.70 – 5	1.497	7.33 – 2
2.5 – 2	0.196	2.26 – 4	1.35 – 5	0.749	3.71 – 2
1.25 – 2	9.82 – 2	1.01 – 4	6.72 – 6	0.376	1.93 – 2
6.25 – 3	4.97 – 2	5.05 – 5	3.37 – 6	0.189	9.8 – 3
order	0.99	0.99	0.99	0.99	0.97

Table 8.2: Orders of Convergence for the Chorin-Uzawa Scheme ($\alpha = 0.8$)

cretization error). These facts lead to improved estimates for the velocity field, measured in the norm $l^2(0, T; \mathbf{H}_0^1)$.

Of course, the Chorin-Uzawa method $CU_k\{u^{m+1}, p^{m+1}; u^0, p^0\}$ is no pressure stabilization scheme for the quantities $\{\tilde{u}^{m+1}, p^{m+1}\}$ any more. Moreover, the scheme has been developed in order to alleviate the disadvantages of the Chorin scheme. Now, if we further think of a stable spatial discretization via finite elements of high accuracy, here Q1/Q1, we need another modification, which is stable for the full discretization, $\overline{CU_{k,h}}\{u_h^{m+1}, p_h^{m+1}; u_h^0, p_h^0\}$. Therefore, we have to apply the following modification to the last step (8.5),

$$(1 - h^2\Delta_h)p_h^{m+1} = -\alpha\nu\mathrm{div}_h\tilde{u}_h^{m+1} + p_h^m,$$
$$\partial_{n,h}p_h^{m+1}|_{\partial\Omega} = 0.$$
(8.61)

This gives a stable procedure for the Q1/Q1 finite elements. The conditioning of problem (8.61) is good, due to the normalizing factor. Therefore, this equation is easily solvable.

The test calculations that have been done for our model problem $\{T, \gamma, \nu\} = \{2, 1, 1\}$ are summarized in Table 8.2.

If we compare the results with the corresponding errors for the basic and the modified Chorin scheme of Timmermans et al., we recognize a drastic improvement with respect to the errors. There have been done several test calculations for the Chorin-Uzawa scheme, leading to errors that are only of magnitude 1% through 10% of those produced by the other two schemes in all

$\alpha =$	u_2	div$u$ bef. proj.	div$u$ after proj.	∇u_2	p
0.2	0.314	$3.24 - 4$	$2.16 - 5$	1.195	0.143
0.4	0.157	$1.62 - 4$	$1.07 - 5$	0.599	$6.76 - 2$
0.6	0.105	$1.08 - 4$	$7.16 - 6$	0.400	$4.25 - 2$
0.8	$7.9 - 2$	$8.11 - 5$	$5.38 - 6$	0.302	$3.01 - 2$
1.0	$6.33 - 2$	$6.48 - 5$	$4.31 - 6$	0.246	$2.28 - 2$
2.0	$3.16 - 2$	$3.25 - 5$	$2.17 - 6$	0.125	$1.15 - 2$
2.2	$2.88 - 2$	$2.95 - 5$	$1.97 - 6$	0.115	$1.18 - 2$
2.4	$2.64 - 2$	$2.7 - 5$	$1.81 - 6$	0.106	$1.11 - 2$
2.6	$7.91 - 2$	$8.1 - 5$	$5.38 - 6$	0.302	$3.00 - 2$
3.2	DIV	DIV	DIV	DIV	DIV

Table 8.3: Errors for the Chorin-Uzawa Scheme (k = 0.01)

norms considered. Thus, these numerical experiments support the theoretical results and again illustrate the dominant impact of the boundary layers in case of the Chorin scheme on the global quality of the approximation.

The objective of the remaining investigation is now to find out the dependence of the stability on the chosen parameter α. The results of these studies are given in Table 8.3.

We recognize that the restriction $\alpha < 1$ is sufficient for optimal convergence behavior. Further, it is apparent that the errors decrease for increasing values of α, until a critical value is reached, α_{crit}; if we choose a higher value, the scheme diverges ("DIV").

Let us conclude this section and chapter with a comparison of the *three proposed projection methods of first order* that are analyzed in this book. The error functions that are given in Figures 8.2 through 8.4 (with $k = 0.01$) for the pressure function and the divergence of the computed velocity field illustrate the arising error mechanisms once more: The dominant boundary layers can be clearly seen, whereas the Chorin-Uzawa method gives good approximations for pressure function and the divergence of $\tilde{u}^{m+1}$ up to the boundary. In this point, we would like to stress that the latter in Figure 8.4 is scaled with a factor of 10 in order to actually make any structure of the error visible.

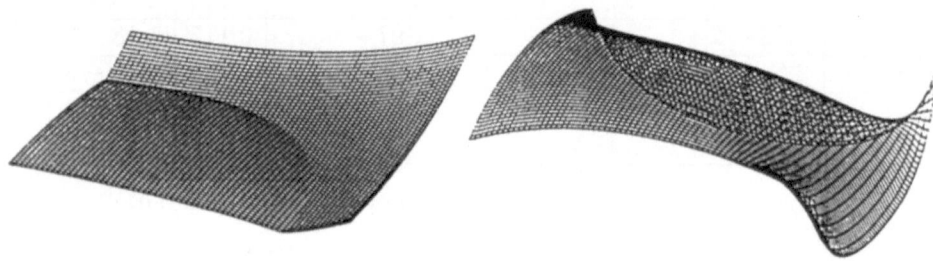

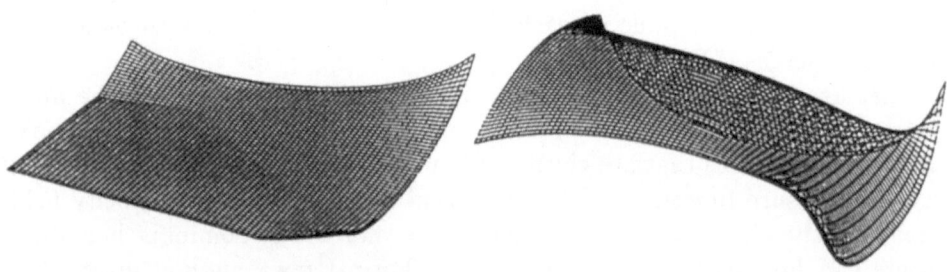

Figure 8.3: modified Chorin method: pressure error (left) and $\operatorname{div}\tilde{u}_{Cho}^{M+1}$

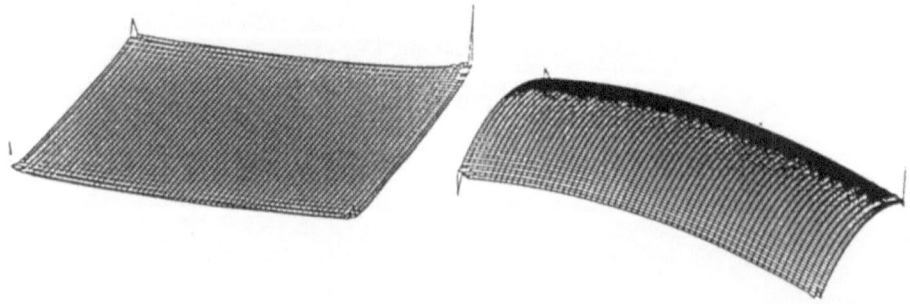

Figure 8.4: Chorin-Uzawa method: pressure error (left) and $\mathrm{div}\tilde{u}_{Cho}^{M+1}$ (right, scaled with 10)

Chapter 9

Multi-Component Schemes

9.1 Overview

The subject of Chapter 7 has been the separation and investigation of the diverse error sources that act on the approximation quality in case of the Van Kan scheme (with $\beta \geq \frac{1}{2}$). Especially, we treated the consistency error independently of those errors that stem from the quasi-compressibility constraint. Finally, we dealt with the particular problems that arise in the case of incompatibly posed initial data, both, from the theoretical and the computational point of view. As a result, we furnished that the assumptions given in *Postulate B_2* are necessary to secure a second order convergence rate, otherwise there might be a global reduction of order of convergence for the approximation $\tilde{u}^{m+1}$ from $\mathcal{O}(k^2 \log \frac{1}{k})$ to $\mathcal{O}(k)$.

Thus, the objective of the first part of the present chapter is to propose another more stable splitting ansatz of second order, based on the idea of the Van Kan scheme, that *does not need the additional assumptions of Postulate B_2 for an optimal convergence behavior.* Because of that, the first part of this chapter will be concerned with the problematic nature of achieving the "environment" of approximations, necessary for the initial data of the Van Kan scheme, compare (7.4). The following catalog of requirements should be satisfied by an improved second order projection scheme:

1. The starting algorithm (which we call the predictor) that is employed in a multi-component ansatz has smoothing properties. Especially, this scheme is expected to work in an optimal way for all solutions of the

incompressible Stokes system (2.1), only assuming (A1) and (A2).

2. The predictor is supposed to cause no higher numerical effort compared to the previously proposed projection schemes.

3. The assumptions (7.4) are satisfied at a time $t_{m_0} = \mathcal{O}(1)$ to guarantee continuation of the (long-time) simulation via the (slightly modified) Van Kan scheme as the corrector.

Evidently, the classical Chorin method satisfies the first requirement, whereas we are forced to take an actual time-step $\tilde{k} = k^2$ in order to guarantee the verification of item 3. (and the preservation of second order accuracy of the scheme). It is clear that the immense numerical effort arising from such a strategy cannot justify this. But surprisingly, it is possible to develop a multi-component scheme in Section 9.2, the *Chorin-/Van Kan scheme*, that satisfies the *whole* catalog of requirements.

The second part of this chapter is concerned with a *modified Chorin-Uzawa scheme*, which is a projection scheme of *first* order. It can be seen as a continuation of material presented in Section 8.3. In that context, we will propose and analyze a modification of the Chorin-Uzawa scheme, that is of the same complexity but now gives *linear order of convergence in k for the pressure error*, measured in the $l^\infty(0, T; L^2(\Omega)/\mathbb{R})$-norm. This can be interpreted as the theoretical foundation for the fact that no global time-induced error sources are present any more that act through the nonstationary character of the quasi-compressibility constraint. Let us recall that this error mechanism prevented a corresponding result for the Chorin-Uzawa scheme. This algorithm, which will be proposed in Section 9.3, is again a multi-component scheme, with the application of the robust Chorin scheme as the predictor up to a "macroscopic" time $t_{m_0} = \mathcal{O}(1)$ and the replacement by the Chorin-Uzawa scheme for times $t_{m+1} \geq t_{m_0}$, but now for the discretely differentiated functions $\{d_t\tilde{u}^{m+1}, d_t p^{m+1}\}$. In order to guarantee the transfer of the right amount of information at the "interface" time t_{m_0}, we have to employ a (stationary) $\mathbf{J}_1(\Omega)$-projection. We refer to Section 9.3 for further details and especially to Theorem 9.2 that incorporates the main results for this multi-component projection scheme.

Both above multi-component projection schemes are parameterized by the time t_{m_0}, where the calculation of the flow is replaced by a more accurate time-discretization scheme of projection type. Of course, this time-point can depend on the given problem parameters (like the Reynolds number Re) of a complex flow in a critical way. For the further studies, we are motivated to apply these numerical models to flows with medium Reynolds numbers, expecting that there is no sensitivity with respect to the choice of t_{m_0} for a large range of applications. We will come back to this again in the subsequent sections.

The proofs of the following theorems employ a huge quantity of auxiliary results that have been achieved in the previous chapter. Again, we will restrict ourselves to the case of the incompressible Stokes equations (2.1). We want to stress that the following analysis for the Chorin-/Van Kan scheme relies heavily on a damping mechanism with respect to perturbations that is inherent to the Stokes equations. This is due to the fact that we start with a first order scheme, looking for a damping of the resulting errors to second order for macroscopic times. Thus, a transfer of the following investigations to arbitrary flows governed by the incompressible Navier-Stokes equations is *not* possible. But, if we are provided with additional information regarding the stability properties of the flow under consideration, this allows the application of the analytical methods which will be given in the present context. In contrast, the applicability of the Chorin-/Chorin Uzawa scheme is not touched by this problematic nature.

9.2 The Chorin-/Van Kan Scheme — Proposal and Analysis

This section is devoted to the proposal and analysis of a scheme of second order that combines both the advantages of a stable projection scheme of first order with the gain of accuracy of a method of higher order, satisfying the three requirements that have been given in the introductory part of the previous section. This allows us to approximate the solution of the Stokes system in an accurate and stable manner, by only requiring the assumptions (A1) and (A2) to hold.

We already mentioned the fact that the "naive" application of the Chorin method as a predictor causes a numerical effort that can not be accepted, if we want to satisfy the three introductory requirements (this requires the choice $\tilde{k} = k^2$). We will now propose another strategy that justifies the application of the Chorin scheme over the time range $[0, t_{m_0}]$ with time-step size k, which we call the *Chorin-/Van Kan scheme*:

1. *predictor method:* Start the simulation with the standard Chorin scheme, using the time-step size k, and calculate the fluid flow up to the time $t_{m_0} = \mathcal{O}(1)$.

2. $\mathbf{J_1}(\Omega)$-*projection:* Let $\{U^{m_0}, P^{m_0}\} \in \mathbf{J_1}(\Omega) \cap \mathbf{H}^2(\Omega) \times H^1(\Omega)/\mathbb{R}$ be the solution of the incompressible, stationary Stokes equations,

$$- \Delta U^{m_0} + \nabla P^{m_0} = -d_t \tilde{u}^{m_0}_{Cho} + f^{m_0}, \qquad (9.1)$$

with $d_t \tilde{u}^{m_0}_{Cho} \in \mathbf{H}^1_0(\Omega)$ the approximation that has been established by means of the Chorin method. Of course, the difference operator is related to the functions with tilde notation, i.e., $d_t \tilde{u}^{m_0}_{Cho} \equiv \frac{1}{k} \{ \tilde{u}^{m_0}_{Cho} - \tilde{u}^{m_0-1}_{Cho} \}$.

3. *corrector method:* Together with the initial functions $\{U^{m_0}, P^{m_0}\}$, we apply the *implicitly shifted Van Kan scheme* (with $\beta \geq \frac{1}{2}$). This differs from the original scheme in the way that the viscous part of the operator is shifted implicitly, i.e., it is replaced by the terms $-\Delta \{ \frac{1+k}{2} \tilde{u}^{m+1} + \frac{1-k}{2} \tilde{u}^m \}$.

Remark 9.1   *1. The application of the $\mathbf{J_1}(\Omega)$-projection is essential in the above algorithm. At a certain amount, it serves as a compensation of the inaccurate Chorin scheme, which approximates higher solution frequencies only in a very modest way, even in the case of the pressure approximation. On the other hand, the $\mathbf{J_1}(\Omega)$-projection can only be applied successfully in this framework, owing to the stability properties of the Chorin scheme, see below the current analysis.*

2. The introduction of the implicit shift of the Van Kan scheme is motivated by related studies of Rannacher in [31], where this idea is proposed in the case of the Crank-Nicolson method, applied to the (linear) heat transfer equation. Owing to that modification, Rannacher derived the following classical "damping property", which is giving a statement with respect to the qualitative behavior of initially introduced perturbations,

$$\|v^{m+1}\| \leq \exp(-\lambda t_{m+1}) \|v_0\| + \frac{1}{\sqrt{\lambda}} \sup_{0 \leq m \leq M} \|f^{m+1}\|_{-1}. \qquad (9.2)$$

Here, we used v^{m+1} as the $(m+1)$th iterate of the modified Crank-Nicolson method, applied to the heat equation. Additionally, the value $\lambda \equiv \lambda(-\Delta_D(\Omega))$ denotes the minimal eigenvalue of the Laplacian. The necessity of such a modification of the original Crank-Nicolson scheme is given owing to the fact that such a statement does not hold for the original scheme. Let us recall, that in the latter case a corresponding result can only be verified in the context of a full discretization and the coupling condition $k \approx h$ (for a quasi-uniform triangulation). In the present work, we will only restrict on semi-discretization and its effects, further, we intend to avoid step-size restrictions of type $F(k,h) > 0$. Therefore, we will prefer the introduction of the modified Van Kan scheme in the above algorithm.

Now, we will discuss and analyze the Chorin-/Van Kan scheme, aiming at optimal convergence results for this multi-component scheme. The analysis is decomposed in three steps, corresponding to the components of the scheme.

1. *The Chorin scheme as a predictor method:* Based on the investigation of the Chorin method in Chapter 6, it is known that this projection method has a *smoothing property*. This property is of crucial importance in order to understand the multi-component ansatz proposed above. In particular, we have

$$\max_{0 \leq m \leq M} \left\{ \tau_{m+1}^{3/2} \| u_t(t_{m+1}) - d_t \tilde{u}_{Cho}^{m+1} \| \right\} \leq Ck, \tag{9.3}$$

with u the solution of (2.1). The verification of this estimate will be given in the contents of the proof of Theorem 9.1.

2. *The $J_1(\Omega)$-projection at time t_{m_0}:* In order to apply the Van Kan scheme in an optimal manner, we need to satisfy (7.4) for the initial data of pressure and velocity. Therefore, we employ a $J_1(\Omega)$-projection at time t_{m_0}, as given in (9.1). Because of estimate (9.3) and by means of standard a-priori bounds for the stationary Stokes equations we can now give an error estimate for these projected functions. Thus, if we subtract the first equation in (2.1) — evaluated at time t_{m_0} — from the first one in (9.1), we obtain

$$\begin{aligned} \| U^{m_0} - u(t_{m_0}) \|_2 + \| P^{m_0} - p(t_{m_0}) \|_1 \\ \leq C \| d_t \tilde{u}_{Cho}^{m_0} - u_t(t_{m_0}) \| \leq Ck. \end{aligned} \tag{9.4}$$

Note that this pointwise $\mathbf{J}_1(\Omega)$-projection (9.1) gives a transfer of order with respect to the regularity of the approximating solution. Especially, the solution of (9.1) *does not suffer from any numerically induced boundary layers any more*, owing to the strong character of the projection. This gives an improvement of the order of convergence by half an order. Let us mention the fact that the history of the fluid flow is solely represented by the velocity function on the right hand side of (9.1), which is sufficient to gain super-convergence results of the above sort. Due to (9.4), we get a first initial approximation result for the Van Kan scheme, namely for the pressure function. Unfortunately, this transfer of order through the $\mathbf{J}_1(\Omega)$-projection does not provide us with an initial velocity function that satisfies (7.4). This problematic nature will be mastered via the damping character of this implicitly shifted numerical model (see the discussion above). This is the topic of the subsequent point.

3. *The implicitly shifted Van Kan scheme as the corrector:* In the preceding step, we had to cope with the problem of how to improve the velocity approximates at time t_{m_0}, which are of course only of first order. In order to continue with the Van Kan scheme providing us with iterates $\{u^{m+1}, p^{m+1}\}|_{m \geq m_0}$ that converge of second order to the actual solution $\{u(t_{m+1}), p(t_{m+1})\}|_{m \geq m_0}$, we make use of the damping property of the applied discretization scheme that even holds true for the implicitly shifted version of the Van Kan scheme introduced above. We will focus on the mechanisms inherent to that scheme in the proof of the subsequent theorem, which gives the statements of convergence for the solution of the Chorin-/Van Kan scheme.

The main results of the present section are collected in the subsequent theorem.

Theorem 9.1 *Assume the given data of problem (2.1) satisfy the basic assumptions (A1), (A2), with $\{u, p\}$ being the solution. Let $\{\tilde{u}^{m+1}, p^{m+1}\}$ be the solution of the Chorin-/Van Kan scheme (with $\beta = 1/2$). We introduce the notation $\hat{\tau}_{m+1} := \tau(t_{m+1} - t_{m_0})$. Then, the following error statements are valid, for times $t_{m_0} = \mathcal{O}(1)$,*

   i)   *provided that $t_{m+1} \leq t_{m_0}$:*

$$\max_{0 \leq m \leq m_0} \left\{ \|u(t_{m+1}) - \tilde{u}^{m+1}\| + \sqrt{k}\|u(t_{m+1}) - \tilde{u}^{m+1}\|_1 \right\} \leq Ck,$$

$$\max_{0 \leq m \leq m_0} \sqrt{\tau_{m+1}} \left\{ \sqrt{\tau_{m+1}}\|p(t_{m+1}) - p^{m+1}\|_{-1} \right.$$
$$\left. + \sqrt{k}\|p(t_{m+1}) - p^{m+1}\| \right\} \leq Ck,$$

ii) *provided that* $t_{m+1} > t_{m_0}$:

$$\max_{0 \leq m \leq m_0} \left\{ \sqrt{\tilde{\tau}_{m+1/2}}\|u(t_m) - \bar{\bar{u}}^m\| + k\|u(t_m) - \bar{\bar{u}}^m\|_1 \right\}$$
$$\leq Ck^2(1 + \log\tfrac{1}{k}) + C_{m_0}k\exp\left(-\lambda(t_{m+1} - t_{m_0})\right),$$

$$\max_{0 \leq m \leq m_0} \left\{ \sqrt{\tilde{\tau}_{m+1/2}}\|p(t_m) - \bar{p}^m\| \right\}$$
$$\leq Ck(1 + \log\tfrac{1}{k}) + C_{m_0}\sqrt{k}\exp\left(-\lambda(t_{m+1} - t_{m_0})\right),$$

and $0 < k < 1$. *Here, we used the notation, which has been introduced in Chapter 7. The sub-index at the constant* C_{m_0} *is to indicate that it is not growing in time but only depends on the selection of* t_{m_0}. *The constant* $\lambda \equiv \lambda(A)$ *is the minimal eigenvalue of the Stokes operator* A.

Remark 9.2   *1. In this section, we restrict ourselves to the case of the original Van Kan scheme, i.e.,* $\beta = \tfrac{1}{2}$. *The formulation of sharper statements in the case* $\beta > \tfrac{1}{2}$ *is analogous by making use of the results in Chapter 7.*

2. *In the case of weighting the Stokes operator* A *in (2.1) with the viscosity parameter* ν *we have to multiply the eigenvalue in ii) with* ν *such that* $\lambda_\nu \equiv \nu\lambda$. *Evidently, the resulting estimate gives only striking validity for moderate values of* ν. *Thus, this multi-component projection scheme seems to be limited to certain problem classes. Nevertheless, this scheme can also be profitably applied to problems with higher Reynolds numbers and incompatibly posed initial data, if we are concerned with long-time studies of fluid flows. Of course, the accuracy for the velocity field of the scheme is limited for times* $t_{m+1} - t_{m_0} << \tfrac{1}{\nu}$ *by* Ck — *in the same way as the original Van Kan scheme. But, by contrast, for increasing times* t_{m+1} *the order of convergence climbs up to second order (for* $t_{m+1} - t_{m_0} > C\tfrac{1}{\nu}$), *whereas the order stagnates for the original scheme in cases where the conditions of Postulate* B_2 *are not satisfied.*

By means of the multi-component strategy we get rid of the problematic nature of missing initial compatibility, which is otherwise causing a global reduction of order of convergence. Let us mention the fact that again there appear logarithms in the estimates of Theorem 9.1 caused by arising incompatibility properties.

Proof:

We can restrict on the verification of part ii). This requires results taken from previous chapters and [31]. We will start with a sketch of the validity of statement (9.3).

a) Verification of estimate (9.3): We will employ the results that have been verified in Chapter 6, where we have shown optimal results of convergence for the Chorin scheme. The way of reaching this aim was the separate treatment of consistency and perturbation errors and those that stem from the semi-explicitness of the projection scheme. We will take the same idea in order to verify statement (9.3):

a_1) Estimation of the consistency error caused by the implicit Euler discretization: If we set $e^{m+1} := u(t_{m+1}) - u^{m+1}$, $\eta^{m+1} := p(t_{m+1}) - p^{m+1}$ and the integral residual

$$R^{m+1} = \frac{1}{k} \int_{t_m}^{t_{m+1}} (s - t_m) u_{tt}(s) ds,$$

the corresponding error equations are as follows,

$$
\begin{aligned}
d_t e^{m+1} - \Delta e^{m+1} + \nabla \eta^{m+1} &= R^{m+1}, \\
\operatorname{div} e^{m+1} = 0, \qquad e^{m+1}|_{\partial\Omega} &= 0, \qquad e^0 \equiv 0.
\end{aligned}
\tag{9.5}
$$

Application of d_t to the first equation of (9.5) and multiplication with the test function $\tau_m^3 d_t e^{m+1}$ gives, after summation over all iteration steps,

$$
\begin{aligned}
\frac{1}{2} \tau_{M+1}^3 \| d_t e^{M+1} \|^2 &+ k \sum_{m=0}^{M} \tau_{m+1}^3 \| \nabla d_t e^{m+1} \|^2 \\
&\leq Ck \sum_{m=0}^{M} \tau_{m+1}^2 \| d_t e^{m+1} \|^2 + Ck \sum_{m=0}^{M} \tau_{m+1}^3 \| d_t R^{m+1} \|_{-1}^2.
\end{aligned}
\tag{9.6}
$$

In order to control the last term on the right hand side of (9.6), we use the identity

$$d_t R^{m+1} = \frac{1}{k}\{R^{m+1} - R^m\}$$
$$= \frac{1}{k^2}\{\int_{t_m}^{t_{m+1}} (s - t_m)u_{tt}(s)ds - \int_{t_{m-1}}^{t_m} (s - t_{m-1})u_{tt}(s)ds\}. \quad (9.7)$$

We can continue, using the mean-value rule of the calculus of integration,

$$= \frac{1}{k^2}\{u_{tt}(\xi)|_{\xi \in [t_m, t_{m+1}]} \int_{t_m}^{t_{m+1}} (s - t_m)\, ds$$
$$- u_{tt}(\xi')|_{\xi \in [t_{m-1}, t_m]} \int_{t_{m-1}}^{t_m} (s - t_{m-1})\, ds\}$$
$$= \frac{1}{2}\{u_{tt}(\xi)|_{\xi \in [t_m, t_{m+1}]} - u_{tt}(\xi')|_{\xi' \in [t_{m-1}, t_m]}\} \quad (9.8)$$
$$= \frac{1}{2}(\xi - \xi')u_{ttt}(\xi'')|_{\xi'' \in [t_{m-1}, t_{m+1}]},$$

thanks to the mean-value theorem. Owing to the bound $|\xi - \xi'| < 2k$ and some well-known regularity results, we can now control the last term on the right hand side of (9.6) by Ck^2. Let us mention that solely at this point we need the stronger time-weight τ_{m+1}^3 instead of τ_{m+1}^2, which is sufficient else. This gives rise to the conjecture that the above estimate (9.6) is too strong in that respect. Nevertheless, we are interested in simulating the fluid flow with the Chorin scheme up to a time $t = t_{m_0} = \mathcal{O}(1)$, before we switch to the Van Kan scheme, so we do not have to bother too much concerning that question. — We will skip the easy considerations that bound the first term on the right hand side of (9.6) through Ck^2. By means of a regularity statement for the velocity function of the incompressible Stokes equations (2.1), we get

$$\tau_{m+1}^{3/2}\|u_t(t_{m+1}) - d_t u^{m+1}\| \leq \tau_{m+1}^{3/2}\|d_t\{u(t_{m+1}) - u^{m+1}\}\|$$
$$+ \tau_{m+1}^{3/2}\|d_t u(t_{m+1}) - u_t(t_{m+1})\| \quad (9.9)$$
$$\leq Ck + k\tau_{m+1}^{3/2}\{\|u_{tt}(t_{m+1})\| + \max_{t_m \leq s \leq t_{m+1}} \|u_{tt}(s)\|\} \leq Ck.$$

Here, we used the result (9.6) and twice employed the Taylor evolution formula.

a_2) Transfer of estimate (9.9) to the Chorin scheme: In order to generalize (9.9) to the Chorin scheme, we can make use of the auxiliary results (6.38) and (6.64) in Chapter 6. This gives the estimate (9.3).

With inequality (9.4), we have specified the regularizing effect of the $J_1(\Omega)$-projection at time t_{m_0} on the error function. It now remains to show that these initial errors are sufficient to reach second order convergence results for times $t_{m+1} > t_{m_0}$ to the solution of (2.1).

b) *Proof of second order accuracy for the Chorin-/Van Kan scheme:* The verification of the estimates given in Theorem 9.1, *ii*) is again rather techni-cal. In order to keep the essential steps of the proof clear, let us introduce notations that are operator-oriented. Thus, $CONT_{u_0}\{u, p; u(t_{m_0}), p(t_{m_0})\}$ represents the incompressible Stokes equations (2.1) with the tuple $\{u, p\}$ as the solution at times $t_{m+1} > t_{m_0}$, and the solution $\{u(t_{m_0}), p(t_{m_0})\}$ at time $t = t_{m_0} = \mathcal{O}(1)$. The subindex u_0 is to refer to the fact that this dynamical system is already defined for times $t \geq 0$. - We use $CN - I_{u_0}\{u^{m+1}, p^{m+1}; u^1_{m_0}, p^1_{m_0}\}$ to denote the implicitly shifted Crank-Nicolson scheme, possessing a pre-history, which means, that it is defined for times $t \geq 0$, but is here restricted to times $t \geq t_{m_0}$. In contrast to the subsequent semi-discretization, the tilde sign is to emphasize that the pressure gradient in the momentum equation is treated in an implicit way. Further, $CN - I\{\tilde{u}^{m+1}, \tilde{p}^{m+1}; u^2_{m_0}, p^2_{m_0}\}$ is to de-note the discrete dynamical system of the implicitly shifted Crank-Nicolson scheme, together with the initial data $\{u^2_{m_0}, p^2_{m_0}\}$. The pressure gradients are dealt with in the trapezoidal sense, compare formula (7.11). We stress the fact that this system is now defined only for times $t_{m+1} \geq t_{m_0}$. Let us remark that the pressure function $p^2_{m_0}$ is prescribed in a way that is anal-ogous to the procedure in Chapter 7: $p_{m_0} = \bar{p}^{m_0+1/2}$. Furthermore, the Stokes operator on pressure correction basis, whose action is described by system (7.26) but with the modification of an implicit shift of the viscos-ity part like in the case of the Crank-Nicolson scheme, will be denoted as $PC - I\{u^{m+1}_\varepsilon, p^{m+1}_\varepsilon; u^3_{m_0}, p^3_{m_0}\}$. Finally, $VK - M\{u^{m+1}_{VK}, p^{m+1}_{VK}; u^4_{m_0}, p^4_{m_0}\}$ rep-resents the modified Van Kan scheme (with implicitly shifted diffusive part). Now, we can make use of these notations in order to gather the structure of the subsequent proof in Figure 9.1. (The symbol " $\longrightarrow$ " is used to denote that the solutions of the related equations have to be compared in the following.)

The first two flows that are given in Figure 9.1 possess "full" compatibility

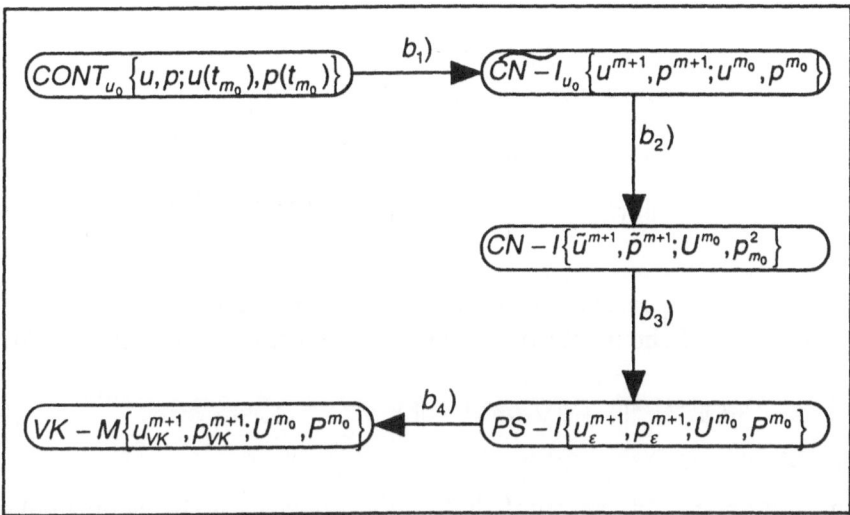

Figure 9.1: Sketch of proof for Theorem 9.1

properties, for times $t_{m+1} \geq t_{m_0}$, i.e.: The time derivatives (continuous and discrete) of velocity and pressure function of arbitrary order are controllable. We refer to the Lemmata 2.5 and 7.1 in this context, and we will take benefit from this later on.

The proof of statement ii) in Theorem 9.1, as it is given in Figure 9.1, splits into two parts: The first step is to cope with the transitions b_1) and b_2). For that, the choice of the intermitted auxiliary problems and their related initial data will be crucial. In the framework of this analysis we have to check the following "invariants of the errors", for $m \geq m_0$:

1. We have to affirm that the initial error for the pressure gradient is convergent of first order, i.e., we have

$$\|p(t_{m_0}) - p^{m_0}\|_1 + \|p^{m_0} - p^2_{m_0}\|_1 \leq Ck. \tag{9.10}$$

2. We have to guarantee the global conservation as a semi-discretization of second order. Here, we will be able to prove the sharper statements,

for times $t_{m+1} \geq t_{m_0}$,

$$\|u(t_{m+1}) - u^{m+1}\| + \|u^{m+1} - \tilde{u}^{m+1}\|$$
$$\leq Ck^2 + C_{t_{m_0}} k \exp\left(-\lambda(t_{m+1} - t_{m_0})\right),$$
$$\|p(t_{m+1}) - p^{m+1}\| + \|\bar{p}^{m+1/2} - \tilde{\bar{p}}^{m+1/2}\| \tag{9.11}$$
$$+ \|u(t_{m+1/2}) - \bar{u}^{m+1/2}\|_1 + \|\bar{u}^{m+1/2} - \tilde{\bar{u}}^{m+1/2}\|_1 \leq Ck.$$

3. We need to guarantee a certain amount of "regularity transfer" of the solutions of the involved auxiliary problems, i.e., there must hold:

$$\|d_t \tilde{u}^{m+1/2}\|_2 + \|\nabla d_t \tilde{p}^{m+1/2}\| \leq C, \qquad \text{for } m \geq m_0. \tag{9.12}$$

In the second step of the proof that is concerned with the error analysis of the transitions $b_3)$ and $b_4)$, we have to employ the arguments that have been presented in Chapter 7. Additionally, we have to cope with the technical difficulties stemming from the implicit shift of the viscous part of the Stokes operator. — In the following notation, we will refer to the items given in Figure 9.1.

$b_1)$ *consistency error analysis:* Subject of this step is the comparison of the flows that are governed by the continuous Stokes equations and the implicitly shifted Crank-Nicolson ansatz, which are both compatible in the above sense, for times $t_{m+1} \geq t_{m_0}$. The validity of inequality (9.11) is verified in [31]. In order to ensure that (9.10) is satisfied, we use the error notations $e^{m+1} := u(t_{m+1}) - u^{m+1}$ and $\eta^{m+1} := p(t_{m+1}) - p^{m+1}$, which lead us to

$$d_t\{e^{m+1} - \frac{1}{2}k^2 \Delta e^{m+1}\} - \Delta \bar{e}^{m+1/2} + \nabla \eta^{m+1} = \tilde{R}^{m+1}(u), \tag{9.13}$$

with the term $\tilde{R}^{m+1}(u) = R^{m+1}(u) - \frac{1}{2}k^2 \Delta d_t u_t(t_{m+1})$. Here, the integral residual $R^{m+1}(u)$ is given in Chapter 7, following (7.12). Owing to the requirement $m \geq m_0$, we can bound the second term of this difference in the $l^\infty(0, T; \mathbf{L}^2(\Omega))$-norm by $\mathcal{O}(k^2)$, thus it is a perturbation of the residual $R^{m+1}(u)$ which maintains the order. If we apply the averaging operator to (9.13) and test this identity with $\tau_m^r A d_t \bar{e}^{m+1/2}$, with $r \geq 2$ to be specified later on, summation over all time-steps $m \geq 0$ gives

$$k \sum_{m=1}^{M} \tau_{m+1}^r \left\{ \|\nabla d_t \bar{e}^{m+1/2}\|^2 + \frac{1}{2}k^2 \|\Delta d_t \bar{e}^{m+1/2}\|^2 \right\} + \tau_{M+1}^r \|\Delta \bar{e}^{M+1/2}\|^2$$

$$\leq C \frac{1}{k} \sum_{m=1}^{M} \tau_{m+1}^r \|\overline{\tilde{R}}^{m+1/2}(u)\|^2 \tag{9.14}$$

$$+ rk \sum_{m=1}^{M} \tau_{m+1}^{r-1} \|\Delta \bar{e}^{m+1/2}\|^2 + Ck^2.$$

In this inequality, we did take profit from the stabilizing effect of the second term on the left hand side in this relation. The initial term related to the third one on the left hand side of (9.14) is represented through the last term on the right hand side. The first term on the right hand side of this inequality can immediately be controlled by Ck^2, whereas the second sum on this side requires another consideration. For that, we test (9.13) with $\tau_m^{r-1} A \bar{e}^{m+1/2}$ and sum over all iteration steps,

$$\tau_{M+1}^{r-1} \|\nabla e^{M+1}\|^2 + \frac{1}{2}k^2 \tau_{M+1}^{r-1} \|\Delta e^{M+1}\|^2 + k \sum_{m=0}^{M} \tau_{m+1}^{r-1} \|\Delta \bar{e}^{m+1/2}\|^2$$

$$\leq Ck \sum_{m=0}^{M} \tau_{m+1}^{r-1} \|\tilde{R}^{m+1}(u)\|^2 \tag{9.15}$$

$$+ C(r-1)k \sum_{m=0}^{M} \tau_{m+1}^{r-2} \left\{ \|\nabla e^{m+1}\|^2 + \frac{1}{2}k^2 \|\Delta e^{m+1}\|^2 \right\}.$$

We refer to [31], in order to bound the second sum on the right hand side of (9.15) by Ck^2, for $r \geq 4$. The same bound can also be provided for the last term in (9.15), by applying another result derived in [31]. Thanks to (9.15), we can now bound the right hand side of (9.14) through Ck^2 and therefore arrive at a pointwise statement for the velocity error in the $l^\infty(0, T; \mathbf{H}_0^1 \cap \mathbf{H}^2)$-norm. This renders a pointwise statement for the error in the pressure gradient possible: Testing of the first equation in (9.13) with $\nabla \eta^{m_0}$, for $m = m_0 - 1$, gives result (9.10), due to (9.11), (9.14) and (9.15).

b_2) *Perturbation analysis:* The error equation that has to be analyzed in this step reads, with the notations $\bar{e}^{m+1} := u^{m+1} - \tilde{u}^{m+1}$ and $\bar{\eta}^{m+1} :=$

$p^{m+1} - \bar{p}^{m+1},$

$$d_t\{\bar{e}^{m+1} - \frac{1}{2}k^2\Delta\bar{e}^{m+1}\} - \Delta\bar{e}^{m+1/2} + \nabla\bar{\eta}^{m+1/2} = \frac{1}{2}k\nabla d_t p^{m+1}.$$

$$(9.16)$$

We will start with the second part of inequality (9.10). This requires the provision of the following inequality,

$$\|\bar{e}^{m_0+1/2}\|_2 + \|\bar{e}^{m_0+3/2}\|_2 \leq Ck. \tag{9.17}$$

In order to derive (9.17), let us look at the error identities that we obtain if we compare the systems $CONT_{u_0}\{u,p;u(t_{m_0}),p(t_{m_0})\}$ and $CN - I\{\tilde{u}^{m+1}, \bar{p}^{m+1}; U^{m_0}, p^2_{m_0}\}$ with each other. For $\hat{e}^{m+1} := u(t_{m+1}) - \tilde{u}^{m+1}$ and $\hat{\eta}^{m+1} := p(t_{m+1}) - \bar{p}^{m+1}$, the evolution of the error is prescribed by the following equations,

$$d_t\{\hat{e}^{m+1} - \frac{1}{2}k^2\Delta\hat{e}^{m+1}\} - \Delta\bar{e}^{m+1/2} + \nabla\bar{\eta}^{m+1/2} = \tilde{R}_1^{m+1}(u) + R_2^{m+1}(\nabla p).$$

$$(9.18)$$

Here, $\tilde{R}_1^{m+1}$ is identical to the function $\tilde{R}^{m+1}$, which is given in (9.13), whereas the residual for the pressure function, $R_2^{m+1}(\nabla p)$, is given in (7.19). Now, making use of the statement (9.4) for the starting velocity approximation $U^{m_0} \in \mathbf{J}_1 \cap \mathbf{H}^2$ by setting $m = m_0$ in (9.18) and testing this identity with $kAd_t\hat{e}^{m_0+1}$, an easy (iterated) calculation leads to

$$\|\hat{e}^{m_0}\|_2 + \|\hat{e}^{m_0+1}\|_2 + \|\hat{e}^{m_0+2}\|_2 \leq Ck. \tag{9.19}$$

We can combine the results (9.14) and (9.19), which give the desired result (9.17). Now, we are able to measure the error at initial time. To proceed, we perform an easy reformulation — based on the choice of the starting data for the pressure function of system $CN - I\{\tilde{u}^{m+1}, \bar{p}^{m+1}; U^{m_0}, p^2_{m_0}\}$,

$$\begin{aligned}
\nabla\bar{\eta}^{m_0+1/2} &\equiv \frac{1}{2}\nabla(p^{m_0} + p^{m_0+1}) - \frac{1}{2}\nabla(\bar{p}^{m_0} + \bar{p}^{m_0+1}) \\
&= \frac{1}{2}\nabla(p^{m_0} + p^{m_0+1}) - \nabla p^2_{m_0} \\
&= \nabla(p^{m_0} - p^2_{m_0}) + \frac{1}{2}k\nabla d_t p^{m_0+1}.
\end{aligned} \tag{9.20}$$

Owing to this, we can write:

$$\|\nabla(p^{m_0} - p_{m_0}^2)\| \leq \|\nabla\overline{\tilde{\eta}}^{m_0+1/2}\| + \frac{1}{2}k\|\nabla d_t p^{m_0+1}\|$$
$$\leq \|\overline{\tilde{e}}^{m_0+1/2}\|_2 + Ck\|\nabla d_t p^{m_0+1}\| \leq Ck. \tag{9.21}$$

The second inequality benefits from an a-priori statement for the pressure gradient in (9.16), which is immediate. (9.21) proves the validity of inequality (9.10). In order to finish this part of the proof, we have to show (9.11) and (9.12). For the verification of the error statements given in (9.11) for the transition b_2), we employ the damping property of the modified Crank-Nicolson method. This mechanism of the operator gives the following estimate for perturbed initial data, for steps $m \geq m_0$ (compare (9.2) and [31]),

$$\|\tilde{e}^{m+1}\| \leq \exp\left(-\lambda(\Omega)(t_{m+1} - t_{m_0})\right)\|u^{m_0} - U^{m_0}\|$$
$$\leq C_{t_{m_0}}\exp\left(-\lambda(\Omega)(t_{m+1} - t_{m_0})\right)k. \tag{9.22}$$

The verification of an optimal estimate for the perturbation, measured in the Dirichlet-norm, is easy, owing to the regularity shift (9.4) of the $J_1(\Omega)$-projection (9.1). This gives the following estimate for the pressure error,

$$\|\overline{\tilde{\eta}}^{m+1/2}\| \leq C\|\nabla\overline{\tilde{e}}^{m+1/2}\| + k\|d_t p^{m+1}\|$$
$$+ \|d_t\tilde{e}^{m+1}\| + k^2\|d_t\tilde{e}^{m+1}\|_1 \leq Ck. \tag{9.23}$$

This verifies that the statements given in (9.11) hold true. In order to show the compatibility estimate (9.12) for system $CN - I\{\tilde{u}^{m+1}, \tilde{p}^{m+1}; U^{m_0}, p_{m_0}^2\}$, we again start from (9.16): Employing the average- and the difference operator d_t, testing with $Ad_t^2\overline{\tilde{e}}^{m+1/2}$ and summing over all iteration steps, $m \geq m_0$, we end up with

$$k\sum_{m=m_0+2}^{M}\left\{\|\nabla d_t^2\overline{\tilde{e}}^{m+1/2}\|^2 + \frac{1}{2}k^2\|\Delta d_t^2\overline{\tilde{e}}^{m+1/2}\|^2\right\} + \|\Delta d_t\overline{\tilde{e}}^{M+1/2}\|^2$$
$$\leq \|\Delta d_t\overline{\tilde{e}}^{m_0+3/2}\|^2 \leq \frac{1}{k^2}\left\{\|\Delta\overline{\tilde{e}}^{m_0+3/2}\|^2 + \|\Delta\overline{\tilde{e}}^{m_0+1/2}\|^2\right\} \leq C. \tag{9.24}$$

The last inequality is owing to (9.17). The compatibility of the flows of $CONT_{u_0}\{u, p; u(t_{m_0}), p(t_{m_0})\}$ and $CN - I_{u_0}\{u^{m+1}, p^{m+1}; u^{m_0}, p^{m_0}\}$ and further standard arguments thus give (9.12).

The contents of the items b_1), b_2) has been the study of the nonstationary stability properties of the implicitly shifted Crank-Nicolson scheme, which are also inherent to the Van Kan scheme. Thanks to the damping property it will be possible in the subsequent pressure correction analysis to satisfy the necessary requirement (7.4).

b_3) *Pressure stabilization analysis:* In this part of the error analysis, we have the homogeneity of the initial velocity error. Owing to the statements (9.4), (9.10) we have the following estimate for the pressure error at time $t = t_{m_0}$,

$$\|P^{m_0} - p^2_{m_0}\|_1 \le Ck. \tag{9.25}$$

Together with the statements (9.12) for system $CN - I\{\tilde{u}^{m+1}, \tilde{p}^{m+1}; U^{m_0}, p^2_{m_0}\}$ the assumptions are satisfied, such that the pressure correction ansatz (7.26) in its implicitly shifted version gives approximations of second order accuracy. Despite of the fact that we only have investigated the "equilibrated", i.e., original version of the second order time-discretization up to now, let us mention that the present analysis is quite the same. Thus, we restrict on a short sketch of the proof. — The first error identity in (7.27) will now be replaced by the following one, using the notation that has been introduced in Chapter 7,

$$\{1 - \frac{1}{2}k^2\Delta\}d_t e^{m+1} - \Delta \bar{e}^{m+1/2} + \nabla \bar{\eta}^{m+1/2} = 0, \tag{9.26}$$

with the quasi-compressibility constraint remaining unchanged in (7.27). The analysis is then corresponding to the one in Section 7.2. Especially, we can couple inequality (7.63) with the estimation in Lemma 7.4,

$$\left(k \sum_{m=0}^{M} \left\{\tau_{m+1}\|d_t E^{m+1}\|^2 + k^2\|\nabla d_t E^{m+1}\|^2\right\}\right)^{1/2} \le Ck^2. \tag{9.27}$$

Therefore, it is sufficient to analyze the error equations (7.38), together with the following replacement of the first equation,

$$\{1 - \frac{1}{2}k^2\Delta\}d_t e^{m+1}_\varepsilon - \Delta \bar{e}^{m+1/2}_\varepsilon + \nabla \bar{\eta}^{m+1/2}_\varepsilon = \{1 - \frac{1}{2}k^2\Delta\}d_t E^{m+1}, \tag{9.28}$$

compare with Subsection 7.2.2. — We can proceed like in the derivation of error estimate (7.39), using statement (9.27). In order to control the averaged error functions $k \sum_{m=2}^{M} \{\|\bar{e}_\varepsilon^{m+1/2}\|^2 + \frac{1}{2}k^2\|\nabla\bar{e}_\varepsilon^{m+1/2}\|^2\}$ and $\varepsilon k \sum_{m=2}^{M} \|\nabla\bar{\bar{\eta}}^m\|^2$, we again employ dual auxiliary problems that are also implicitly shifted. Thus, in order to bound the first sums, the auxiliary problem (7.41) is modified as follows, for times $t_{M+1} > t_{m_0}$,

$$\{1 + \frac{1}{2}k^2\Delta\}d_t w^{m+1} + \Delta\overline{w}^{m+1/2} - \nabla\overline{q}^{m+1/2} = \overline{g}^{m+1/2},$$

$$\mathrm{div}w^m = 0, \qquad w^{M+1} = 0, \qquad q^{M+1} = 0. \tag{9.29}$$

Setting $\overline{g}^{m+1/2} \equiv \overline{e}^{m+1/2}$, testing the first equation in (9.29) with $\overline{e}^{m+1/2}$ and proceeding correspondingly (9.26) with $\overline{w}^{m+1/2}$, subsequent summation of the arising identities gives

$$\|\bar{e}^{m+1/2}\|^2 = d_t(w^{m+1}, e^{m+1}) - \frac{1}{2}k^2\{(d_t\nabla w^{m+1}, \nabla\bar{e}^{m+1/2})$$

$$- (d_t\nabla e^{m+1}, \nabla\overline{w}^{m+1/2})\} + \varepsilon(\nabla\overline{q}^{m+1/2}, \nabla d_t\overline{p}_\varepsilon^{m+1/2}). \tag{9.30}$$

Owing to the identity

$$(d_t\phi^{m+1}, \overline{\psi}^{m+1/2}) - (d_t\psi^{m+1}, \overline{\phi}^{m+1/2})$$

$$= 2(d_t\phi^{m+1}, \overline{\psi}^{m+1/2}) - d_t(\phi^{m+1}, \psi^{m+1}), \tag{9.31}$$

we can change the right hand side of equality (9.30) to

$$= d_t(w^{m+1}, e^{m+1}) + \frac{1}{2}k^2 d_t(\nabla w^{m+1}, \nabla e^{m+1})$$

$$- k^2(\nabla d_t w^{m+1}, \nabla\bar{e}^{m+1/2}) + \varepsilon(\nabla\overline{q}^{m+1/2}, \nabla d_t\overline{p}_\varepsilon^{m+1/2}). \tag{9.32}$$

We can apply an a-priori statement for the dual problem (9.29), that can be found in [31],

$$k^3 \sum_{m=0}^{M} \|\nabla d_t w^{m+1}\|^2 + k \sum_{m=0}^{M} \|\nabla\overline{q}^{m+1/2}\|^2 \leq Ck \sum_{m=0}^{M} \|\overline{g}^{m+1/2}\|^2. \tag{9.33}$$

Thus, owing to (9.33) and the homogeneity of the initial data of the dual problem, we obtain from (9.30)/(9.32),

$$k \sum_{m=0}^{M} \|\bar{e}^{m+1/2}\|^2 \leq Ck^3 \sum_{m=0}^{M} \|\nabla\bar{e}^{m+1/2}\|^2 + C\varepsilon^2 k \sum_{m=0}^{M} \|\nabla d_t\overline{p}_\varepsilon^{m+1/2}\|^2. \tag{9.34}$$

The first term on the right hand side of this inequality can easily be bounded by Ck^4. Therefore, we test (9.26) with $\bar{e}^{m+1/2}$ and sum over all iteration steps. We leave the details to the reader. — In order to bound the second term in (9.34) by Ck^4, setting $\varepsilon \equiv k^2$, we use an a-priori estimate for the averaged pressure gradient that is differentiated in time, of problem $PC - I\{u_\varepsilon^{m+1}, u_\varepsilon^{m+1}; U^{m_0}, P^{m_0}\}$. In that respect, we refer to an analogous argumentation in Subsection 7.2.2, compare estimate (7.43).

In order to complete the error analysis in this part, we need an optimal convergence result for the term $\varepsilon k \sum_{m=2}^M \|\nabla \bar{\bar{\eta}}_\varepsilon^m\|^2$. Therefore, we formulate the following equations as an auxiliary problem, corresponding to (7.44): For given $t_{M+1} > t_{m_0}$, find the solution $\{\kappa^m, \rho^m\}$ of the system:

$$\{1 + \frac{1}{2}k^2\Delta\}d_t\bar{\kappa}^{m+1/2} + \Delta\bar{\kappa}^m - \nabla\bar{\rho}^m = 0,$$

$$\mathrm{div}\bar{\kappa}^{m-1/2} + \varepsilon\Delta d_t\bar{\rho}^{m+1/2} = -\varepsilon\Delta\bar{\eta}_\varepsilon^{m-1/2}, \qquad \varepsilon \equiv \mathcal{O}(k^2),$$

$$\partial_n d_t\bar{\rho}^{m+1/2}|_{\partial\Omega} = -\partial_n\bar{\eta}_\varepsilon^{m-1/2}|_{\partial\Omega}, \qquad \kappa^m|_{\partial\Omega} = 0,$$

$$\bar{\kappa}^{M+1/2} = 0, \qquad \bar{\rho}^{M+1/2} = 0.$$

A procedure, as it is carried out in Subsection 7.2.2 is then leading to an inequality that corresponds to the result (7.52). In order to estimate the arising sum with the both terms in the brackets, we have to apply the first error identity, here equation (9.28): Thus, after averaging this relation, we test with $\bar{\bar{e}}_\varepsilon^m$ and proceed correspondingly with (9.28), testing it with $\bar{\bar{\kappa}}_\varepsilon^m$, and we obtain the following identity after summation,

$$d_t(\bar{e}_\varepsilon^{m+1/2}, \bar{\kappa}^{m+1/2}) + \frac{1}{2}k^2\{(\nabla d_t\bar{e}_\varepsilon^{m+1/2}, \nabla\bar{\kappa}^m) - (\nabla d_t\bar{\kappa}^{m+1/2}, \nabla\bar{e}_\varepsilon^m)\}$$

$$+ \{(\nabla\bar{\eta}_\varepsilon^m, \bar{\kappa}^m) - (\bar{e}_\varepsilon^m, \nabla\bar{\rho}^m)\}$$

$$= (\{1 - \frac{1}{2}k^2\Delta\}d_t\overline{E}^{m+1/2}, \bar{\kappa}^m), \qquad (9.35)$$

instead of equation (7.53). By means of this equation we intend to derive an estimation for the second bracket term on the left side of this relation. Owing to homogeneous data for the initial functions of the dual problems for the velocities and the statement (9.27), we can proceed as in (7.54) or (7.55). It only remains to treat the first bracket expression on the left side of (9.35).

This difference can be easily reformulated, thanks to (9.31),

$$(\nabla d_t \overline{e}_\varepsilon^{m+1/2}, \nabla \overline{\kappa}^m) - (\nabla d_t \overline{\kappa}^{m+1/2}, \nabla \overline{e}_\varepsilon^m)$$
$$= 2(\nabla d_t \overline{e}_\varepsilon^{m+1/2}, \nabla \overline{\kappa}^m) - d_t(\nabla \overline{e}_\varepsilon^{m+1/2}, \nabla \overline{\kappa}^{m+1/2}). \tag{9.36}$$

If we sum (9.35) over all iteration steps, we can write, owing to (9.36),

$$\sum_{m=m_0}^{M} \left\{ (\nabla \overline{\eta}_\varepsilon^m, \overline{\kappa}^m) - (\overline{e}_\varepsilon^m, \nabla \overline{\rho}^m) \right\} \le Ck^4 + k \sum_{m=m_0}^{M} \left\{ \frac{1}{4\delta_2} + \frac{1}{4\delta_3} \right\} \|\overline{\kappa}^m\|^2$$

$$+ k \sum_{m=m_0}^{M} \delta_2 \|d_t \overline{E}^{m+1/2}\|^2$$

$$+ \delta_3 k^5 k \sum_{m=m_0}^{M} \left\{ \|\nabla d_t \overline{e}_\varepsilon^{m+1/2}\|^2 + \|\nabla d_t \overline{E}^{m+1/2}\|^2 \right\}, \tag{9.37}$$

with $\delta_2, \delta_3 > 0$. The constant δ_2 will be chosen like in Subsection 7.2.2, compare with what follows (7.54). The boundedness of the last term in (9.37) is given, owing to the argument that has been employed in what follows (9.34). For constants $\tilde{\delta}_2, \delta_3 > 0$, compare Subsection 7.2.2, and after insertion of (9.37) in a relation that is corresponding to (7.52), these terms can be absorbed by the corresponding terms on the left hand side, owing to an a-priori result which is analogous to (7.48), whereas the remaining terms on the right hand side of this relation can be bounded by Ck^4. Thus, we presented all necessary arguments for the proof that ensure the conservation of order for this transition.

b_4) *Projection analysis:* This part of the transition needs no further new considerations and follows immediately from the corresponding procedure in Section 7.3.

This establishes the proof of Theorem 9.1.   □

Remark 9.3 *Corresponding to the estimates in (7.33) and (7.40) for the "pointwise" error evolution in the first initial steps we need similar statements for the auxiliary problems (9.26) and (9.28) that can be gained in an analogous fashion. We remark, that the first difference quotient of the iterates for the pressure gradients disappears, because of the prescription of the initial pressure function at time $t_{m+1} = t_{m_0}$ of problem $CN-I\{\tilde{u}^{m+1}, \tilde{p}^{m+1}; U^{m_0}, p_{m_0}^2\}$ This leads to the disappearance of the related terms in (7.33) and (7.40).*

9.3 Proposal and Analysis of a Modified Chorin-Uzawa Scheme

The objective of this section is to present another projection method which is a further development of the Chorin-Uzawa scheme that does not cause numerical boundary layers. Moreover, we intend to switch off the error source which is owing to the nonstationary character of the quasi-compressibility condition. In order to succeed in this, we construct another multi-component scheme that is parameterized in time $t_{m_0} = \mathcal{O}(1)$,

1. *Original Chorin scheme:* Start with the basis Chorin scheme, for times $t_{m+1} \leq t_{m_0} = \mathcal{O}(1)$.

2. $\mathbf{J}_1(\Omega) - projection:$ At time $t = t_{m_0}$, we will invert the following Stokes problem, for $\{U^{m_0}, P^{m_0}\} \in \mathbf{J}_1(\Omega) \cap \mathbf{H}^2(\Omega) \times H^1(\Omega)/\mathbb{R}$, with the given function $d_t \tilde{u}^{m_0}_{Cho} \in \mathbf{H}^1_0$ from the previous item,

$$-\Delta U^{m_0} + \nabla P^{m_0} = -d_t \tilde{u}^{m_0}_{Cho} + f^{m_0}. \tag{9.38}$$

3. *The modified Chorin-Uzawa scheme:*

   (a) *Choice of the initial data:* We set

   $$\mu^0 := d_t \tilde{u}^{m_0}_{Cho}, \qquad \phi^0 := d_t p^{m_0}_{Cho} \in L^2_0(\Omega), \qquad \tilde{\phi}^0 := 0,$$

   and

   $$\tilde{u}^{m_0} := U^{m_0}, \qquad p^{m_0} := P^{m_0} \in L^2_0(\Omega).$$

   (b) *Iteration step:* For $m \geq 0$, we will pursue the following four steps, with the given functions $\{\mu^m, \phi^m \tilde{\phi}^m; \tilde{u}^m, p^m\}$.

   i. Determine the auxiliary function $\tilde{\mu}^{m+1} \in \mathbf{H}^1_0$ through inversion of the Helmholtz equation

   $$\frac{1}{k}\{\tilde{\mu}^{m+1} - \mu^{m+1}\} - \Delta\tilde{\mu}^{m+1} + \nabla\{\phi^m - \tilde{\phi}^m\} = d_t f^{m+1}. \tag{9.39}$$

   ii. Determine $\mu^{m+1} = P_{\mathbf{J}_0}\tilde{\mu}^{m+1}$ as the solution of the subsequent equations,

   $$\frac{1}{k}\{\mu^{m+1} - \mu^m\} + \nabla\tilde{\phi}^{m+1} = 0,$$
   $$\mathrm{div}\,\mu^{m+1} = 0, \qquad \mu^{m+1}|_{\partial\Omega} \cdot n = 0. \tag{9.40}$$

iii. The pressure function ϕ^{m+1} will be determined to be

$$\phi^{m+1} = \phi^m - \alpha \operatorname{div} \tilde{\mu}^{m+1}, \qquad \alpha < 1. \tag{9.41}$$

iv. Set

$$\tilde{u}^{m+1} := k \tilde{\mu}^{m+1} + \tilde{u}^m, \qquad p^{m+1} := k \phi^{m+1} + p^m. \tag{9.42}$$

Remark 9.4 *The formulation of this multi-component scheme in case of the full Navier-Stokes equations is immediate, owing to the comments in the preceding chapters. Therefore, we only have to keep in mind that step (9.41) is then as follows, using the viscosity parameter ν,*

$$\phi^{m+1} = \phi^m - \alpha \nu \operatorname{div} \tilde{\mu}^{m+1}, \qquad \alpha < 1.$$

Owing to the fact that this is the formulation of the Chorin-Uzawa scheme for the sizes that are (discretely) differentiated in time, the resulting quasi-compressibility constraint is now a regular perturbation for the solutions of the incompressible Stokes equations that obey the conditions (A1) and (A2). Let us emphasize, that this character of a *regular* perturbation is given in the analysis of the artificial compressibility method in Chapter 4, formula (4.1), provided that the solution to be approximated satisfies *Postulate B_1*. — The next theorem deals with the quantification of the error influences that are immanent to this algorithm.

Theorem 9.2 *Assume that the given data of problem (2.1) satisfy the conditions (A1) and (A2). Let $\{\tilde{u}^{m+1}, p^{m+1}\} \in H_0^1 \times L_0^2(\Omega)$ be the solution of the modified Chorin-Uzawa scheme, as it is given above in (1. - 3.). Then, the following error statements hold, for given times $t_{m_0} = \mathcal{O}(1)$,*

i) *provided that $t_{m+1} \leq t_{m_0}$:*

$$\max_{0 \leq m \leq m_0} \left\{ \|u(t_{m+1}) - \tilde{u}^{m+1}\| + \sqrt{k}\|u(t_{m+1}) - \tilde{u}^{m+1}\|_1 \right\} \leq Ck,$$

$$\max_{0 \leq m \leq m_0} \sqrt{\tau_{m+1}} \left\{ \sqrt{\tau_{m+1}} \|p(t_{m+1}) - p^{m+1}\|_{-1} \right.$$
$$\left. + \sqrt{k}\|p(t_{m+1}) - p^{m+1}\| \right\} \leq Ck.$$

ii) *provided that $t_{m+1} > t_{m_0}$:*

$$\max_{0 \leq m \leq m_0} \left\{ \|u(t_{m+1}) - \tilde{u}^{m+1}\|_1 + \|p(t_{m+1}) - p^{m+1}\| \right\} \leq (C + C_{m_0})k.$$

The subindex at the constant C_{m_0} is to indicate that this constant is not growing in time but only depends on the choice of t_{m_0}. Further, the employed constant C is only depending on the given data of the continuous problem (2.1).

Proof:
We do only have to verify part *ii)* of Theorem 9.2. The proof is split into several steps.

a) Let us start with the verification of several error estimates for the functions that are discretely differentiated, using the *smoothing property* of the Chorin scheme. We recall the following results that have already been achieved, for times $t_{m_0} \in \mathcal{O}(1)$,

$$\|d_t \tilde{u}_{Cho}^{m_0} - u_t(t_{m_0})\| + \sqrt{k}\|d_t p_{Cho}^{m_0} - p_t(t_{m_0})\| \leq Ck, \tag{9.43}$$

and

$$\|U^{m_0} - u(t_{m_0})\|_2 + \|P^{m_0} - p(t_{m_0})\|_1 \leq Ck, \tag{9.44}$$

see (9.3), (9.4), which proves inequality (9.44) immediately. Correspondingly, we can verify estimate (9.43) for the first term on the left hand side because of (9.3) and the subsequent argumentation, and it remains to give a verification for the other estimate in (9.43) for the pressure. In order to do that, we distinguish between different error contributions, simultaneously introducing the notation for the further studies,

$$\begin{aligned}\|d_t p_{Cho}^{m_0} - p_t(t_{m_0})\| &\leq \|d_t p_{Eu}^{m_0} - p_t(t_{m_0})\| \\ &\quad + \|d_t p_{Eu}^{m_0} - d_t p_{PS}^{m_0}\| + \|d_t p_{PS}^{m_0} - d_t p_{Cho}^{m_0}\|.\end{aligned} \tag{9.45}$$

Let the functions with subindex "Eu" denote the solutions of the implicit Euler scheme, that has already been introduced in (6.12), with "PS" we will denote the pressure-stabilized Stokes equations (6.11) that are semi-discretized in time (via implicit Euler). Finally, with "Cho" we will denote the solution of the Chorin scheme (6.6) (omitting the nonlinear term).

a_1) We will start with an estimation of the first term on the right hand side of (9.45), that measures the consistency error contribution of the Chorin scheme. We proceed by splitting the analysis in several steps: If we employ the difference operator d_t onto the first error equation of (9.5) that is describing the error evolution of $e_1^{m+1} := u(t_{m+1}) - u_{Eu}^{m+1}$ and $\eta_1^{m+1} := p(t_{m+1}) - p_{Eu}^{m+1}$ and finally test with $\tau_m^5 A d_t e_1^{m+1}$, we get after summation,

$$
\frac{1}{2} \tau_{M+1}^5 \| \nabla d_t e_1^{M+1} \|^2 + k \sum_{m=2}^{M} \tau_{m+1}^5 \| A d_t e_1^{m+1} \|^2
$$
$$
\leq k \sum_{m=2}^{M} \tau_{m+1}^5 \| d_t R^{m+1} \|^2 + Ck \sum_{m=2}^{M} \tau_{m+1}^4 \| \nabla d_t e_1^{m+1} \|^2.
$$
(9.46)

The argumentation that is needed in order to bound the first term on the right hand side of this inequality is corresponding to (9.8). In order to control the second norm on the right hand side of (9.46), we proceed testing the first identity in (9.5) with $\tau_m d_t e_1^{m+1}$, after having employed the difference operator d_t. Summation gives

$$
\frac{1}{2} \tau_{M+1}^3 \| d_t e_1^{M+1} \|^2 + k \sum_{m=1}^{M} \tau_{m+1}^3 \| \nabla d_t e_1^{m+1} \|^2
$$
$$
\leq k \sum_{m=1}^{M} \tau_{m+1}^3 \| d_t R^{m+1} \|_{-1}^2 + Ck \sum_{m=1}^{M} \tau_{m+1}^2 \| d_t e_1^{m+1} \|^2.
$$
(9.47)

It is now easy to bound the terms on the right hand side by Ck^2, and we leave the further details to the reader. Therefore, we can bound the right hand side of (9.46) by Ck^2.

Owing to estimate (9.46), we can now proceed as follows,

$$
\| d_t \eta^{m+1} \| \leq C \| \nabla d_t e^{m+1} \| + \| d_t^2 e^{m+1} \|_{-1} + \| d_t R^{m+1} \|
$$
$$
\leq C \tau_{m+1}^{-5/2} \sqrt{k}.
$$
(9.48)

The last inequality is thanks to Lemma 2.3 that enables an estimation for the second term on the right hand side of the first inequality sign through the remaining two. Thus, we can control the first term on the right hand side of (9.45) as follows,

$$
\| d_t p_{Eu}^{m_0} - p_t(t_{m_0}) \| \leq \| d_t p_{Eu}^{m_0} - d_t p(t_{m_0}) \|
$$
$$
+ \| d_t p(t_{m_0}) - p_t(t_{m_0}) \| \leq C \{ \sqrt{k} + k \}.
$$
(9.49)

a_2) In order to estimate the second error contribution on the right hand side of (9.45), we employ another result from Chapter 6. We can benefit from

(6.22)/(6.23) and (6.38) to get

$$\|d_t u_{Eu}^{mo} - d_t u_{PS}^{mo}\|_1 \leq C\sqrt{k}, \qquad t_{mo} = \mathcal{O}(1). \tag{9.50}$$

Further, we need the verification of the following "suboptimal" error bound,

$$\|d_t^2 u_{Eu}^{mo} - d_t^2 u_{PS}^{mo}\| \leq C\sqrt{k}. \tag{9.51}$$

This requires the error identities (6.18), with the following notation instead, $e_2^{m+1} := u_{Eu}^{m+1} - u_{PS}^{m+1}$ and $\eta_2^{m+1} := p_{Eu}^{m+1} - p_{PS}^{m+1}$. Now, we will present a "generating mechanism" of two error equalities that give the justification for (9.51). If we employ the difference operator d_t^r, with $r \geq 0$ onto the momentum equation and test with $d_t^r e_2^{m+1}$, we arrive at

$$d_t\|d_t^r e_2^{m+1}\|^2 + \|\nabla d_t^r e_2^{m+1}\|^2 + k\|\nabla d_t^r \eta_2^{m+1}\|^2$$
$$\leq Ck\|\nabla d_t^r p^{m+1}\|^2, \qquad r \geq 0. \tag{9.52}$$

We obtain another relation through application of the difference operator d_t^r onto the first relation in (6.18) and subsequent testing with $d_t^{r+1} e_2^{m+1}$,

$$\|d_t^{r+1} e_2^{m+1}\|^2 + d_t\|\nabla d_t^r e_2^{m+1}\|^2 + kd_t\|\nabla d_t^r \eta_2^{m+1}\|^2$$
$$\leq Ck\|\nabla d_t^{r+1} p^{m+1}\|^2, \qquad r \geq 0. \tag{9.53}$$

Then, the following estimate is valid, setting $r = 2$ in (9.52),

$$\tau_{m+1}^2\|d_t^2 e_2^{m+1}\| \leq C\sqrt{k}. \tag{9.54}$$

This gives assertion (9.51). Together with (9.50) we get, thanks to a stability statement for the divergence operator,

$$\|d_t p_{Eu}^{mo} - d_t p_{PS}^{mo}\| \leq C\|d_t^2 e_2^{mo}\| + \|\nabla d_t e_2^{mo}\| \leq C\sqrt{k}. \tag{9.55}$$

a_3) It remains to derive an upper bound for the last term in the inequality (9.45), i.e., $\|d_t p_{PS}^{mo} - d_t p_{Cho}^{mo}\|$. An error analysis for the transition from the pressure stabilization ansatz to the Chorin scheme is given in Chapter 6 (for the Navier-Stokes case). We generalize them for the discretely differentiated physical quantities. The verification of a corresponding error estimate uses several steps: At first, the following inequality can be shown, compare with (9.59) below, with $r = 1$,

$$\|d_t u_{PS}^{mo} - d_t u_{Cho}^{mo}\|_1 \leq C\sqrt{k}. \tag{9.56}$$

In order to derive the error statement that is additionally necessary,

$$\|d_t^2 u_{PS}^{mo} - d_t^2 u_{Cho}^{mo}\| \leq C\sqrt{k}, \tag{9.57}$$

we make use of the notations $e_3^{m+1} := u_{PS}^{m+1} - u_{Cho}^{m+1}$ and $\eta_3^{m+1} := p_{PS}^{m+1} - p_{Cho}^{m+1}$ for the error functions, that are the solution of system (6.65), omitting the arising convective part. The details of the following statements are given in Chapter 6 (for the Navier-Stokes case). — Again, for these equations we have a "generating mechanism" that leads us to the identities for the error,

$$\frac{1}{2} d_t \|d_t^r e_3^{m+1}\|^2 + \frac{1}{2} k \|d_t^{r+1} e_3^{m+1}\|^2 + \|\nabla d_t^r e_3^{m+1}\|^2 + k\|\nabla d_t^r \eta_3^{m+1}\|^2$$
$$= k^2 (\nabla d_t^r \eta_3^{m+1}, \nabla d_t^{r+1} \eta_3^{m+1})$$
$$- k(\nabla d_t^{r+1} p_{PS}^{m+1}, d_t^r e_3^{m+1}), \qquad r \geq 0. \tag{9.58}$$

Owing to

$$k^2 (\nabla d_t^r \eta_3^{m+1}, \nabla d_t^{r+1} \eta_3^{m+1}) = k(d_t^{r+1} e_3^{m+1}, \nabla d_t^r \eta_3^{m+1}), \qquad r \geq 0,$$

we can absorb the first term on the right hand side of (9.58) through the left hand side and get

$$\frac{1}{2} d_t \|d_t^r e_3^{m+1}\|^2 + \|\nabla d_t^r e_3^{m+1}\|^2 + k\|\nabla d_t^r \eta_3^{m+1}\|^2$$
$$\leq Ck^2 \|\nabla d_t^{r+1} p_{PS}^{m+1}\|^2, \qquad r \geq 0. \tag{9.59}$$

We obtain another relation if we apply the difference operator d_t^r, $r \geq 0$, onto the first equation in (6.65), which is now formulated for the functions $\{e_3^{m+1}, \eta_3^{m+1}\}$, setting $\mathcal{Q} \equiv 0$, and finally test with $d_t^{r+1} e_3^{m+1}$. This leads us to

$$\|d_t^{r+1} e_3^{m+1}\|^2 + d_t \|\nabla d_t^r e_3^{m+1}\|^2 + kd_t \|\nabla d_t^r \eta_3^{m+1}\|^2$$
$$\leq 2k^2 \|\nabla d_t^{r+1} \eta_3^{m+1}\|^2 + k^2 \|\nabla d_t^{r+1} p_{PS}^{m+1}\|^2, \quad r \geq 0. \tag{9.60}$$

Mutual application of the results (9.59), (9.60) give, for $r = 2$, the following error bound,

$$\tau_{m+1}^2 \|d_t^2 e_3^{m+1}\| \leq Ck. \tag{9.61}$$

Especially, this proves (9.57). Now, an easy consideration leads us to the desired estimate for the discrete time derivative of the pressure error, owing to (9.56) and (9.57) and the relation (9.53) that can be interpreted as an a-priori bound for $d_t p_{PS}^{mo}$,

$$\|d_t p_{Eu}^{mo} - d_t p_{PS}^{mo}\| \leq C\|d_t^2 e_3^{mo}\| + \|\nabla d_t e_3^{mo}\| + k\|d_t^2 \eta_3^{mo}\| \leq Ck.$$
(9.62)

With that result, we can come back to (9.45). We can apply the results (9.49), (9.55) and (9.62) to verify estimate (9.43).

b) We can reformulate the equations (9.39) to (9.42) as follows,

$$d_t^2 \tilde{u}^{m+1} - \Delta d_t \tilde{u}^{m+1} + \nabla d_t p^m = d_t f^{m+1},$$
$$\mathrm{div} d_t \tilde{u}^{m+1} + \alpha k d_t^2 p^{m+1} = 0, \qquad \alpha < 1.$$
(9.63)

This is a system of equations that requires the following quadruple of data in order to be well-defined,

$$\left\{ \tilde{u}^{mo}, d_t \tilde{u}^{mo}; \tilde{p}^{mo}, d_t \tilde{p}^{mo} \right\}.$$
(9.64)

These data satisfy the approximation properties which are given in (9.43), (9.44). Owing to the results for the Chorin-Uzawa scheme for times $t_{M+1} \geq t_{mo}$, it immediately follows from Theorem 9.2,

$$\|d_t \tilde{u}^{M+1} - u_t(t_{M+1})\|^2 + k \sum_{m=mo}^{M} \|d_t \tilde{u}^{m+1} - u_t(t_{m+1})\|_1^2 \leq Ck^2.$$
(9.65)

By means of the Fundamental Theorem of the calculus of integration and the approximation property for the velocity gradient, given in (9.44) we obtain

$$\|\tilde{u}^{M+1} - u(t_{M+1})\|_1 \leq C\sqrt{t_{M+1}}k + \|\tilde{u}^{mo} - u(t_{mo})\|_1 \leq Ck.$$
(9.66)

Now, integration of the error identity that corresponds to the first equation in (9.63) and taking care of the compatibility of the equation at time t_{mo}, see the inequalities (9.43), (9.44), leads to the following result for the pressure approximation,

$$\|p^{M+1} - p(t_{M+1})\| \leq Ck.$$
(9.67)

We skip the presentation of more technical details here. — This completes the proof of Theorem 9.2.   □

Chapter 10

Time Discretization on Time-Grids with Structure — from Euler and Trapezoidal Method to Revised Projection Schemes

10.1 Introduction

The analyses of projection schemes of higher order in the preceding chapters have highlighted the difficulties of these numerical schemes to cope with highly incompatible flows. As results, we proved an optimal behavior of convergence for the classical Van Kan scheme for problem constellations that satisfy the regularity assumptions demanded in Postulate B_2, cf. page 145. The sharpness of this restrictive limitation of applicability is shown by numerical experiments leading to a global reduction of the convergence order in the case of violation, i.e., in case of incompatible flows.

A first idea to manage an accurate numerical modeling of incompatible flows on the basis of projection schemes has been proposed in Chapter 9 with methods called *multi-component schemes*. These composed algorithms inherit the robustness properties of the Chorin scheme during the initial phase and the higher accuracy of the Van Kan or the Chorin-Uzawa scheme for "macroscopic" times. In order to link both components successfully to obtain an algorithm which gives optimal approximations for a general range of problem classes, we have to adjust the computed data stemming from the Chorin-method at a chosen time $t_{m_0} = \mathcal{O}(1)$ by means of a $\mathbf{J_1}$-projection, which is not always practicable in existing computational environments.

The objective of this chapter is the presentation of new higher order projection schemes *that are applicable to general fluid flows*, using solely Burgers- and Laplace-solver components. In contrast to the multi-component schemes that need additional effort due to the adaptation process (i.e., the $\mathbf{J_1}$-projection), these new schemes to be presented below are motivated by another idea to strengthen the robustness of the original Chorin-Uzawa and the Van Kan scheme. This idea can be summarized as follows: Firstly, the underlying equi-distant time-grid is replaced by a problem-adapted structure, using fine time-steps at the beginning, which grow larger for increasing times. Secondly, the character of the (semi-explicit) nonstationary quasi-compressibility methods is relaxed in a way made precise below. This leads to an increase of robustness of the related projection-methods which are now capable of modeling *general* — especially incompatible — flows with optimal order of convergence.

The idea of using special a-priori time-grid structures to improve the stability seems to be new, even in the context of time-discretization schemes for parabolic equations. As a matter of fact, it is well-known that the Euler-method suffers from a lack of accuracy in the neighborhood of the starting time, i.e., for times $t_{m+1} = \mathcal{O}(k^\alpha), 0 \leq \alpha < 1$, provided the initial data are rough, i.e., u_0 in $H^1(\Omega)$ or, even worse, in $L^2(\Omega)$. In case of the trapezoidal method, the same phenomenon holds true also for smooth initial data, i.e., $u_0 \in H^2(\Omega)$, whereas the order of convergence is *globally in time* reduced to $o(k)$ or $o(1)$ for u_0 in $H^1(\Omega)$ or $L^2(\Omega)$, respectively. Therefore, the subject of Section 10.2 is to propose new time-grids adapted to the dynamical behavior of the solution in the limiting case $t \to 0$, which cause *no* significant additional effort. As a result, we obtain a first order convergence rate for the Euler-method for problems with $u_0 \in L^2(\Omega)$, even for times that are "close" to $t = 0$. Correspondingly, second order accuracy is preserved for the trapezoidal method also in case of rough initial data, i.e., $u_0 \in H^1$, *uniformly* on the whole time interval $[k, 1]$. The results of this section are intended to provide a general framework for applications in time-discretization methods of evolutionary problems.

Apart from the general applicability of time-grids with stabilizing structures to other evolution problems, they will be successfully applied to the
projection schemes in the subsequent sections, adding an additionally needed
amount of stabilization to them. As main results, it will be possible to construct a *revised Chorin-Uzawa* and a *revised Van Kan scheme* which are both
applicable to general fluid flows. *More precisely, slight modifications of these
original schemes on stretched time-grids will give schemes of optimal accuracy for general fluid flows and without the restrictive necessity of prescribing
sufficiently accurate data for the initial pressure function.*

Section 10.3 will be concerned with the proposal as well as the analysis of
a revised Chorin-Uzawa method based on time-grid structures. Apart from
the proposed scheme, the main results are presented in Theorem 10.3. As
results, we prove rates of convergence that are comparable to those gained
for the original Chorin-Uzawa scheme (which is defined on the equidistant
time-grid and requires good initial data for the pressure as well as a sufficiently regular function, i.e., $p_t \in L^\infty(0, t_{M+1}; L^2/\mathbb{R})$). Unfortunately, we are
not able to extend these convergence rates to estimates for the pressure error
in the norm $l^\infty(0, t_{M+1}; L^2/\mathbb{R})$. It is suspected that this is a consequence
of the fact that (pressure) correction type projection schemes are not capable of approximating the time-derivative $d_t \tilde{u}^{m+1}$ in a sufficiently accurate
manner leading to sub-optimal error statements for the pressure function in
the desired norm. This lack of quality of the approximation is well-known
from the previously discussed original Chorin-Uzawa scheme (i.e., those on
equidistant grids) and it seems that this cannot be overcome by the structured time-grids. For this purpose, we have to apply the strategy of multi-
component schemes to assure first order convergence for the pressure in the
norm $l^\infty(t_{m_0}, t_{M+1}; L^2/\mathbb{R})$ with $t_{m_0} = \mathcal{O}(1)$. In that respect, we refer to
Chapter 9 — see the Chorin-/Chorin-Uzawa scheme — for further details.

In order to obtain first order accuracy for the pressure in the demanded
pointwise norm for general fluid flows, we will propose a revised Van Kan
scheme on structured time-grids. Subsequently, Section 10.4 is devoted to
the proposal as well as study of this higher order scheme on structured time-
grids. The manifestation of an optimal behavior of convergence is given in
Theorem 10.4.

Finally, let us stress the fact that the contents of Chapter 4 is very
much connected to the present one. In the former analysis of the continuous damped nonstationary quasi-compressibility method we obtained an
operator with strong damping properties of initially rough relevant functions

governing the flow. This results in a reduction of perturbations that appear during the initial phase, which would otherwise have been amplified by the nonstationary quasi-compressibility method. We already mentioned the fact that this strategy cannot be successfully transfered to the projection schemes, owing to the explicit treatment of the pressure function in the corresponding quasi-compressibility reformulation. Now, the coupling with problem-adapted time-steps is to shift the splitting of the projection-schemes to a more implicit form, introducing damping facilities to the scheme that are able to cope with the rough initial perturbations.

10.2   The Euler and Trapezoidal Method on Stabilizing Time-Grids

The necessity of numerical modeling of time evolution processes is arising in diverse disciplines like biology, chemistry, physics and engineering. The choice of a certain discretization scheme has to be adjusted to the stability features of the solution to be approximated. Further, after the selection of any standard numerical algorithm it has to be adapted to the behavior of the solution in question in order to guarantee the stability of the resulting scheme for a wide range of applicability.

This stand-point will be outlined for the non-homogeneous linear evolution equation in the present section, to be considered on the time interval $t \in [0, T]$,

$$u_t + A(t)u = f(t), \qquad u|_{\partial\Omega} = 0, \qquad u(0) \equiv u_0, \tag{10.1}$$

which is defined on a Hilbert space H. The operator $A(t)$ has to satisfy certain assumptions that are given below. Essentially, it is the sum of a positive definite self-adjoint operator A and a lower order operator $B(t)$ at each time $t \in [0, T]$. Note that this class of problems to be studied incorporates many interesting evolutionary, nonlinear problems that are linearized, for instance convection-diffusion problems or the Navier-Stokes equations. For such a problem constellation, one-step methods for semi-discretization in time are often in use and are subject of a vast quantity of scientific papers. Let us recall the now "classical" results for the Euler- and the trapezoidal method, starting with the former one.

At first, in order to clarify the notations, an equi-distributed time-grid is parametrized by the mesh-size $k_0 \equiv k_{m+1} := t_{m+1} - t_m$, $0 \le m \le M$, with

$T = t_{M+1}$. Subsequently, this time-grid will be denoted by $\mathcal{G}_0(k_{m+1})$. —
Then, the implicit Euler scheme reads as follows,

$$d_t u^{m+1} + A(t_{m+1})u^{m+1} = f^{m+1}, \tag{10.2}$$

with the abbreviative notation $d_t u^{m+1} = \frac{1}{k_{m+1}}\{u^{m+1} - u^m\}$. If the given force
$f(t)$ of (10.1) is sufficiently regular, we obtain the following *smoothing* error
estimates, with certain time-weights that reflect the roughness of the initial
data $u_0 \in H$,

$$\|u(t_{m+1}) - u^{m+1}\|_H \leq k_0 E(t_{m+1}; i)\{\|A^{i/2}u_0\|_H + 1\}, \qquad i \in \{0, 1, 2\}, \tag{10.3}$$

with $E(s; i) = Cs^{-1+i/2}$, and C a generic constant that is depending on the
domain Ω, the right hand side f and the interval length T. Depending on
the degree of roughness of the initial data, these results have been proven by
several authors. We refer here to the papers of Luskin & Rannacher, [23],
[24] and the book of Thomee [43]. Provided smooth initial data are given,
for instance $u_0 \in D(A)$, first order convergence is ensured *uniformly* over the
whole interval $[0, T]$. On the other hand, an increasing degree of roughness
of the initial data leads to an *anisotropic* convergence behavior over the
time range. In particular, for the first $\mathcal{O}(k_0^{-\alpha})$ time-steps, with $\alpha \in [0, 1)$,
we do only obtain approximations that are rather crude. In the worst case
($M = \mathcal{O}(1)$ and $u_0 \in H$) no convergence statement can be extracted from
(10.3).

For the trapezoidal method, the situation is even worse. This method,
applied to the equation (10.1), can be formulated in some modifications.
Without restrictions, we prefer the version

$$d_t u^{m+1} + A_{m+1/2}\overline{u}^{m+1/2} = f^{m+1/2}, \tag{10.4}$$

with $\overline{u}^{m+1/2} \equiv \frac{1}{2}\{u^{m+1} + u^m\}$. For $u_0 \in D(A)$, the following estimate has
been proven by Luskin & Rannacher in [24], see also Rannacher [31],

$$\|u(t_{m+1}) - u^{m+1}\|_H \leq C\frac{k_0^2}{t_{m+1}}\{\|Au_0\|_H + 1\}. \tag{10.5}$$

Again, this estimate shows a suboptimal efficiency of the trapezoidal method
in the vicinity of the starting point $t = 0$. — Moreover, for decreasing

regularities of the initial function the *global* order of convergence is reduced to $o(k)$ (if $u_0 \in D(A^{1/2})$) and $o(1)$ (if $u_0 \in H$), which is rather disappointing. This reflects the property of the trapezoidal method of not being able to cope with rough perturbations. The only strategy to enhance the stability of the trapezoidal method in case of H-initial data known to the author has been proposed by Luskin & Rannacher, see [24]. These authors use a couple of Euler steps in order to determine the discrete dynamical system, thus getting the following result,

$$\|u(t_{m+1}) - u^{m+1}\|_H \leq C \left(\frac{k_0}{t_{m+1}} \right)^2 \left\{ \|u_0\|_H + 1 \right\}. \tag{10.6}$$

This assures the preservation of second order convergence rate for times $t_{m+1} = \mathcal{O}(1)$. On the other side, apart from the strong time-weight necessary in this error estimate this approach via inserted Euler-steps suffers from a severe drawback: The numerical algorithm employed *is dependent on where one is interested in optimal error results across the transient phase*, and there is no possibility of controlling the dynamical behavior of the approximation features during the initial time phase $[0, 1]$ for every time-step.

To the author's opinion, in order to qualify any competitive time-discretization scheme for any evolutionary problem class as being of βth order convergent, there must hold

$$\|u(t_{m+1}) - u^{m+1}\|_H \leq Ck^\beta, \tag{10.7}$$

uniformly along the range $[0, T]$. So, to adjust a selected discretization scheme to a specific problem class, we sometimes also have to deal with the underlying time-grid topology as an essential part of well-defining the scheme in the latter sense. The objective of the present section is to propose a-priori time-grid strategies that lead to uniform error estimates of type (10.7) for both, the Euler- and the trapezoidal method, applied to (10.1). They are to enhance the stability of both discretization methods *without increasing the computational costs significantly*. Depending on the degree of roughness of the initial given function, different adjusted grid topologies are necessary in order to warrant the transfer of the roughness information along the grid structure for Euler's scheme. Corresponding results will be obtained for the trapezoidal-method, provided $u_0 \in D(A)$ or $u_0 \in D(A^{1/2})$.

The case of initial data of very poor smoothness, i.e., $u_0 \in H$, causes much more difficulties for the trapezoidal scheme, and we have not just reached a satisfying grid generation technique to secure a uniform convergence behavior on the whole time range. Nevertheless, let us mention the fact that we are able to verify (improved) smoothing-type error statements that hold, *provided* we use time-grids that start with step sizes $\mathcal{O}(k_0^{10})$, with k_0 being the reference discretization size, see below. Obviously, this restriction can not be accepted for practical calculations, owing to permanently acting truncation errors. This is the reason why we skip the treatment of this problem scenario.

10.2.1 Preliminaries and Notations

The study in the subsections 10.2.1 through 10.2.5 can be considered independently from the special focus on the instationary (Navier-)Stokes equations in this monograph. Owing to this, we will propose the results in a general framework.

Let V, H be separable Hilbert spaces such that V is dense in H and the injection $V \hookrightarrow H$ being compact. From this, we have a Gelfand triple $V \subset H \subset V'$, with V' being the dual space of V.

Let $A(t) : V \to V'$ be a linear continuous operator that is defined for all $t \in (0, T]$. We can associate with it a bilinear form function $t \mapsto a(t; \cdot, \cdot)$, $\forall t \in (0, T]$, on V in the way

$$a(t; u, v) = (A(t)u, v), \qquad \forall u, v \in V.$$

Here and throughout the subsections 10.2.1 through 10.2.5 we employ the notation for the inner product in H, $(\cdot, \cdot)$; this notation will also be used for the duality product between V and V'.

For the applications in mind, we suppose $a(t; \cdot, \cdot)$ to have the following form,

$$a(t; \cdot, \cdot) = a(\cdot, \cdot) + b(t; \cdot, \cdot). \tag{10.8}$$

The leading part $a(\cdot, \cdot)$ is supposed to be symmetric and coercive, i.e., there exists a positive constant $C > 0$ such that

$$a(v, v) \geq C\|v\|_V^2, \qquad \forall v \in V. \tag{10.9}$$

From the Riesz representation theorem, this determines a positive definite, self-adjoint, inverse-compact operator A in H given by the following action:

$$(Au, v) = a(u, v), \qquad \forall u, v \in V.$$

The spectral theorem then allows to define the powers A^s of A, for $s \in \mathbb{R}$. In the following, we will make frequent use of the set

$$D(A) = \{v \in V, \; Av \in H\}.$$

In particular, we have $V = D(A^{1/2})$ and $H = D(A^0)$.

Finally, the "lower order" part in (10.8) has to be continuous, for all admissible times t,

$$|b(t, u, v)| \leq C \|u\|_H \|v\|_V, \qquad \forall u, v \in V. \tag{10.10}$$

The same property is assumed to hold for the bilinear forms b_t and b_{tt}.

Again, we assume $\Omega \subset \mathbb{R}^d$, $d \in \{2, 3\}$ to be a bounded polygonal domain and the given right hand side $f(t)$ to be sufficiently smooth with respect to time derivatives, i.e., there holds for a given $T > 0$,

$$f, f_t, f_{tt} \in L^\infty(0, T; H). \tag{10.11}$$

For this class of problems, we have the following a-priori results, that provide a statement of the solution's behavior in time, depending on the degree of roughness of the initial function u_0. The verification of it is quite standard, so we omit the proof (cf. [46], e.g.).

Lemma 10.1 *Suppose that V, H, $a(t; \cdot, \cdot)$, f satisfy the assumptions (10.8) through (10.11). Then, for any $u_0 \in V \cap D(A^{\kappa/2})$, $\kappa \in \{0, 1, 2\}$, there exists a unique solution u of problem (10.1),*

$$u \in L^2(0, T; V) \cap C(0, T; H) \quad \text{and} \quad u_t \in L^2(0, T; V'),$$

and the following a-priori estimates are satisfied,

$$\tau^{\ell_1 + \frac{1}{2}(\ell_2 - \kappa)}(t) \|A^{\ell_2/2} \partial_t^{\ell_1} u(t)\|_H +$$

$$+ \int_0^t \tau^{2\ell_1 + \ell_2 - \kappa - 1}(s') \|A^{\ell_2/2} \partial_t^{\ell_1} u(s')\|_H^2 \, ds' \leq C, \qquad \forall t \in (0, T],$$

with $\ell_1 \geq 1$, $0 \leq \ell_2 \leq 2$, and $2\ell_1 + \ell_2 \leq 5$. This a-priori statement is also valid for the parameter range $-2 \leq \ell_2 \leq 0$, provided $2\ell_1 + \ell_2 - \kappa - 1 \geq 0$. – The above bound C is bounded uniformly, for times $[0, T]$, and is depending on the given data of the problem.

The next section is devoted to the proposal of different time-grid topologies that are to enhance the stability of the Euler- and the trapezoidal method in case of rough initial functions.

10.2.2 Time Cascades and Main Results

In the first section we have already pointed out that the choice of an equidistant grid $\mathcal{G}_0(k_{m+1})$ leads to reductions with respect to the order of convergence in the vicinity of the starting time, (at least) for non-smooth initial data. This phenomenon arises for the Euler- as well as the trapezoidal scheme. The reason for this is the fact that the information related to rough initial data cannot be transported across the time-grid. The corresponding smoothing-type estimates then reflect the regularizing property initiated by the dissipative main part $a(\cdot, \cdot)$. In order to guarantee the transport of information across the underlying time-grids we introduce two new grid strategies, with their application depending on the degree of roughness of the initial data u_0.

The grid topology $\mathcal{G}_2(k_{m+1})$: The grid function k_{m+1} is given in the following way,

$$k : m \mapsto k_{m+1} \equiv \begin{cases} (m+1)k_0^2, & \text{for } 0 \leq t_{m+1} \leq 1, \\ \gamma k_0, & \text{for } t_{m+1} \geq 1, \end{cases} \qquad (10.12)$$

with k_0 the basic grid size and $\gamma = \mathcal{O}(1)$. Obviously, this grid structure is very fine near the origin, with increasing mesh-size for increasing times. As will be pointed out in the following, this grid is well-adapted to the Euler-method, provided $u_0 \in V$, and the trapezoidal scheme, for initial data $u_0 \in D(A)$. In order to get striking error results for the remaining scenarios of rough initial data, we need another grid topology that is more refined in the neighborhood of the origin. This third grid is composed of several cascades and is thus capable to transporting the whole information related to rough initial data.

The grid topology $\mathcal{G}_3(k_{m+1})$: Let $S \equiv \{s_i\}|_{i\in N^+}$ be the sequence

$$s_i = 2 + \frac{1}{2^{i-2}}, \qquad (10.13)$$

then, with $[\cdot]_G$ denoting the Gauss-bracket, let there be

$$k_m = m_i k_0^{s_i}, \qquad \text{with} \qquad m_i = m - \frac{1}{k_0}[m \cdot k_0]_G. \qquad (10.14)$$

As we see, the coupling between the grid points is even stronger in the beginning, compared to $\mathcal{G}_2(k_{m+1})$.

If we look at the computational effort E that is caused by the diverse grid structures $\mathcal{G}_\ell(k_{m+1})$, $\ell \in \{1, 2, 3\}$, in order to bridge over the time interval $(0, 1]$, we obtain the following asymptotic results,

- $E\big(\mathcal{G}_1(k_{m+1})\big) = \frac{1}{k_0}$,

- $E\big(\mathcal{G}_2(k_{m+1})\big) = \sqrt{2}E\big(\mathcal{G}_1(k_{m+1})\big)$. This asymptotic result can be easily verified: we are looking for the number Z, such that

$$\sum_{m=0}^{Z} k_{m+1} = 1 \qquad \Rightarrow \qquad \frac{Z(Z+1)}{2} = \frac{1}{k_0^2}. \qquad (10.15)$$

An elementary calculation gives the result.

- $E\big(\mathcal{G}_3(k_{m+1})\big) = \infty$. This result follows immediately from the construction of this grid.

Owing to the last item, the grid structure $\mathcal{G}_3(k_{m+1})$ is not practicable that way of course, and we are compelled to stop the cascade generation at a sequence number i_0. As a consequence, this will lead to a loss of information and therefore a slight reduction of the convergence rate, which will be discussed below. At this point, we will propose the following modification of $\mathcal{G}_3(k_{m+1})$:

The grid topology $\mathcal{G}_3^{i_0}(k_{m+1})$: We apply the grid construction $\mathcal{G}_3(k_{m+1})$ up to the i_0th cascade, then replaced by the structure $\mathcal{G}_2(k_{m+1})$. Therefore, the grid function k_{m+1} reads as follows,

$$k : m \mapsto k_m \equiv \begin{cases} m_i k_0^{s_i}, & \text{for } i \leq i_0 \text{ and } \{m_i, s_i\} \text{ as in (10.13), (10.14),} \\ m_{i_0+1} k_0^2, & \text{for } \delta k_0^{s_{i_0}-2} \leq t_{m+1} \leq 1, \\ \gamma k_0, & \text{for } t_{m+1} \geq 1, \end{cases}$$

with a suitable parameter $\delta > 0$. Now, an easy calculation shows the computational effort of this scheme to be

$$E\big(\mathcal{G}_3^{i_0}(k_{m+1})\big) = (\sqrt{2} + i_0)E\big(\mathcal{G}_1(k_{m+1})\big). \qquad (10.16)$$

It will be justified by subsequent error statements that a moderate choice for the parameter i_0 is already justified to secure an quasi-optimal convergence behavior of the discretization, uniformly in time.

We are now in a position to state the main results. The following theorem describes the error behavior of the approximation u^{m+1} achieved by the Euler-method, depending on the degree of initial regularity and the grid topology that will be employed.

The subsequent theorem applies to the general type of equation (10.1), (10.8). However, for simplicity, the verification of these statements will only be given for the case $A(t) = A$, with the operator being (purely) positive definite type, selfadjoint and inverse compact, to demonstrate the ongoing mechanism of improved accuracy of the iterates, pointwise in time. However, the idea of the proof can also be applied to the general case (10.8).

Theorem 10.1 *Let $u_0 \in V \cap D(A^{\kappa/2})$, $\kappa \in \{0, 1, 2\}$ be the initial function of the evolution equation (10.1), and the assumptions (10.8) through (10.11) be satisfied. Let u^{m+1} denote the approximation that is determined by using the Euler method (10.2). Finally, let $\mathcal{G}_\ell(k_{m+1})$, $\ell \in \{1, 2\}$ and $\mathcal{G}_3^{i_0}(k_{m+1})$, with $i_0 \geq 0$ be the time-grid topologies that have been introduced above, with i_0 governing the breaking off of the cascade generation of grids. Then, depending on the respective regularity of the initial data the following error statements are valid,*

i) *for $u_0 \in D(A)$, on $\mathcal{G}_1(k_{m+1})$*: $\max_{0 \leq m \leq M} \|u(t_{m+1}) - u^{m+1}\|_H \leq Ck_0$.

ii) *for $u_0 \in V$, depending on the selected grid structure,*

    ii_1) *on $\mathcal{G}_1(k_{m+1})$*: $\max_{0 \leq m \leq M} \{\sqrt{t_{m+1}}\|u(t_{m+1}) - u^{m+1}\|_H\} \leq Ck_0$,

    ii_2) *on $\mathcal{G}_2(k_{m+1})$*: $\max_{0 \leq m \leq M} \|u(t_{m+1}) - u^{m+1}\|_H \leq Ck_0$.

iii) *for $u_0 \in H$, depending on the selected grid structure,*

    iii_1) *on $\mathcal{G}_1(k_{m+1})$*: $\max_{0 \leq m \leq M} \{t_{m+1}\|u(t_{m+1}) - u^{m+1}\|_H\} \leq Ck_0$,

    iii_2) *on $\mathcal{G}_2(k_{m+1})$*: $\max_{0 \leq m \leq M} \{\sqrt{t_{m+1}}\|u(t_{m+1}) - u^{m+1}\|_H\} \leq Ck_0$,

    iii_3) *on $\mathcal{G}_3^{i_0}(k_{m+1})$, for $m \geq \frac{1}{k_0}$*:

$$\|u(t_{m+1}) - u^{m+1}\|_H \leq C\{\chi_{i_0,m} + k_0\},$$

with $\chi_{i_0,m} = 0$, provided $m \cdot k_0 \leq i_0$, and else

$$\chi_{i_0,m} = \min\Big\{ k_0^{1-2^{1-i_0}}, \frac{1}{\sqrt{t_{m+1}}} k_0 \Big\}.$$

The constant C applied in these inequalities reflects the long-time approximation properties of the Euler scheme. Especially, it is uniformly bounded in the vicinity of the origin, i.e., for times $t_M = \mathcal{O}(1)$.

The proof of this theorem will be given in the Section 10.2.3.

Remark 10.1 *The results in this theorem for the grid structure $\mathcal{G}_1(k_{m+1})$ are more or less well-known. The verification of statement iii_1) is due to Rannacher, cf. [31].*

The next theorem is related to the trapezoidal method. Especially, it establishes an optimal, uniform convergence behavior of this method in the case of $u_0 \in D(A)$, or even $u_0 \in V$, right up to the beginning of the evolutionary process, on stretched time-grid structures $\mathcal{G}_\ell(k_{m+1})$, $\ell = \{2, 3\}$. Again, the proof will be given for the case $A(t) = A$.

Theorem 10.2 *Suppose that the hypotheses of Theorem 10.1 regarding the data u_0, f, f_t and Ω are satisfied. If we denote by u^{m+1} the approximation via the trapezoidal method (10.4), using the grid structures $\mathcal{G}_\ell(k_{m+1})$, $\ell \in \{1, 2\}$ and $\mathcal{G}_3^{i_0}(k_{m+1})$, with $i_0 \geq 0$, we can distinguish the following error statements that reflect the varying stability behavior dependent on the roughness of the initial function u_0,*

i)   *for $u_0 \in D(A)$, depending on the selected grid structure,*

    i_1) *on $\mathcal{G}_1(k_{m+1})$: $\max_{0 \leq m \leq M} \big\{ t_{m+1} \| u(t_{m+1}) - u^{m+1} \|_H \big\} \leq C k_0^2$.*

    i_2) *on $\mathcal{G}_2(k_{m+1})$: $\max_{0 \leq m \leq M} \| u(t_{m+1}) - u^{m+1} \|_H \leq C k_0^2$.*

ii) *for $u_0 \in V$, depending on the selected grid structure,*

    ii_1) *on $\mathcal{G}_1(k_{m+1})$: $\max_{0 \leq m \leq M} \| u(t_{m+1}) - u^{m+1} \|_H = o(k_0)$,*

    ii_2) *on $\mathcal{G}_2(k_{m+1})$, starting with $\frac{2}{k_0}$ Euler steps on $\mathcal{G}_3(k_{m+1})$:*

$$\max_{2/k_0 \le m \le M} \left\{ \sqrt{t_{m+1}} \| u(t_{m+1}) - u^{m+1} \|_H \right\} \le C k_0^2,$$

$ii_3)$ on $\mathcal{G}_3^{i_0}(k_{m+1})$, starting with one Euler step:

$$\| u(t_{m+1}) - u^{m+1} \|_H \le C \left\{ \chi_{i_0,m}^* + k_0^2 \right\}.$$

The constant C applied in these inequalities possesses the same properties as in Theorem 10.1, and we further employed the notation $\chi_{i_0,m}^*$, which is zero for $m \cdot k_0 \le i_0$, and else

$$\chi_{i_0,m}^* = \min \left\{ k_0^{2-2^{1-i_0}}, \frac{1}{\sqrt{t_{m+1}}} k_0^2 \right\}.$$

This theorem will be verified in Section 10.2.4.

Remark 10.2   1. *Again, the results outlined in this theorem for the grid structure $\mathcal{G}_1(k_{m+1})$ are well-known. We refer to the works of [43] and [31] for corresponding proofs.*

2. *In order to verify statement ii_2), $\frac{2}{k_0}$ initial Euler steps are needed to bridge the time interval $[k_0^4, k_0^1]$. It is conjectured that one cascade (on the time interval $[k_0^4, k_0^2]$) is already sufficient to secure second order conservation for the trapezoidal method.*

10.2.3   Proof of Theorem 10.1 for the Euler-method

Owing to Remark 10.1, we can focus on the verification of the statements ii_2), iii_2) and iii_3). — At first, let us introduce the notations used throughout this section. The error that is committed if we intend to approximate (10.1) via (10.2) will be abbreviated e^{m+1}. Further, the integral residuum $r^{m+1}(u)$ has the topic,

$$r^{m+1}(u) = -\frac{1}{k_{m+1}} \int_{t_m}^{t_{m+1}} (s - t_m) u_{tt}(s) \, ds. \tag{10.17}$$

Then the error equation is given as follows:

$$d_t e^{m+1} + A e^{m+1} = r^{m+1}(u). \tag{10.18}$$

We will now study the impact of the chosen time-grid structures on the stability and convergence properties of the Euler method, proving the named items of Theorem 10.1.

ii_2) Testing (10.18) with e^{m+1} and summation over the iteration steps give immediately

$$\|e^{M+1}\|_H^2 + \sum_{m=0}^{M} k_{m+1}\|e^{m+1}\|_V^2 \leq C \sum_{m=0}^{M} k_{m+1}\|r^{m+1}(u)\|_{V'}^2. \tag{10.19}$$

We will now treat the right hand side independently, by taking benefit from the specific structure of the underlying time-grid $\mathcal{G}_2(k_{m+1})$. Before proceeding that way, let us bear in mind the well-known result ii_1), which gives for the first iteration step on the grid $\mathcal{G}_2(k_{m+1})$,

$$\|e^1\|_H \leq Ck_0. \tag{10.20}$$

Thus, without any restrictions, we can start the summation index with $m = 1$ in the following.

$$\sum_{m=1}^{M} k_{m+1}\|r^{m+1}\|_{V'}^2 = \sum_{m=1}^{M} \frac{1}{k_{m+1}} \left\| \int_{t_m}^{t_{m+1}} (s - t_m) u_{tt}(s) \, ds \right\|_{V'}^2$$

$$\leq \sum_{m=1}^{M} \frac{1}{k_{m+1}} \int_{t_m}^{t_{m+1}} \frac{1}{s}(s - t_m)^2 \, ds \int_{t_m}^{t_{m+1}} s\|u_{tt}(s)\|_{V'}^2 \, ds$$

$$\leq C \max_{1 \leq m \leq M} \left\{ \frac{1}{k_{m+1}} \int_{t_m}^{t_{m+1}} \frac{1}{s}(s - t_m)^2 \, ds \right\}$$

$$\leq C \max_{1 \leq m \leq M} \left\{ \frac{1}{t_m} k_m^2 \right\}$$

$$\leq Ck_0^4 \max_{1 \leq m \leq M} \left\{ \frac{1}{t_m}(m + 1)^2 \right\} \leq Ck_0^2.$$

The last inequality is owing to the fact, that we have an auxiliary result relating the iteration number $m + 1$ with the time t_{m+1},

$$\sum_{m=0}^{M} k_{m+1} = t_{M+1}, \tag{10.21}$$

which gives after a short calculation, owing to the choice of k_{m+1},

$$m + 1 = -\frac{1}{2} + \sqrt{\frac{1}{4} + \frac{t_{m+1}}{k_0^2}} = \mathcal{O}\left(\frac{\sqrt{t_{m+1}}}{k_0}\right) \tag{10.22}$$

Therefore, this part of the theorem is verified.

iii_2) Again, we start from (10.18), testing this relation with $t_m e^{m+1}$ and summing over the iteration steps $1 \leq m \leq M$. This leads us to

$$t_{M+1}\|e^{M+1}\|_H^2 + \sum_{m=1}^{M} k_{m+1} t_m \|e^{m+1}\|_V^2$$

$$\leq Ck_0^2 + \sum_{m=1}^{M} k_{m+1}\|e^{m+1}\|_H^2 + \sum_{m=1}^{M} k_{m+1} t_m \|r^{m+1}(u)\|_{V'}^2. \qquad (10.23)$$

The first term stems from the time weighted error, evaluated at time $k_1 = k_0^2$. This result follows, analogously to (10.20), from the smoothing type estimate iii_1) which is valid due to the smallness of the initial time step. The third term on the right side can be controlled by Ck_0^2. This is justified by an argument as it is given in part ii_2). The additional time-weight here is now because of the rougher initial data. In order to bound the second term on the right hand side of (10.23), we test (10.18) with $A^{-1}e^{m+1}$ and sum over all iteration steps,

$$\|e^{M+1}\|_{V'}^2 + \sum_{m=1}^{M} k_{m+1}\|e^{m+1}\|_H^2$$

$$\leq \|e^1\|_{V'}^2 + \sum_{m=1}^{M} k_{m+1}\|A^{-1}r^{m+1}(u)\|_H^2 \leq Ck_0^2.$$

Again, the verification of the last estimation employs arguments that have been outlined in the previous part of the proof. We leave the details to the reader.

iii_3) For the first step in the cascade, i.e., the interval $[k_0^4, k_0^2]$, we can entirely follow the idea presented in the previous item, leading to

$$\|e^{\frac{1}{k_0}+1}\|_{V'} + k_0\|e^{\frac{1}{k_0}+1}\|_H \leq Ck_0^2. \qquad (10.24)$$

This argument can now be easily be applied to the general case, and we obtain in the i-th component of the cascade:

$$\|e^{m+1}\|_H^2 + \frac{1}{t_{m+1}}\|e^{m+1}\|_{V'}^2 \leq \frac{1}{t_{m+1}}\mathcal{L}_{m+1}(i), \qquad (10.25)$$

with the notation

$$\mathcal{L}_{m+1}(i) = t_{m-m_i+1}\|e^{m-m_i+1}\|_H^2 + \|e^{m-m_i+1}\|_{V'}^2$$

$$+ Ck_0^{2s_i} \max_{1 \leq m_i \leq 1/k_0} \left\{ \frac{1}{t_m}(m_i+1)^2 \right\}.$$

The subsequent analysis is split into two steps. The first is devoted to the study of the error at the beginning of each cascade, whereas the second deals with the error in a certain cascade.

1st step: "error along the cascades" This part is devoted to the investigation of the case $[m \cdot k_0]_G = m \cdot k_0 \equiv i+1 \geq 2$. We will start with the treatment of the last term in $\frac{1}{t_{m+1}} \mathcal{L}_{m+1}(i)$, using the identity $t_{m+1} = \mathcal{O}(k_0^{s_i-2})$. We will take benefit from

$$m_i = \mathcal{O}\left(\frac{\sqrt{t_{m_i+1}}}{k_0^{s_i/2}}\right). \tag{10.26}$$

Then, we can control the third part on the right hand side in (10.25) as follows, using (10.26),

$$
\begin{aligned}
k_0^{2s_i-s_i+2} & \max_{1 \leq m_i \leq 1/k_0} \left\{ \frac{(m_i+1)^2}{t_m} \right\} \\
& \leq k_0^2 \max_{1 \leq m_i \leq 1/k_0} \frac{t_{m_i+1}}{t_{m_i+1}\{1 + \frac{t_{m-m_i+1}}{t_{m_i+1}}\}} \leq C k_0^2.
\end{aligned} \tag{10.27}
$$

In the following, we will use the following abbreviative notation for the error,

$$
\begin{aligned}
\mathcal{E}_i^{m+1} &:= \|e^{m+1}\|_H^2 + \frac{1}{t_{m+1}} \|e^{m+1}\|_{V'}^2 \\
&= \|e^{m+1}\|_H^2 + k_0^{2-s_i} \|e^{m+1}\|_{V'}^2.
\end{aligned} \tag{10.28}
$$

Now, result (10.27) can be used to treat the inequality (10.25),

$$
\begin{aligned}
\mathcal{E}_i^{m+1} &\leq C k_0^2 + C k_0^{s_i-(s_i-2)} \|e^{m-m_i+1}\|_H^2 + k_0^{2-s_i} \|e^{m-m_i+1}\|_{V'}^2 \\
&= C k_0^2 + k_0^{s_i-1-s_i} \mathcal{E}_{i-1}^{m-m_i+1} \\
&= C k_0^2 + k_0^{2-i+2} \mathcal{E}_{i-1}^{m-m_i+1}.
\end{aligned} \tag{10.29}
$$

This can be spooled back to the start and we finally obtain the inequality:

$$
\begin{aligned}
\mathcal{E}_i^{m+1} &\leq C k_0^2 \left\{ 1 + \sum_{j=2}^{i-1} \prod_{l=2}^{j} k_0^{2-i+l} \right\} + \prod_{j=2}^{i} k_0^{2-i+j} \mathcal{E}_1^{\frac{1}{k_0}+1} \\
&= C k_0^2 \left\{ 1 + \sum_{j=2}^{i-1} k_0^{\sum_{l=2}^{j} 2-i+l} \right\} + k_0^{\sum_{j=2}^{i} 2-i+j} \mathcal{E}_1^{\frac{1}{k_0}+1}.
\end{aligned} \tag{10.30}
$$

Now, in order to bound the terms on the right hand side, we make use of the following results for geometric series,

$$\sum_{l=2}^{j} 2^{-i+l} = 2^{-i+j}\left(2 - 2^{2-j}\right) \qquad \text{and} \qquad \sum_{j=2}^{i} 2^{-i+j} = 2 - 2^{2-i}.$$

$$(10.31)$$

We can insert this result in the corresponding terms of the right hand side of (10.30), which gives a first upper bound for the last sum, by using (10.24). We complete the analysis of the terms that are denoted on the right hand side of (10.30) with the subsequent observation:

$$\sum_{j=2}^{i-1} k_0^{2-i+j} \leq C \equiv C(i; k_0).$$

These results can be inserted into (10.30), which finally gives, for indices $[m \cdot k_0]_G = m \cdot k_0 \equiv i + 1 \geq 2$,

$$\|e^{m+1}\|_H \leq C k_0. \qquad (10.32)$$

2nd step: "error in one cascade element" It is sufficient to show the following inequality, for an arbitrary fix value $i_* \geq 2$,

$$\|e^{m+1}\|_H^2 \leq \frac{1}{t_{m+1}} \mathcal{L}_{m-m_{i_*}-1/k_0+1}(i_*-1)$$

$$+ \frac{1}{t_{m+1}} \sum_{n=m-m_{i_*}-1/k_0}^{m-m_{i_*}} k_{n+1}\left\{ t_{n+1}\|r^{n+1}(u)\|_{V'}^2 + \|A^{-1}r^{n+1}(u)\|_H^2 \right\}$$

$$(10.33)$$

$$+ \frac{1}{t_{m+1}} \sum_{n=m-m_{i_*}}^{m} k_{n+1}\left\{ t_{n+1}\|r^{n+1}(u)\|_{V'}^2 + \|A^{-1}r^{n+1}(u)\|_H^2 \right\}.$$

Note that this gap permits the upper bound

$$k_0^{2-s_{i_*}}\|e^{m-m_{i_*}-1/k_0+1}\|_{V'}^2 \leq C k_0^{s_{i_*}-1+2-s_{i_*}} \leq C k_0^2.$$

The other terms that arise on the right hand side can all easily be estimated by $C k_0^2$, apart from the last sum. In order to give an optimal upper bound for it, we proceed as follows,

$$\frac{1}{t_{m+1}} \sum_{n=m-m_{i_*}}^{m} k_{n+1}\left\{ t_{n-1}\|r^{n+1}(u)\|_{V'}^2 + \|A^{-1}r^{n+1}(u)\|_H^2 \right\}$$

$$\leq \frac{1}{t_{m+1}} \sum_{n=m-m_{i_*}}^{m} \frac{1}{k_{n+1}} \int_{t_n}^{t_{n+1}} \frac{1}{s}(s-t_n)^2 \, ds \times \qquad (10.34)$$

$$\times \int_{t_n}^{t_{n+1}} s\left\{ s\|u_{tt}(s)\|_{V'}^2 + \|A^{-1}u_{tt}(s)\|_H^2 \right\} ds.$$

We can use the following consideration

$$
\int_{t_m-m_{i_*}}^{t_m} s\{s\|u_{tt}(s)\|_{V'}^2 + \|A^{-1}u_{tt}(s)\|_H^2\}\, ds
$$

$$
\leq \frac{1}{k_0} k_0^{s_{i_*}} \max_{t_m-m_{i_*} \leq s \leq t_{m+1}} \{s^2\|u_{tt}(s)\|_{V'}^2 + s\|A^{-1}u_{tt}(s)\|_H^2\}
$$

$$
\leq C k_0^{1+s_{i_*}-s_{i_*}-1} \times
$$

$$
\times \left(k_0^{s_{i_*}-1-2} \max_{t_m-m_{i_*} \leq s \leq t_{m+1}} \{s^2\|u_{tt}(s)\|_{V'}^2 + s\|A^{-1}u_{tt}(s)\|_H^2\} \right).
$$

The term in the brackets is now bounded, and we can continue in (10.34), obtaining

$$
\leq C k_0^{2-s_{i_*}-1+2+1+s_{i_*}-s_{i_*}-1} \leq C k_0^2. \tag{10.35}
$$

After this, let us return back to (10.33), and we obtain

$$
\|e^{m+1}\|_H \leq C k_0. \tag{10.36}
$$

So, the error is equal to first order in k_0 on the grid $\mathcal{G}_3(k_{m+1})$. If we deviate from it at a time point $t_{m+1} \geq t_{m-m_{i_0}+1} \equiv \mathcal{O}(k_0^{s_{i_0}-2})$ by continuation on the structure $\mathcal{G}_2(k_{m+1})$, we commit another error that is of magnitude $\mathcal{O}(k_0^{1-2^{1-i_0}})$. — This completes the proof of Theorem 10.1.

10.2.4 Proof of Theorem 10.2 for the Trapezoidal Method

This section deals with the verification of the statements of Theorem 10.2. Again, we restrict ourselves to the new results, omitting the proof of the parts i_1) and ii_1). The analyses of the remaining items partly use corresponding pattern of arguments that we have employed in the previous subsection. Nevertheless, the situation is actually more complicated, owing to the fact, that the integral stability term in our error analysis carries another time-derivative, causing additional problems. It is also important to preserve convergence properties in the initial step of the first cascade or the whole one, respectively. In the beginning, let us state the governing error identity for the trapezoidal method,

$$
d_t e^{m+1} + A\bar{e}^{m+1/2} = R^{m+1}(u). \tag{10.37}
$$

Again, we use the abbreviation e^{m+1} for the error $u(t_{m+1}) - u^{m+1}$, with $u(t_{m+1})$ and u^{m+1} being the solutions of (10.1) and (10.4), respectively. Furthermore, the residual term $R^{m+1}(u)$ is given by

$$R^{m+1}(u) = R_1^{m+1}(u) + R_2^{m+1}(u),$$

where

$$R_1^{m+1}(u) \equiv \frac{1}{k_{m+1}} \int_{t_m}^{t_{m+1}} \beta(s) u_{ttt}(s) ds,$$

$$R_2^{m+1}(u) \equiv + \frac{1}{k_{m+1}} \int_{t_m}^{t_{m+1}} \{\alpha - \beta\}(s) A u_{tt}(s) ds,$$

with the functions

$$\alpha(s) \equiv \frac{1}{12}(t_{m+1} - s)(s - t_m),$$

$$\beta(s) \equiv \frac{1}{2} \min\{(t_{m+1} - s)^2, (s - t_m)^2\}.$$

For the further investigations, we restrict on the treatment of the first term of $R^{m+1}(u)$. The term $R_2^{m+1}(u)$ can be analyzed in an analogous way. After having established the basic framework necessary for the error analysis, we can proceed with the verification of the statements that are given in Theorem 10.2.

i_2) Testing (10.37) with $\bar{e}^{m+1/2}$ and final summation over the first M iteration steps gives

$$\|e^{M+1}\|_H^2 + \sum_{m=0}^{M} k_{m+1} \|\bar{e}^{m+1/2}\|_V^2 \leq C \sum_{m=0}^{M} k_{m+1} \|R_1^{m+1}(u)\|_{V'}^2. \tag{10.38}$$

Thus, we have to bound the sum over the residuals in the following. To do so, we use the enhanced stability of the trapezoidal method that stems from the dynamic structure of the time-grid,

$$\sum_{m=0}^{M} k_{m+1} \|R_1^{m+1}(u)\|_{V'}^2 = \sum_{m=0}^{M} \frac{1}{k_{m+1}} \left\| \int_{t_m}^{t_{m+1}} \beta(s) u_{ttt}(s) \, ds \right\|_{V'}^2. \tag{10.39}$$

In order to avoid technical difficulties, we will treat the case $m = 0$ in the sum in an independent way, giving a sketchy argument that preserves second order accuracy for the approximation $u^1 \approx u(t_1)$, which is owing to the step size $t_1 = k_0^2$. In this case, we have $\beta(s) \leq sk_0^2$. Thus, we get

$$\frac{1}{k_0^2}\left(\int_0^{k_0^2} sk_0^2\|u_{ttt}(s)\|_{V'}\ ds\right)^2 \leq Ck_0^4 \int_0^{k_0^2} s^2\|u_{ttt}(s)\|_{V'}^2\ ds \leq Ck_0^4.$$

The last inequality is owing to the boundedness of the third factor of the previous expression and the interval length. Thanks to this leading the way investigation we continue our analyses of the sum (10.39), but now with the retracted summation over the range $1 \leq m \leq M$. This allows not to bother too much of whether we are concerned with t_m or t_{m+1}. So we continue:

$$\sum_{m=1}^{M} \frac{1}{k_{m+1}}\left(\int_{t_m}^{t_{m+1}} (s-t_m)^2\|u_{ttt}(s)\|_{V'}\ ds\right)^2$$

$$\leq \sum_{m=1}^{M} \frac{1}{k_{m+1}}\left(\int_{t_m}^{t_{m+1}} \frac{1}{t_m^2}(s-t_m)^4\ ds\right)\left(\int_{t_m}^{t_{m+1}} s^2\|u_{ttt}(s)\|_{V'}^2\ ds\right)$$

$$\leq C \max_{1\leq m\leq M}\left\{\frac{1}{k_{m+1}}\int_{t_m}^{t_{m+1}} \frac{1}{t_m^2}(s-t_m)^4\ ds\right\}.$$

Here, we have used an a-priori result that bounds the integral term for the velocity, three times differentiated in time. In order to control the maximum term, we continue as follows,

$$\leq C \max_{1\leq m\leq M}\left\{k_{m+1}^3 \int_{t_m}^{t_{m+1}} \frac{1}{t_m^2}\ ds\right\}$$

$$\leq Ck_0^6 \max_{1\leq m\leq M}\left\{(m+1)^3 k_{m+1}\frac{1}{t_m^2}\right\}.$$

To proceed further, let us recall result (10.22) from the previous section. So, we can continue,

$$\leq C\ k_0^6 \max_{1\leq m\leq M}\left\{\frac{t_{m+1}^{3/2}}{k_0^3}(m+1)k_0^2\frac{1}{t_m^2}\right\}$$

$$\leq C\ k_0^5 \max_{1\leq m\leq M}\left\{t_{m+1}^{-1/2}(m+1)\right\}.$$

If we apply the relationship (10.22) once more, we finally arrive at

$$\sum_{m=0}^{M} k_{m+1}\|R_1^{m+1}(u)\|_{V'}^2 \leq Ck_0^4, \tag{10.40}$$

which proves statement i_2) of Theorem 10.2.

ii_2) In this case, we start with $\frac{1}{k_0}$ Euler steps on the time range $[k_0^4, k_0^2]$, with actual step size $k_{m_1+1} = (m_1 + 1)k_0^4$ and $0 \le m_1 \le \frac{1}{k_0}$. This leads to

$$\|e^{\frac{1}{k_0}+1}\|_H + k_0\|e^{\frac{1}{k_0}+1}\|_V \le Ck_0^2,$$

what is resulting from the studies presented above. Further, on the subsequent time cascade $[k_0^2, k_0]$, we choose the size $k_{m_2+1} = (m_2 + 1)k_0^3$, for $0 \le m_2 \le \frac{1}{k_0}$. Again, we can copy the chain of inequalities (10.21), leading to

$$\sum_{m_2=1}^{1/k_0} k_{m_2+1} \|A^{-1}r^{m_2+1}\|_H^2$$

$$\le \sum_{m_2=1}^{1/k_0} \frac{1}{k_{m_2+1}} \int_{t_{m_2}}^{t_{m_2+1}} (s - t_{m_2})^2 \, ds \int_{t_{m_2}}^{t_{m_2+1}} \|A^{-1}u_{tt}(s)\|_H^2 \, ds$$

$$\le C \max_{1\le m_2 \le 1/k_0} \left\{ \frac{1}{k_{m_2+1}} \int_{t_{m_2}}^{t_{m_2+1}} (s - t_{m_2})^2 \, ds \right\}$$

$$\le C \max_{1\le m_2 \le 1/k_0} k_{m_2}^2 \le Ck_0^4$$

on this cascade. This is the integral part for an error estimate in a negative norm. Therefore, we finally arrive at,

$$\|e^{\frac{2}{k_0}+1}\|_{V'} + \sqrt{k_0}\|e^{\frac{2}{k_0}+1}\|_H + k_0\|e^{\frac{2}{k_0}+1}\|_V \le Ck_0^2. \tag{10.41}$$

The verification of the second statement in the latter estimate will be omitted.- These quantitative bounds describing the actual error behavior will be sufficient to start with the trapezoidal method right from this time point $t_{\frac{2}{k_0}+1}$, on a displaced time mesh with structure $\mathcal{G}_2(t_{m+1})$. We stress the fact that the subsequent counting neglects the first $\frac{2}{k_0}$ Euler-steps, in order to avoid difficulties that could stem from the "opening"-strategy of the widening grid. We start as usual, testing (10.37) with $t_m \bar{e}^{m+1/2}$ and summing over all iteration steps $1 \le m \le M$,

$$t_{M+1}\|e^{M+1}\|_H^2 + \sum_{m=1}^{M} k_{m+1}t_{m+1}\|\bar{e}^{m+1/2}\|_V^2$$

$$\le Ck_0^4 + \sum_{m=1}^{M} k_{m+1}t_{m+1}\|R_1^{m+1}(u)\|_{V'}^2 + \sum_{m=1}^{M} k_{m+1}\|e^{m+1}\|_H^2. \tag{10.42}$$

The bound for the error after the first iteration is taken from (10.41). The second term on the right hand side can now easily be bounded, using an analogous argument to the one that is given to treat the term (10.39) in the preceding case. Note that no further complications arise in treating this term, because we are strictly away from the case $t = 0$. We leave the easy details to the reader. In order to control the remaining term in (10.42), we employ two other considerations: At first, testing (10.37) with $d_t e^{m+1}$ and observing (10.41) leads to

$$\sum_{m=1}^{M} k_{m+1} \| d_t e^{m+1} \|_H^2 + \| e^{M+1} \|_V^2 \leq C k_0^2 + \sum_{m=1}^{M} k_{m+1} \| R_1^{m+1}(u) \|_H^2. \tag{10.43}$$

The critical part in the following is now to control the sum for the term $R_1^{m+1}(u)$. We would like to emphasize the fact that *no* additional order of convergence can be extracted from this term in the elected problem scenarios on equi-distributed grids.

The analysis of this sum relies on a study corresponding to (10.40):

$$\sum_{m=1}^{M} \frac{1}{k_{m+1}} \left(\int_{t_m}^{t_{m+1}} (s - t_m)^2 \| u_{ttt}(s) \|_H \, ds \right)^2$$

$$\leq \sum_{m=1}^{M} \frac{1}{k_{m+1}} \left(\int_{t_m}^{t_{m+1}} \frac{1}{t_m^4} (s - t_m)^4 \, ds \right) \left(\int_{t_m}^{t_{m+1}} s^4 \| u_{ttt}(s) \|_H^2 \, ds \right)$$

$$\leq C \max_{1 \leq m \leq M} \left\{ \frac{1}{k_{m+1}} \int_{t_m}^{t_{m+1}} \frac{1}{t_m^4} (s - t_m)^4 \, ds \right\}$$

$$\leq C \max_{1 \leq m \leq M} \left\{ k_{m+1}^4 \frac{1}{t_m^4} \right\} \leq C k_0^4 \max_{1 \leq m \leq M} \frac{1}{t_m^2}.$$

Owing to the starting procedure with Euler's scheme over the range $[0, k_0]$ we can bound the last term by $C k_0^2$, and we finally obtain

$$\sum_{m=1}^{M} k_{m+1} \| R_1^{m+1}(u) \|_H^2 \leq C k_0^6. \tag{10.44}$$

This completes the first part. — Secondly, if we test (10.37) with $A^{-1} \bar{e}^{m+1/2}$, before summing over the whole range of iterations, we can use (10.41) to get

$$\| e^{M+1} \|_{V'}^2 + \sum_{m=0}^{M} k_{m+1} \| \bar{e}^{m+1/2} \|_H^2$$

$$\leq C k_0^4 + \sum_{m=0}^{M} k_{m+1} \| A^{-1} R^{m+1}(u) \|_H^2 \leq C k_0^4. \tag{10.45}$$

Again, we will not go into details here. — Now, we can use (10.45) and the combination (10.44), (10.45) to control the outstanding term on the right hand side of (10.42), via the parallelogram identity, and we are done.

ii_3) The derivation of a uniform, non-smoothing type error estimate is now possible for the trapezoidal method on the grid structure $\mathcal{G}_3^{i_0}(k_{m+1})$. Let us note the fact that the application of one Euler-step in the beginning is necessary to ensure the stability. From the mathematical point of view, this enables us not to worry too much about whether k_m or k_{m+1} is to be used in the following.

Thus, if we look at the error identity for the Euler method, it is clear that we get the following result on this grid for the approximation u^1, $\|e^1\|_H \leq Ck_0^2$. Thus, the transfer of order of convergence is ensured for the first step. The remainder of the proof is devoted to the analysis of the stabilizing effects that arise from the cascade of grids. Again, we will restrict our investigations on the residual $R_1^{m+1}(u)$. Owing to formula (10.38), we proceed as follows,

$$\sum_{m=1}^{M} \frac{1}{k_{m+1}} \left(\int_{t_m}^{t_{m+1}} (s - t_m)^2 \|u_{ttt}(s)\|_{V'} \, ds \right)^2$$

$$\leq \sum_{m=1}^{M} \frac{1}{k_{m+1}} \left(\int_{t_m}^{t_{m+1}} \frac{1}{t_m^3} (s - t_m)^4 \, ds \right) \left(\int_{t_m}^{t_{m+1}} s^3 \|u_{ttt}(s)\|_{V'}^2 \, ds \right)$$

$$\leq C \max_{1 \leq m \leq M} \left\{ \frac{1}{k_{m+1}} \int_{t_m}^{t_{m+1}} \frac{1}{t_m^3} (s - t_m)^4 \, ds \right\}.$$

$$\leq C \max_{1 \leq m \leq M} \left\{ k_{m+1}^4 \frac{1}{t_m^3} \right\}.$$

We can continue as follows,

$$\leq C \max_{0 \leq i \leq i_0} \left\{ k_0^{4s_i} \max_{1 \leq m_i \leq 1/k_0} \left\{ (m+1)_i^4 \frac{1}{t_m^3} \right\} \right\} \tag{10.46}$$

for $i_0 > 1$. Again, the sizes indexed with i are locally in the cascade of grids, whereas m is globally measuring the numbers of iterations. Therefore, we get

$$\leq C \max_{0 \leq i \leq i_0} \left\{ k_0^{2s_i} \max_{1 \leq m_i \leq 1/k_0} \left\{ (t_m - t_{m-m_i})^2 \frac{1}{t_m^3} \right\} \right\}$$

$$\leq C \max_{0 \leq i \leq i_0} \left\{ k_0^{2s_i} \max_{1 \leq m_i \leq 1/k_0} \frac{1}{t_m} \right\} \leq C \max_{0 \leq i \leq i_0} k_0^{2s_i - 2(s_i - 2)} = Ck_0^4,$$

which holds for each $i \leq i_0$, owing to an inequality that is similar to (10.40).

Now, if we neglect the cascade structure at a number i_0, given from the sequence $\{s_i\}|_{0 \leq i \leq i_0}$, by continuing with the grid $\mathcal{G}_2(k_{m+1})$, we globally loose order of convergence. Again, this can be interpreted as a loss of information due to an problem-incompatible time-grid. We will not draw the details here that essentially employ the same arguments as they are given in the proof of the preceding theorem. — Therefore, Theorem 10.2 is proved entirely.

10.2.5   Comments and Outlook

As has been pointed out, the analysis of time-adapted strategies to construct the underlying grid is different from "standard" investigations for equi-distributed time-grid points in the case of rough initial data. Former investigations take advantage of the smoothing property of the dynamical system relying on the properties of the dissipative generator A (as the main part of $A(t)$), *without* adapting the numerical model to the inherent problems that are known from an a-priori standpoint. The grid adjustments presented above seem to be efficient in order to enhance the stability properties of our discretization approaches. We believe that corresponding (a-priori) adaptation processes can be successfully employed to different (more complicated) dynamical systems.

Finally, we would like to consider the complexity of the time cascade model (10.14) and (possible future) modifications causing minor numerical effort. This above introduced time-grid strategy gives rise to discontinuities in the grid size between the end of one and the beginning of the next time cascade. This could be interpreted as a sign that the numerical complexity of this approach allows minimization. In order to look for a corresponding strategy we would like to draw the reader's attention to the fact that there are (at least) two effects competing with each other in that respect: On the one hand, we are compelled to choose small time-steps at the beginning, the degree of smallness being related to the degree of roughness of the given initial data of the problem. On the other hand, the grid structure has to be able to bridge the incompatibility phase by keeping the numerical effort to a moderate amount. These two conditions exclude rather common approaches and require more sophisticated considerations.

10.3 A Revised Chorin-Uzawa Scheme on the Time-Grid $\mathcal{G}_2(k_{m+1})$

The main subject of this section is the construction of a revised projection scheme of first order with enhanced stability properties to cope with incompatibly posed (Navier-)Stokes flows. The basic idea is to employ different scalings of the evolutionary effects in the scheme in combination with a stretched time-grid (i.e., the time-grid $\mathcal{G}_2(k_{m+1})$ introduced in section 10.2.2).

In order to make clear the acting perturbation sources in these projection schemes, we will start with the investigation of the nonstationary quasi-com-pressibility schemes as related candidates of *modifications of the Chorin-Uzawa scheme* on the grid structure $\mathcal{G}_2(k_{m+1})$. Again, we consider the Stokes problem (2.1) in order to avoid technical difficulties stemming from the non-linearity.

A first candidate to construct a Chorin-Uzawa type scheme on structured grids is the following,

$$d_t \tilde{u}^{m+1} - \Delta \tilde{u}^{m+1} + \nabla p^m = f^{m+1},$$

$$\operatorname{div}\tilde{u}^{m+1} + \frac{1}{\alpha} k_0 \tilde{d}_t \big((m+1)k_0 p^{m+1}\big) = 0, \qquad \alpha < 1, \tag{10.47}$$

together with $u^0 = u_0$, $p^0 = 0$ and the underlying time-grid structure $\mathcal{G}_2(k_{m+1})$, i.e., $k_{m+1} = (m+1)k_0^2$. Here, there are two facts to be emphasized: Once, the nonstationary quasi-compressibility constraint is now relaxed, using the notations

$$\tilde{d}_t \phi^{m+1} = \frac{1}{k_0}\{\phi^{m+1} - \phi^m\} \qquad \text{and} \qquad d_t \phi^{m+1} = \frac{1}{k_{m+1}}\{\phi^{m+1} - \phi^m\}.$$

Therefore, we do not have the original time-derivative any more in the second equation in (10.47), thanks to the underlying time-grid. Secondly, only homogeneous initial data for the pressure function p^0 are prescribed, which makes the scheme attractive in comparison to the original Chorin-Uzawa scheme. Unfortunately, this scheme does not work in an optimal way, owing to a lack of stability that is caused by the explicit treatment of the pressure in the momentum equation. The basic mechanisms that prevent an optimal behavior of convergence will be demonstrated below.

Remark 10.3 *An optimal convergence behavior for the solution can be proven for the fully implicit version of system (10.47), i.e., with p^m replaced by*

p^{m+1} in the first equation. This implicit version then provides a robust optimal discretization scheme, i.e., the following error estimates can be proved
for the solution $\{u^{m+1}_{impl}, p^{m+1}_{impl}\}$, for $0 < k_0 < 1$,

$$\max_{0 \leq m \leq M} \{\|u(t_{m+1}) - u^{m+1}_{impl}\| + \sqrt{k_0}\|p(t_{m+1}) - p^{m+1}_{impl}\|\}$$

$$+ \left(k_0 \sum_{m=0}^{M} \|u(t_{m+1}) - u^{m+1}_{impl}\|^2_1\right)^{1/2} \leq C(1 + \log\frac{1}{k_0})k_0,$$

without demanding additional regularity of the solution $\{u, p\}$ of (2.1) in the
limit case $t \to 0$. We omit the proof of these results, referring to corresponding studies for (10.48) below.

The studies of (10.47) indicate that this quasi-compressibility constraint is
not sufficient in order to guarantee stability. In order to arrive at a stable
one, we will insert another perturbation term of higher order in the second
equation, leading to the following scheme,

$$d_t \tilde{u}^{m+1} - \Delta \tilde{u}^{m+1} + \nabla\left(2\frac{m}{m+1}p^m - \frac{m-1}{m+1}p^{m-1}\right) = f^{m+1},$$

$$\text{div}\tilde{u}^{m+1} + \frac{1}{\alpha}k_0\left(\text{Id} - k_0\Delta\right)\tilde{d}_t\left((m+1)k_0 p^{m+1}\right) = 0,$$

$$\partial_n \tilde{d}_t\left((m+1)k_0 p^{m+1}\right)|_{\partial\Omega} = 0, \qquad \alpha < 1,$$

(10.48)

together with $u^0 = u_0$ and $p^{-1} = p^0 = 0$. As we will see below, this provides a stable time-discretization of the Stokes equations, giving first order
of convergence for the velocity field, see Theorem 10.3 below.

The *revised Chorin-Uzawa scheme* on the time-grid $\mathcal{G}_2(k_{m+1})$ is now as
follows:

1. Start with initial data $u^0 = u_0$ (or an accurate approximation) and
   $p^0 \equiv \bar{p}^0 \equiv 0$.

2. Compute $\tilde{u}^{m+1} \in \mathbf{H}^1_0(\Omega)$ as the solution of

$$\frac{1}{k_{m+1}}\left\{\tilde{u}^{m+1} - u^m\right\} - \Delta\tilde{u}^{m+1} + \nabla\left\{2\frac{m}{m+1}p^m - \bar{p}^m\right\} = f^{m+1}.$$

(10.49)

3. Projection-step: Find the solution $\{u^{m+1}, \bar{p}^{m+1}\}$ of

$$\frac{1}{k_{m+2}}\{u^{m+1} - \tilde{u}^{m+1}\} + \nabla\{\bar{p}^{m+1} - \frac{m}{m+2}p^m\} = 0,$$ (10.50)

$$\mathrm{div}\, u^{m+1} = 0,\, u^{m+1}|_{\partial\Omega} \cdot n = 0.$$

4. Computation of the pressure p^{m+1}: Taking $\alpha < 1$, solve

$$\left(\mathrm{Id} - k_0\Delta\right)\mathcal{Q}^{m+1} = \left(\mathrm{Id} - k_0\Delta\right)\mathcal{Q}^m - \alpha\mathrm{div}\tilde{u}^{m+1},$$ (10.51)

$$\partial_n d_t\mathcal{Q}^{m+1}|_{\partial\Omega} = 0.$$

Then, the pressure p^{m+1} is given to be

$$p^{m+1} = \frac{1}{(m+1)k_0}\mathcal{Q}^{m+1}.$$ (10.52)

The main result of this section now is the verification of optimal error statements, collected in the following theorem.

Theorem 10.3 *Let the tuple* $\{\tilde{u}^{m+1}, p^{m+1}\}$ *be the solution of the revised Chorin-Uzawa scheme (10.49) through (10.52) on the time-grid* $\mathcal{G}_2(k_{m+1})$ *that approximates the solution* $\{u(t_{m+1}), p(t_{m+1})\}$ *of (2.1). The given data of the problem are assumed to satisfy (A1) and (A2). Then, the following error estimates are valid in the "canonical" framework of regularities for the given data, using a constant* C *that only depends on the given data of the problem,*

$$\max_{0\leq m\leq M}\left\{\|u(t_{m+1}) - \tilde{u}^{m+1}\| + \sqrt{k_0}\|p(t_{m+1}) - p^{m+1}\|\right\}$$

$$+ \left(k_0\sum_{m=0}^M \|u(t_{m+1}) - \tilde{u}^{m+1}\|_1^2\right)^{1/2} \leq C(1 + \log\frac{1}{k_0})k_0.$$

The proof of this theorem is rather technical. In particular, the study of the interplay of higher order perturbation of the incompressibility constraint and the semi-explicity of the related quasi-compressibility method on the dynamical time-grid structure are the major parts of the proof, assuring the stability of the projection scheme. It is split in the now "well-known" three parts: The analysis of the time-discretization on the space of divergence-free functions constitutes the first one, succeeded by the investigation of the error mechanisms which are initiated by the perturbation of the incompressibility constraint. Then, the study of the impact of the explicit treatment of the pressure function in (10.48) completes the proof of the theorem.

10.3.1 Stability Statements for the Euler Scheme on the Time-Grid $\mathcal{G}_2(k_{m+1})$

The equations stemming from an implicit Euler ansatz are as follows,

$$d_t u^{m+1} - \Delta u^{m+1} + \nabla p^{m+1} = f^{m+1},$$
$$\operatorname{div} u^{m+1} = 0, \tag{10.53}$$

with an initial function $u^0 = u_0$. In the following, we will again employ the Stokes operator $A = -P_{J_0}\Delta$, leading to

$$d_t u^{m+1} + A u^{m+1} = P_{J_0} f^{m+1}.$$

By a standard procedure, the following is easy to verify,

$$\max_{0 \le m \le M} \left\{ \|\Delta u^{m+1}\| + \|d_t u^{m+1}\| \right\} + \left(\sum_{m=0}^{M} k_{m+1} \|\nabla d_t u^{m+1}\|^2 \right)^{1/2} \le C. \tag{10.54}$$

The subsequent considerations will be started from the equation

$$\tilde{d}_t u^{m+1} + (m+1)k_0 A u^{m+1} = (m+1)k_0 P_{J_0} f^{m+1}, \tag{10.55}$$

and its differentiated form,

$$\tilde{d}_t^2 u^{m+1} + mk_0 A \tilde{d}_t u^{m+1} + A u^{m+1} = \tilde{d}_t \left\{ (m+1)k_0 P_{J_0} f^{m+1} \right\}. \tag{10.56}$$

Let us stress the fact that we have applied $\tilde{d}_t$ here instead of d_t. If we now test the last equation with $A d_t u^{m+1}$ while taking benefit from the inequality (10.54) for the resulting initial data, we arrive at

$$\|\nabla \tilde{d}_t u^{M+1}\|^2 + \sum_{m=0}^{M} k_{m+1} \|A \tilde{d}_t u^{m+1}\|^2 \le C. \tag{10.57}$$

Another nontrivial result on these time-grids can be deduced from (10.56), after having employed the differencing operator $\tilde{d}_t$. Thanks to (10.54) and (10.57), testing with $\tilde{d}_t^2 u^{m+1}$ leads to

$$\|\tilde{d}_t^2 u^{M+1}\|^2 + \sum_{m=1}^{M} k_{m+1} \|\nabla \tilde{d}_t^2 u^{m+1}\|^2 \le C. \tag{10.58}$$

Now, we are able to establish the following striking a-priori bounds,

$$\max_{0 \le m \le M} \left\{ \| \Delta \tilde{d}_t \big((m+1) k_0 u^{m+1} \big) \| + \| \nabla \tilde{d}_t \mathcal{P}^{m+1} \| \right\} \le C, \tag{10.59}$$

with the abbreviative notation $\mathcal{P}^{m+1} := (m+1) k_0 p^{m+1}$. Moreover, the following bounds can be verified by means of a canonical proceeding,

$$\max_{1 \le m \le M} \left\{ \| \nabla \tilde{d}_t^2 \big((m+1) k_0 u^{m+1} \big) \| + \| \tilde{d}_t^2 \mathcal{P}^{m+1} \| \right\}$$
$$+ \left(\sum_{m=1}^{M} k_{m+1} \tau_{m+1} \| \Delta \tilde{d}_t^2 u^{m+1} \|^2 \right)^{1/2} + \left(\sum_{m=1}^{M} k_{m+1} \| \nabla \tilde{d}_t^2 \mathcal{P}^{m+1} \|^2 \right)^{1/2} \tag{10.60}$$
$$+ \max_{1 \le m \le M} \left\{ \sqrt{\tau_{m+1}} \| \nabla \tilde{d}_t^2 \mathcal{P}^{m+1} \| \right\} \le C.$$

The details for checking these statements will be omitted. — Note that the a-priori results presented with the latter two inequalities reflect the improved stability of the final projection method which is owing to the time-grid structure $\mathcal{G}_2(k_{m+1})$. This allows to distinguish between the time-grid function $t_{m+1} \mapsto k_{m+1} \equiv (m+1) k_0^2$ and the nonstationary perturbation of the quasi-compressibility constraint which is in powers of k_0. This observation is essential for the further analyses showing that the perturbations are controllable all over the time-range for general fluid flow problems.

10.3.2 Analysis of a nonstationary Quasi-Compressibility Method on the Time-Grid $\mathcal{G}_2(k_{m+1})$

This part of the proof of Theorem 10.3 is devoted to the study of the error impacts that are stemming from the perturbation of the incompressibility constraint. Thus, the object of study below is the system of equations for the functions $\{u_\varepsilon^{m+1}, p_\varepsilon^{m+1}\}$,

$$d_t u_\varepsilon^{m+1} - \Delta u_\varepsilon^{m+1} + \nabla p_\varepsilon^{m+1} = f^{m+1},$$
$$\operatorname{div} u_\varepsilon^{m+1} + \varepsilon (\operatorname{Id} - \varepsilon \Delta) \tilde{d}_t \mathcal{P}_\varepsilon^{m+1} = 0, \tag{10.61}$$
$$\partial_n \tilde{d}_t \mathcal{P}_\varepsilon^{m+1} |_{\partial \Omega} = 0,$$

and its stability and approximation features. For the initial data, we set $u_\varepsilon^0 = u_0$ and $p_\varepsilon^0 = 0$. Again, a new notation for the relaxed pressure function is used, $\mathcal{P}_\varepsilon^{m+1} := (m+1) k_0 p_\varepsilon^{m+1}$. We will now proceed in a way that is analogous to earlier considerations which have been done for projection schemes on equi-distributed time-grids. Subsequently, we start with an analysis of the error effects which are initiated by the perturbation of the incompressibility constraint at some instant, whereas later studies are referred to the quantification of resulting error propagations that are caused by the nonstationary momentum equation.

10.3.3   Introduction of an Auxiliary Problem on the Time-Grid $\mathcal{G}_2(k_{m+1})$

The auxiliary problem to be formulated for that purpose is as follows: Find a pair of solutions $\{U_\varepsilon^{m+1}, B_\varepsilon^{m+1}\}$ satisfying the equations

$$
\begin{aligned}
&- (m+1)k_0 \Delta U_\varepsilon^{m+1} + \nabla \mathcal{B}_\varepsilon^{m+1} = (m+1)k_0 f^{m+1} - \tilde{d}_t u^{m+1}, \\
&\mathrm{div} U_\varepsilon^{m+1} + \varepsilon \left(\mathrm{Id} - \varepsilon \Delta \right) \tilde{d}_t \mathcal{B}_\varepsilon^{m+1} = 0, \\
&\partial_n \tilde{d}_t \mathcal{B}_\varepsilon^{m+1}|_{\partial\Omega} = 0.
\end{aligned}
\tag{10.62}
$$

Again, we have used an abbreviative notation, $\mathcal{B}_\varepsilon^{m+1} := (m+1)k_0 B_\varepsilon^{m+1}$. For further studies, we use the error notations

$$
E^{m+1} := u^{m+1} - U_\varepsilon^{m+1} \qquad \text{and} \qquad \mathcal{D}^{m+1} := \mathcal{P}^{m+1} - \mathcal{B}_\varepsilon^{m+1},
$$

with the governing error identities

$$
\begin{aligned}
&- (m+1)k_0 \Delta E^{m+1} + \nabla \mathcal{D}^{m+1} = 0, \\
&\mathrm{div} E^{m+1} + \varepsilon \left(\mathrm{Id} - \varepsilon \Delta \right) \tilde{d}_t \mathcal{D}^{m+1} = \varepsilon \left(\mathrm{Id} - \varepsilon \Delta \right) \tilde{d}_t \mathcal{P}^{m+1}, \\
&\partial_n \tilde{d}_t \mathcal{D}^{m+1}|_{\partial\Omega} = \partial_n \tilde{d}_t \mathcal{P}^{m+1}|_{\partial\Omega},
\end{aligned}
\tag{10.63}
$$

and $\mathcal{D}^0 \equiv 0$.

The main results of the present subsection are collected in the subsequent lemma, providing striking stability statements for the solution of (10.62) as well as optimal results of convergence.

Lemma 10.2 *Assume $\{U_\varepsilon^{m+1}, B_\varepsilon^{m+1}\}$ to be the solution of (10.62), whereas the tuple $\{u^{m+1}, p^{m+1}\}$ is determined by the system of equations (10.53). Concerning the given data of the problem, we assume (A1) and (A2) to be valid. Then, the following a-priori bounds hold true, with a constant C that* only depends on the given data of the problem, for $\varepsilon = \mathcal{O}(k_0)$,

$$\max_{0 \leq m \leq M} \left\{ (1 + \log \frac{1}{k_0}) \| \tilde{d}_t \mathcal{B}_\varepsilon^{m+1} \| + \| \nabla \tilde{d}_t \mathcal{B}_\varepsilon^{m+1} \| \right\}$$

$$+ \sqrt{k_0} \left(\sum_{m=1}^{M} \| \tilde{d}_t^2 \mathcal{B}_\varepsilon^{m+1} \| \right)^{1/2}$$

$$+ k_0 \left(\sum_{m=1}^{M} \| \nabla \tilde{d}_t^2 \mathcal{B}_\varepsilon^{m+1} \| \right)^{1/2} \leq C(1 + \log \frac{1}{k_0}).$$

Further, the following error estimates are satisfied,

$$\max_{0 \leq m \leq M} \left\{ \sqrt{\tau_{m+1}} \| u^{m+1} - U_\varepsilon^{m+1} \|_1 + \| \mathcal{P}^{m+1} - \mathcal{B}_\varepsilon^{m+1} \| \right\}$$

$$+ \left(\sum_{m=0}^{M} k_0 \| \tilde{d}_t(u^{m+1} - U_\varepsilon^{m+1}) \|^2 \right)^{1/2} \leq C\varepsilon(1 + \log \frac{1}{k_0}).$$

Remark 10.4 *The results in this lemma can be sharpened in the way that the appearing logarithmic terms can be canceled in another estimate for the velocity field, measured in the norm $l^\infty(0, t_{M+1}; \mathbf{L}^2)$. Nevertheless, the verification of Theorem 10.3 requires the consideration of further auxiliary problems below that will involve additional terms of that kind, so we will be content with it right here.*

Proof:
Before starting the actual proof, let us consider the initial errors that are caused by scheme (10.63). If we set $m = 0$, keeping in mind the vanishing initial errors in the weighted pressure approximation $\mathcal{B}_\varepsilon^0$, we are led to consider

$$- \Delta E^1 + \nabla \tilde{d}_t \mathcal{D}^1 = 0,$$
$$\operatorname{div} E^1 + \varepsilon \left(\operatorname{Id} - \varepsilon \Delta \right) \tilde{d}_t \mathcal{D}^1 = \varepsilon \left(\operatorname{Id} - \varepsilon \Delta \right) \tilde{d}_t \mathcal{P}^1, \qquad (10.64)$$
$$\partial_n \tilde{d}_t \mathcal{D}^1 |_{\partial\Omega} = \partial_n \tilde{d}_t \mathcal{P}^1 |_{\partial\Omega}.$$

As an auxiliary result, we obtain the following inequality from the first equation of the latter system,

$$\| \tilde{d}_t \mathcal{D}^1 \| \leq C \| \nabla E^1 \|,$$

which gives rise to the statement

$$\|\nabla E^1\| + \|\tilde{d}_t \mathcal{D}^1\| + \varepsilon\|\nabla \tilde{d}_t \mathcal{D}^1\| \le C\varepsilon. \tag{10.65}$$

This result will be employed frequently in our subsequent studies.

After these preliminary considerations, a first sharp error statement can be achieved from (10.63), testing it with $\frac{1}{(m+1)^2 k_0^2} E^{m+1}$. After summation we arrive at

$$\sum_{m=0}^{M} \frac{1}{(m+1)}\|\nabla E^{m+1}\|^2 + \frac{\varepsilon}{\tau_{M+1}}\|\mathcal{D}^{M+1}\|^2$$

$$+ \frac{\varepsilon^2}{\tau_{M+1}}\|\nabla \mathcal{D}^{M+1}\|^2 + \varepsilon k_0 \sum_{m=0}^{M} \tau_{m+1}^{-3/2}\|\mathcal{D}^{m+1}\|^2$$

$$+ \varepsilon^2 k_0 \sum_{m=0}^{M} \tau_{m+1}^{-3/2}\|\nabla \mathcal{D}^{m+1}\|^2 \tag{10.66}$$

$$\le C\Big\{\varepsilon^2 + \varepsilon^2 \sum_{m=0}^{M} k_{m+1}\|\tilde{d}_t \mathcal{P}^{m+1}\|^2 + \varepsilon^2 \sum_{m=0}^{M} k_{m+1}\|\nabla \tilde{d}_t \mathcal{P}^{m+1}\|^2\Big\}.$$

The first term on the right hand side is due to the previous considerations. For the origin of the consequent terms, we have made use of the following stability result,

$$\tau_{m+1}^{-1/2}\|\mathcal{D}^{m+1}\| \le C\|\nabla E^{m+1}\|,$$

in combination with a normalized version of Gronwall's lemma. Now, it is easy to bound the right hand side through $C\varepsilon^2$. This provides the basis for the following considerations. In order to verify a first pointwise error statement for the velocity, we start from (10.63), employing the operator $\tilde{d}_t$ onto the first equation,

$$-mk_0 \Delta \tilde{d}_t E^{m+1} - \Delta E^{m+1} + \nabla \tilde{d}_t \mathcal{D}^{m+1} = 0. \tag{10.67}$$

Again, the stability property of the div-operator gives

$$\|\tilde{d}_t \mathcal{D}^{m+1}\| \le C\big\{\|\nabla E^{m+1}\| + mk_0\|\nabla \tilde{d}_t E^{m+1}\|\big\}. \tag{10.68}$$

If we now test (10.67) with $\tilde{d}_t E^{m+1}$, we obtain

$$
\begin{aligned}
2mk_0\|\nabla\tilde{d}_t E^{m+1}\|^2 + \tilde{d}_t\|\nabla E^{m+1}\|^2 \\
+ \varepsilon\tilde{d}_t\|\tilde{d}_t \mathcal{D}^{m+1}\|^2 + \varepsilon^2\tilde{d}_t\|\nabla\tilde{d}_t\mathcal{D}^{m+1}\|^2 \\
+ \varepsilon k_0\|\tilde{d}_t^2\mathcal{D}^{m+1}\|^2 + \varepsilon^2 k_0\|\nabla\tilde{d}_t^2\mathcal{D}^{m+1}\|^2 \\
= 2\varepsilon(\tilde{d}_t^2\mathcal{P}^{m+1}, \tilde{d}_t\mathcal{D}^{m+1}) + 2\varepsilon^2(\nabla\tilde{d}_t^2\mathcal{P}^{m+1}, \nabla\tilde{d}_t\mathcal{D}^{m+1})
\end{aligned} \tag{10.69}
$$

The first term on the right hand side can be dealt with by means of the stability result (10.68), in combination with (10.66) and the result (10.60). In order to treat the second term, we use the splitting

$$
\begin{aligned}
(\nabla\tilde{d}_t^2\mathcal{P}^{m+1}, \nabla\tilde{d}_t\mathcal{D}^{m+1}) \leq C(1 + t_M + \log\frac{1}{k_0})\tau_{m+1}^{1/2}\|\nabla\tilde{d}_t^2\mathcal{P}^{m+1}\|^2 \\
+ \frac{1}{4}(1 + t_M + \log\frac{1}{k_0})^{-1}\frac{1}{(m+1)k_0}\|\nabla\tilde{d}_t\mathcal{D}^{m+1}\|^2.
\end{aligned} \tag{10.70}
$$

The application of the summation operator is now justified, thanks to (10.65). The normalized Gronwall lemma then leads us to

$$
\begin{aligned}
\sum_{m=1}^{M} k_{m+1}\|\nabla\tilde{d}_t E^{m+1}\|^2 + \|\nabla E^{M+1}\|^2 \\
+ \varepsilon\|\tilde{d}_t\mathcal{D}^{M+1}\|^2 + \varepsilon^2\|\nabla\tilde{d}_t\mathcal{D}^{M+1}\|^2 \\
+ \varepsilon k_0^2\sum_{m=1}^{M}\|\tilde{d}_t^2\mathcal{D}^{m+1}\|^2 + \varepsilon^2 k_0^2\sum_{m=1}^{M}\|\nabla\tilde{d}_t^2\mathcal{D}^{m+1}\|^2 \\
\leq C\varepsilon^2(1 + \log\frac{1}{k_0}).
\end{aligned} \tag{10.71}
$$

We emphasize, that we have applied the stability result (10.68) here to absorb the time-derivative of the pressure error on the right hand side by the velocity term on the left hand side. Moreover, from the first identity in (10.63) we get

$$
\|\mathcal{D}^{m+1}\| + \sqrt{\tau_{m+1}}\|\nabla E^{m+1}\| \leq C\varepsilon(1 + \log\frac{1}{k_0}).
$$

This establishes the first result of convergence presented in the above lemma. As a further result from (10.71) in combination with (10.60), we have verified

the following stability results for the solution of (10.62), which will be shown
to be sufficient for the further proof of Theorem 10.3,

$$\sqrt{\frac{k_0}{\varepsilon}}\Big(k_0 \sum_{m=1}^{M} \|\tilde{d}_t^2 \mathcal{B}_\varepsilon^{m+1}\|^2\Big)^{1/2} + \sqrt{k_0}\Big(k_0 \sum_{m=1}^{M} \|\nabla \tilde{d}_t^2 \mathcal{B}_\varepsilon^{m+1}\|^2\Big)^{1/2}$$

$$\leq C(1 + \log\frac{1}{k_0}). \qquad (10.72)$$

In order to prove a sharp upper bound for the time-derivative of the velocity
error, we have to introduce an auxiliary dual problem. It is given in the form:
Given the triple $\{w^m, q^m, \tilde{d}_t E^{m+1}\}$, the tuple $\{w^{m+1}, q^{m+1}\} \in \mathbf{H}_0^1 \times H^1/R$ is
the solution of

$$-\Delta \tilde{d}_t w^{m+1} + \nabla \tilde{d}_t q^{m+1} = \tilde{d}_t E^{m+1},$$

$$\mathrm{div}\tilde{d}_t w^{m+1} = 0. \qquad (10.73)$$

This equation will be tested with $\tilde{d}_t E^{m+1}$. Note that the related equation is
now (10.67), which possesses a time-dependent operator. Nevertheless, the
duality argument is *scaling invariant*, i.e., from (10.63) we deduce

$$(\nabla E^{m+1}, \nabla \tilde{d}_t w^{m+1}) = 0.$$

Owing to this, the rest of the duality argument is now standard, providing
us with the desired statement

$$\Big(k_0 \sum_{m=0}^{M} \|\tilde{d}_t E^{m+1}\|^2\Big)^{1/2} \leq C\varepsilon(1 + \sqrt{\frac{\varepsilon}{k_0}})(1 + \log\frac{1}{k_0}).$$

This error result shows the importance of adjusting the time-discretization
parameter dependent on the perturbation parameter for getting good approximations of the time-derivatives.   □

As we have seen in this part of the proof, the time-grid strategy employed
gives a problem-adapted splitting of the operator parts "incompressibility"
and "viscosity". This weighted splitting is the reason for a robust modeling of
the relevant solution properties guaranteeing the transport of the necessary
amount of information across the time-grids.

10.3.4 Completion of the Proof of Theorem 10.3

The error equations that will be discussed in this subsection are the following ones,

$$
\begin{aligned}
&\tilde{d}_t e_\varepsilon^{m+1} - (m+1)k_0 \Delta e_\varepsilon^{m+1} + \nabla \mathcal{K}_\varepsilon^{m+1} = \tilde{d}_t E^{m+1}, \\
&\operatorname{div} e_\varepsilon^{m+1} + \varepsilon(\mathrm{Id} - \varepsilon\Delta)\tilde{d}_t \mathcal{K}_\varepsilon^{m+1} = 0, \\
&\partial_n \tilde{d}_t \mathcal{K}_\varepsilon^{m+1}|_{\partial\Omega} = 0,
\end{aligned}
\tag{10.74}
$$

with initial errors $e_\varepsilon^0 \equiv 0$, $\mathcal{K}_\varepsilon^0 = 0$. — Here, we employed the abbreviative notations

$$
e_\varepsilon^{m+1} := u_\varepsilon^{m+1} - U_\varepsilon^{m+1} \qquad \text{and} \qquad \mathcal{K}_\varepsilon^{m+1} := \mathcal{P}_\varepsilon^{m+1} - \mathcal{B}_\varepsilon^{m+1}.
$$

The objective of this section is the verification of striking a-priori bounds for the solution of (10.74). They will be presented in the following lemma.

Lemma 10.3 *For the error functions* $\{e_\varepsilon^{m+1}, \mathcal{K}_\varepsilon^{m+1}\}$, *the following bounds are valid, for* $\varepsilon = \mathcal{O}(k_0)$,

$$
\max_{0 \leq m \leq M} \tau_{m+1}^{-1/4} \left\{ \|e_\varepsilon^{m+1}\| + \sqrt{\varepsilon}\|\mathcal{K}_\varepsilon^{m+1}\| \right\}
$$

$$
+ \left(k_0 \sum_{m=0}^{M} \|\nabla e_\varepsilon^{m+1}\|^2 \right)^{1/2} \leq C(1 + \log\frac{1}{k_0})\varepsilon.
$$

Proof:
In order to verify the error statement for the velocity field, we test the first

equation in (10.74) with $\frac{1}{(m+2)k_0}e_\varepsilon^{m+1}$. We obtain

$$\frac{1}{\sqrt{\tau_{M+1}}}\|e_\varepsilon^{M+1}\|^2 + \sum_{m=0}^{M}\frac{k_0}{\tau_{m+1}}\|e_\varepsilon^{m+1}\|^2$$

$$+ \sum_{m=0}^{M}\frac{k_0^2}{\sqrt{\tau_{m+1}}}\|\tilde{d}_t e_\varepsilon^{m+1}\|^2 + k_0\sum_{m=0}^{M}\|\nabla e_\varepsilon^{m+1}\|^2$$

$$+ \varepsilon\frac{1}{\sqrt{\tau_{M+1}}}\|\mathcal{K}_\varepsilon^{M+1}\|^2 + \varepsilon\sum_{m=0}^{M}\frac{k_0}{\tau_{m+1}}\|\mathcal{K}_\varepsilon^{m+1}\|^2$$

$$+ \varepsilon\sum_{m=0}^{M}\frac{k_0^2}{\sqrt{\tau_{m+1}}}\|\tilde{d}_t\mathcal{K}_\varepsilon^{m+1}\|^2 + \varepsilon^2\frac{1}{\sqrt{\tau_{M+1}}}\|\nabla\mathcal{K}_\varepsilon^{M+1}\|^2 \qquad (10.75)$$

$$+ \varepsilon^2\sum_{m=0}^{M}\frac{k_0}{\tau_{m+1}}\|\nabla\mathcal{K}_\varepsilon^{m+1}\|^2 + \varepsilon^2\sum_{m=0}^{M}\frac{k_0^2}{\sqrt{\tau_{m+1}}}\|\nabla\tilde{d}_t\mathcal{K}_\varepsilon^{m+1}\|^2$$

$$\leq Ck_0\sum_{m=0}^{M}\|\tilde{d}_t E^{m+1}\|^2 \leq C(1+\frac{\varepsilon}{k_0})(1+\log\frac{1}{k_0})\varepsilon^2.$$

This shows conservation of first order accuracy for the employed quasi-compressibility constraint with respect to the velocity components. Again, we see the acting error mechanisms that need to be balanced — perturbation and time-discretization — in order to handle with a stable scheme. Therefore, this already establishes the result of the lemma. □

In order to extend our analysis to the semi-explicit formulation of the related nonstationary quasi-compressibility method, we need another a-priori statement for the pressure function of (10.61). Referring to the known stability behavior of the solution of the "incompressible" equations (10.53), it is easy to verify the following statements,

$$\sqrt{k_0}\left(k_0\sum_{m=1}^{M}\|\tilde{d}_t^2\mathcal{P}_\varepsilon^{m+1}\|^2\right)^{1/2} + \sqrt{\varepsilon}\sqrt{k_0}\left(k_0\sum_{m=1}^{M}\|\nabla\tilde{d}_t^2\mathcal{P}_\varepsilon^{m+1}\|^2\right)^{1/2}$$

$$\leq C(1+\log\frac{1}{k_0}). \qquad (10.76)$$

The details that are necessary to achieve this important result are left to the reader. Even the latter result is to prepare the final step in the error analysis for the modified Chorin-Uzawa scheme on the time-grid structure $\mathcal{G}_2(k_{m+1})$. This will be carried out in the subsequent subsection.

10.3.5   Analysis of the Semi-Explicit System

This final part of the proof of Theorem 10.3 is devoted to the study of the error contribution that is caused by the transition from the fully implicit to the semi-explicit prescription of the perturbed system with respect to the treatment of the pressure function. As we will see below, we will again need a stabilizing parameter $\alpha < 1$ in order to guarantee a stable approximation of the original solution. The equations to be considered here are then as follows,

$$\tilde{d}_t e^{m+1} - (m+1)k_0\Delta e^{m+1} + \nabla \mathcal{Q}^{m+1} = k_0^2 \nabla \tilde{d}_t^2 \mathcal{Q}^{m+1} - k_0^2 \nabla \tilde{d}_t^2 \mathcal{P}_\varepsilon^{m+1},$$

$$\operatorname{div} e^{m+1} + \frac{1}{\alpha} k_0 \big(\operatorname{Id} - k_0\Delta \big) \tilde{d}_t \mathcal{Q}^{m+1} = 0, \qquad \text{with } \alpha < 1,$$

$$\partial_n \tilde{d}_t \mathcal{Q}^{m+1}|_{\partial\Omega} = 0.$$

$$(10.77)$$

The initial data of this problem are homogeneous. Let us recall the error notations valid for the present subsection, $e^{m+1} := u_\varepsilon^{m+1} - u_{ChoUz}^{m+1}$, $\mathcal{Q}^{m+1} := \mathcal{P}_\varepsilon^{m+1} - \mathcal{Q}_{ChoUz}^{m+1}$. Now, optimal error results are easy to obtain, by testing the first equation of the latter system (10.77) with e^{m+1} and taking care of the stabilizing factor α in the second identity of (10.77) in order to absorb certain norms for the pressure function on the left hand side. Provided the above parameter value restriction is satisfied, i.e., $\alpha < 1$, we end up with

$$(1-\alpha)\|e^{M+1}\|^2 + (1-\alpha)k_0^2 \sum_{m=0}^{M} \|\tilde{d}_t e^{m+1}\|^2 + \sum_{m=0}^{M} k_{m+1}\|\nabla e^{m+1}\|^2$$

$$+ k_0 \frac{1}{\alpha}\|\mathcal{Q}^{M+1}\|^2 + (\frac{1}{\alpha} - \alpha k_0)k_0^3 \sum_{m=0}^{M} \|\tilde{d}_t \mathcal{Q}^{m+1}\|^2$$

$$+ \frac{1}{\alpha}k_0^2\|\nabla \mathcal{Q}^{M+1}\|^2 + (\frac{1}{\alpha} - \alpha k_0)k_0^4 \sum_{m=0}^{M} \|\nabla \tilde{d}_t \mathcal{Q}^{m+1}\|^2 \qquad (10.78)$$

$$\leq C k_0^2 (1 + \log\frac{1}{k_0}) + \|e^1\|^2 + \frac{1}{\alpha}\big\{ k_0\|\mathcal{Q}^1\|^2 + k_0^2\|\nabla \mathcal{Q}^1\|^2 \big\}.$$

Therefore, it is sufficient to study the initial error behavior of the modified Chorin-Uzawa scheme on the time-grid $\mathcal{G}_2(k_{m+1})$. In this context, we present the following statements for the error that is committed by the transition from the incompressible to the (implicitly) perturbed case,

$$\frac{1}{k_0}\|u^1 - u_\varepsilon^1\| + \|u^1 - u_\varepsilon^1\|_1 + \|p^1 - p_\varepsilon^1\| + \sqrt{k_0}\|p^1 - p_\varepsilon^1\|_1$$

$$\leq \|p^0\| + \|p^1\| + \sqrt{k_0}\big\{ \|\nabla p^0\| + \|\nabla p^1\| \big\} \leq C. \qquad (10.79)$$

As a consequence, this gives the following a-priori bounds,

$$\|p_\varepsilon^1\| + \sqrt{k_0}\|\nabla p_\varepsilon^1\| \leq C, \tag{10.80}$$

which will now be used in order to qualify the initial error behavior initiated by the transition from the fully implicit to the semi-implicit formulation. The governing equations in this case are

$$\frac{1}{k_0^2}e^1 - \Delta e^1 = -\nabla p_\varepsilon^1,$$
$$\text{div}e^1 + \frac{1}{\alpha}\mathcal{Q}^1 - \frac{1}{\alpha}k_0\Delta\mathcal{Q}^1 = 0. \tag{10.81}$$

Now, the following results can easily be established by means of vanishing initial errors for pressure as well as velocity, using the first equation for the derivation of a statement concerning the velocity error whereas the second identity can be employed to derive a corresponding result for the pressure error,

$$\sqrt{k_0}\|\mathcal{Q}^1\| + k_0\|\nabla\mathcal{Q}^1\| + \|e^1\| \leq Ck_0. \tag{10.82}$$

This can be inserted on the right hand side of (10.78), and the proof is complete.

10.4   A Revised Van Kan Scheme on the Time-Grid structure $\mathcal{G}_2(k_{m+1})$

This section is devoted to a construction as well as analysis of a robust modification of the original Van Kan scheme *on time-grids with structure* $\mathcal{G}_2(k_{m+1})$. As has been pointed out in earlier chapters, the original scheme of Van Kan on equi-distributed time-grids suffers from severe restrictive drawbacks: Firstly, *accurate* initial data for the pressure are needed in order to produce further accurate approximations $\{u^{m+1}, p^{m+1}\}$. Secondly, the fluid flow to be simulated *has* to satisfy nonlocal compatibility conditions in order to guarantee conservation of second order accuracy for the approximation. For general fluid flows, the Van Kan scheme will give approximations that are only of first order accurate over the whole time-interval. Let us recall that it was the motivation for introducing the multi-component Chorin-/Van Kan scheme to abolish these two main drawbacks.

However, also the latter modification to simulate the dynamical behavior of flows governed by the incompressible Navier-Stokes equations suffers from some restrictions. We mention the fact that second order accuracy can only be verified for (exponentially) stable flows, at times t_{m+1} that are of order $\mathcal{O}(Re)$. Moreover, we introduced a $\mathbf{J}_1$-projection at a time $t_{m_0} = \mathcal{O}(1)$, thus increasing the computational tools that are necessary to calculate the approximations $\{u^{m+1}, p^{m+1}\}$.

Now, the objective in this section is to develop a new modification of the Van Kan scheme on the basis of time-grids with structure $\mathcal{G}_2(k_{m+1})$. In a way that is related to the modification of the original Chorin-Uzawa scheme we will combine the ideas of higher order splitting and time-grid structure $\mathcal{G}_2(k_{m+1})$. The *related Van Kan scheme with a chosen parameter* $\beta \geq \frac{1}{2}$ *on* $\mathcal{G}_2(k_{m+1})$ is then as follows,

1. Start with $u^0 \approx u(0)$ and homogeneous pressure, $p^0 \equiv 0$.

2. Find an approximation of the actual velocity field, $\tilde{u}^{m+1} \in \mathbf{H}_0^1$, which satisfies

$$\frac{1}{k_{m+1}}\{\tilde{u}^{m+1} - u^m\} - \Delta\tilde{u}^{m+1/2} + (\frac{3}{2} - \beta)\frac{m}{m+1}\nabla p^m$$
$$+ (\beta - \frac{1}{2})\frac{m-1}{m+1}\nabla p^{m-1} = \overline{f}^{m+1/2}. \quad (10.83)$$

3. Projection-step: Determine $\{u^{m+1}, p^{m+1}\}$ to be the solution of

$$\frac{1}{k_{m+2}}\{u^{m+1} - \tilde{u}^{m+1}\} + \beta\nabla\{\frac{m+1}{m+2}p^{m+1} - \frac{m}{m+2}p^m\} = 0,$$
$$\operatorname{div} u^{m+1} = 0, \qquad u^{m+1}|_{\partial\Omega} \cdot n = 0. \quad (10.84)$$

The revised Van Kan scheme is presented for times $t_{m+1} \ll 1$. Of course, at times $t = \mathcal{O}(1)$ the scheme can be replaced by the original Va n Kan scheme, i.e., the coefficients depending on m can be set equal to 1. We have already discussed the point of how to modify the scheme in the initial step $(m = 0)$ in the case $\beta \neq \frac{1}{2}$, see Section 7.1. For the analytical point of view, this algorithm can be reformulated as a semi-explicit nonstationary quasi-compressibility method,

$$d_t \tilde{u}^{m+1} - \Delta \tilde{\tilde{u}}^{m+1/2} + \frac{1}{2} \nabla \left\{ 3 \frac{m}{m+1} p^m - \frac{m-1}{m+1} p^{m-1} \right\} = \bar{f}^{m+1/2},$$

$$\text{div} \tilde{u}^{m+1} - \beta k_0^2 \Delta \tilde{d}_t \mathcal{P}^{m+1} = 0, \qquad \beta \geq \frac{1}{2},$$

$$\partial_n \tilde{d}_t \mathcal{P}^{m+1}|_{\partial\Omega} = 0, \qquad \text{and} \qquad p^{m+1} = \frac{1}{(m+1)k_0} \mathcal{P}^{m+1}. \tag{10.85}$$

The main result from the analysis of the latter system are the following error statements.

Theorem 10.4 *Given the approximations $\{\tilde{u}^{m+1}, p^{m+1}\}$ as the solution of the modified Van Kan scheme (10.83), (10.84) on the time-grid $\mathcal{G}_2(k_{m+1})$, for $0 < k_0 < 1$. Let $\{u, p\}$ be the solution of the incompressible Stokes equations (2.1), and suppose the assumptions (A1) and (A2) to be satisfied. Then, there exists a constant $C \equiv C(\nu, \Omega, f, u_0, T)$ such that the following error statements are valid for "general fluid flows", in the case $\beta > \frac{1}{2}$,*

$$\max_{1 \leq m \leq M} \left\{ \|u(t_m) - \tilde{\tilde{u}}^m\| + k_0 \tau_{M+1}^{1/4} \|u(t_m) - \tilde{\tilde{u}}^m\|_1 \right.$$

$$\left. + k_0 \tau_{M+1}^{1/2} \|p(t_m) - \bar{\bar{p}}^{m-1/2}\| \right\} \leq C k_0^2 (1 + \log \frac{1}{k_0}).$$

For values $\beta = \frac{1}{2}$, the following error statement for the velocity field is valid, instead of the above one,

$$\max_{0 \leq m \leq M} \|u(t_{m-1/2}) - \tilde{\tilde{u}}^{m-1/2}\| \leq C k_0^2 (1 + \log \frac{1}{k_0}).$$

The remainder of this section is devoted to present a sketch of the proof of this theorem, using ingredients that are already studied in the previous section that is dealing with the revised Chorin-Uzawa method. Therefore, we will concentrate on presenting the main features of inherent error sources, referring to analogous studies that have been executed in the contexts of the original Van Kan scheme. On the other hand, the robustness of the algorithm is not that evident, also not when the grid structure $\mathcal{G}_2(k_{m+1})$ is in use, which is owing to the higher degree of complexity of the algorithm. This is the reason why we have to consider several facts carefully that seem to be evident — owing to previous investigations —, to understand the interplay of the algorithm with the grid structure.

10.4.1   Stability Results for the Incompressible Stokes Problem

Proceeding in the common way, the introductory problem of interest are the equations,

$$d_t u^{m+1} - \Delta \bar{u}^{m+1/2} + \frac{1}{2}\nabla\{p^{m+1} + \frac{m}{m+1}p^m\} = \bar{f}^{m+1/2},$$

$$\text{div} u^{m+1} = 0.$$

(10.86)

Let us start with the derivation of striking a-priori bounds for the related solution. The following bounds are by means of standard arguments,

$$\left(\sum_{m=0}^{M} k_{m+1}\|\nabla d_t u^{m+1}\|^2\right)^{1/2} + \max_{0\le m\le M}\left\{\|\Delta u^{m+1}\| + \|d_t u^{m+1}\|\right\} \le C.$$

(10.87)

For further results, we consider the equations

$$\tilde{d}_t u^{m+1} + (m+1)k_0 A\bar{u}^{m+1/2} = (m+1)k_0 P_{J_0}\bar{f}^{m+1/2}$$

and

$$\tilde{d}_t^2 u^{m+1} + mk_0 A\tilde{d}_t\bar{u}^{m+1/2} + A\bar{u}^{m+1/2}$$
$$= mk_0\tilde{d}_t P_{J_0}\bar{f}^{m+1/2} + P_{J_0}\bar{f}^{m+1/2}.$$

(10.88)

Based on these formulations and another derivation form of it, the following results are valid, setting $\mathcal{P}^{m+1} \equiv (m+1)k_0 p^{m+1}$,

$$\max_{1\le m\le M}\left\{\|\nabla \tilde{d}_t u^{m+1}\| + \|\tilde{d}_t^2 u^{m+1}\| + \|A\tilde{d}_t(\sqrt{\tau_{m+1}}\bar{u}^{m+1/2})\|\right.$$
$$\left. + \|\nabla\tilde{d}_t\overline{\mathcal{P}}^{m+1/2}\|\right\} + \left(\sum_{m=2}^{M} k_{m+1}\|\nabla\tilde{d}_t^2\overline{\mathcal{P}}^m\|^2\right)^{1/2} \le C.$$

(10.89)

Note the necessity of bounding *averaged* functions for higher order time derivatives. This is owing to the stretched nature of the time-grid. We will omit the proof of all statements, only proving the last relation. For subsequent use, let us recall the elementary identity

$$(a+b,a) = \frac{1}{2}\|a+b\|^2 + \frac{1}{2}\{\|a\|^2 - \|b\|^2\}.$$

(10.90)

We are starting from the identity

$$\tilde{d}_t^3 \overline{u}^{m+1/2} + (m-1)k_0 A \tilde{d}_t^2 \overline{\overline{u}}^m + k_0 A \tilde{d}_t^2 \overline{u}^{m+1/2} + 2A\tilde{d}_t \overline{\overline{u}}^m$$

$$= \tilde{d}_t^2 \left((m+1)k_0 P_{\mathbf{J}_0} \overline{f}^{m+1/2}\right) + \tilde{d}_t^2 \left(mk_0 P_{\mathbf{J}_0} \overline{f}^{m-1/2}\right). \tag{10.91}$$

This equation will be tested with $\tau_{m+1} A \tilde{d}_t^2 \overline{\overline{u}}^m$. Using an easy calculation, the result given in (10.89) is furnished. At this place, we omit presenting the technical details.

The stability results presented in this subsection are the basis for the perturbation analysis that is devoted in order to detect the error mechanisms of the revised Van Kan scheme. As the main result of the actual subsection we have verified that the crucial function $m \mapsto \nabla \tilde{d}_t \overline{\mathcal{P}}^{m+1/2}$ that determines the stability features of the projection scheme is now controllable globally and pointwise in time — owing to the time-grid structure $\mathcal{G}_2(k_{m+1})$ employed.

The second part of the verification of Theorem 10.4 is now devoted to the error mechanisms stemming from the nonstationary quasi-compressibility constraint. As a result, we will see that the perturbations given in the algebraic constraint can now be handled with on the employed time-grid. We state the equations that will be considered in the subsequent studies: For the time-grid $\mathcal{G}_2(k_{m+1})$, find the solution $\{u_\varepsilon^{m+1}, p_\varepsilon^{m+1}\}$ of the equations

$$d_t u_\varepsilon^{m+1} - \Delta \overline{u}_\varepsilon^{m+1/2} + \frac{1}{2}\nabla\{p_\varepsilon^{m+1} + \frac{m}{m+1}p_\varepsilon^m\} = \overline{f}^{m+1/2},$$

$$\operatorname{div}\overline{u}_\varepsilon^{m+1} - \varepsilon \Delta \tilde{d}_t \mathcal{P}_\varepsilon^{m+1} = 0, \qquad \varepsilon > 0, \tag{10.92}$$

$$\partial_n \tilde{d}_t \mathcal{P}_\varepsilon^{m+1}|_{\partial\Omega} = 0,$$

with initial data $u_\varepsilon^0 \equiv u_0$ and $p_\varepsilon^0 \equiv 0$. We recall the applied notation for the time-damped pressure function $\mathcal{P}_\varepsilon^{m+1} := (m+1)k_0 p_\varepsilon^{m+1}$. — As in earlier studies, the analysis of this system decouples in certain auxiliary steps that are concerned with the perturbation and the evolutionary error effects of the second and the first equation as well. The subsequent part of the proof is now devoted to the study of "the stationary" auxiliary problem.

10.4.2   Error Statements for an Auxiliary Problem

For the investigation of the error mechanisms caused by the nonstationary quasi-compressibility constraint, let us introduce the following auxiliary problem on the time-grid $\mathcal{G}_2(k_{m+1})$,

$$
\begin{aligned}
& - (m+1)k_0 \Delta \overline{U}_\varepsilon^{m+1/2} + \nabla \overline{\mathcal{B}}_\varepsilon^{m+1/2} = \tilde{d}_t u^{m+1}, \\
& \operatorname{div} U_\varepsilon^{m+1} - \varepsilon \Delta \tilde{d}_t \mathcal{B}_\varepsilon^{m+1} = 0, \\
& \partial_n \tilde{d}_t \mathcal{B}_\varepsilon^{m+1}|_{\partial\Omega} = 0,
\end{aligned}
\tag{10.93}
$$

together with the initial data $U_0^\varepsilon \equiv u_0$ and $P_0^\varepsilon \equiv 0$. Again, we employed the abbreviative notation $\mathcal{B}_\varepsilon^{m+1} \equiv (m+1)k_0 P_\varepsilon^{m+1}$. The study of the errors that are introduced by the perturbation of the incompressibility constraint starts from the following error identities,

$$
\begin{aligned}
& - (m+1)k_0 \Delta \overline{E}^{m+1/2} + \nabla \overline{\mathcal{Q}}^{m+1/2} = 0, \\
& \operatorname{div} E^{m+1} - \varepsilon \Delta \tilde{d}_t \mathcal{Q}^{m+1} = -\varepsilon \Delta \tilde{d}_t \mathcal{P}^{m+1}, \\
& \partial_n \tilde{d}_t \mathcal{Q}^{m+1}|_{\partial\Omega} = \partial_n \tilde{d}_t \mathcal{P}^{m+1}|_{\partial\Omega}.
\end{aligned}
\tag{10.94}
$$

Here, we have used the abbreviative notation $E^{m+1} := u^{m+1} - U_\varepsilon^{m+1}$, $\Pi^{m+1} := p^{m+1} - P_\varepsilon^{m+1}$ and $\mathcal{Q}^{m+1} = (m+1)k_0 \Pi^{m+1}$. Before starting with the main error analysis, we consider the initial error behavior, setting $m = 0$:

$$
\begin{aligned}
& - \Delta E^1 + \nabla \Pi^1 = 0, \\
& \operatorname{div} E^1 - \varepsilon \Delta \Pi^1 = -\varepsilon \Delta p^1, \\
& \partial_n \Pi^1|_{\partial\Omega} = \partial_n p^1|_{\partial\Omega}.
\end{aligned}
\tag{10.95}
$$

This gives

$$
\|E^1\|_1 + \sqrt{\varepsilon}\|\nabla \Pi^1\| \leq \sqrt{\varepsilon}\|\nabla p^1\|.
\tag{10.96}
$$

Apart from the approximation contents, the latter result can be applied in order to verify the bound

$$
\|\nabla P_\varepsilon^1\| \leq C.
\tag{10.97}
$$

Now, we are in a position to verify the following lemma.

Lemma 10.4 Let $\{U_\varepsilon^{m+1}, \mathcal{B}_\varepsilon^{m+1}\}$ be the solution of system (10.93), whereas the tuple $\{u^{m+1}, \mathcal{P}^{m+1}\}$ is determined by the system (10.86). We assume (A1), (A2) to be satisfied, and let $\varepsilon = \mathcal{O}(k_0^2)$. Then, the following error estimates are valid, with a constant C that depends on the given data of the problem (2.1), and $0 < k_0 < 1$,

$$\max_{0 \le m \le M} \{ \| \overline{u}^{m+1/2} - \overline{U}_\varepsilon^{m+1/2} \| + \sqrt{\varepsilon} \tau_{M+1}^{1/4} (\| \overline{u}^{m+1/2} - \overline{U}_\varepsilon^{m+1/2} \|_1$$

$$+ \sqrt{\varepsilon} \| \overline{p}^{m+1/2} - \overline{P}_\varepsilon^{m+1/2} \|) \}$$

$$+ \left(\sum_{m=2}^{M} k_{m+1} \| \tilde{d}_t \{ \overline{\overline{\tilde{u}}}^m - \overline{\overline{U}}_\varepsilon^m \} \|^2 \right)^{1/2} \le C\varepsilon (1 + \log \frac{1}{k_0}).$$

Remark 10.5 *As will be shown in the subsequent proof, the statements for the velocity error, evaluated in the Dirichlet-norm, and the one for the pressure error as well can be sharpened by getting rid of the logarithmic term. Nevertheless, the further analyses on the way to verify Theorem 10.4 will imply logarithmic terms also for these sizes, so we will not focus on this actual improvement here.*

Proof:
From elementary considerations, we are led to the bound

$$k_0 \sum_{m=0}^{M} \| \nabla \overline{E}^{m+1/2} \|^2 + \frac{\varepsilon}{\sqrt{\tau_{m+1}}} \| \nabla \overline{Q}^{M+1/2} \|^2 \le C\varepsilon. \tag{10.98}$$

The bound on the right hand side incorporates initial terms that can be verified to be of the same magnitude. — In order to verify a statement for the velocity error that is pointwise in time we have to use the differentiated form,

$$- m k_0 \Delta \tilde{d}_t \overline{E}^{m+1/2} - \Delta \overline{E}^{m+1/2} + \nabla \tilde{d}_t \overline{Q}^{m+1/2} = 0. \tag{10.99}$$

This gives a pointwise error estimate for the velocity field and a first striking a-priori statement for the pressure function of (10.93),

$$\tau_{M+1}^{1/4} \{ \| \nabla \overline{E}^{M+1/2} \| + \| \overline{\Pi}^{M+1/2} \| + \sqrt{\varepsilon} \| \nabla \tilde{d}_t \overline{Q}^{M+1/2} \| \}$$

$$+ \sqrt{\varepsilon} \left(k_0 \sum_{m=1}^{M} \| \nabla \tilde{d}_t \overline{Q}^{m+1/2} \|^2 \right)^{1/2} \le C\sqrt{\varepsilon}. \tag{10.100}$$

The latter result for the error in the pressure function, $\overline{\Pi}^{M+1/2}$, is by means of the first equation in (10.94), using a stability property of the div-operator. In the following, we intend to improve these a-priori statements for the pressure of (10.93) by annihilation of the time-weight factors. This can be done for the *averaged* quantity at the expense of a logarithmic factor, reflecting initial instability features. Therefore, application of the averaging operator onto the equation (10.99) leads to

$$-(m-1)k_0\Delta\tilde{d_t}\overline{\overline{E}}^m - \frac{1}{2}k_0\Delta\tilde{d_t}\overline{E}^{m+1/2}$$
$$-\Delta\overline{\overline{E}}^m + \nabla\tilde{d_t}\overline{\overline{Q}}^m = 0. \tag{10.101}$$

This identity will be tested with $\tilde{d_t}\overline{\overline{E}}^m$. Owing to (10.90), we can proceed as follows,

$$\sum_{m=0}^{M} k_{m+1}\|\nabla\tilde{d_t}\overline{\overline{E}}^m\|^2 + k_0^2\|\nabla\tilde{d_t}\overline{E}^{M+1/2}\|^2 + \varepsilon\|\nabla\tilde{d_t}\overline{\overline{Q}}^M\|^2$$
$$\leq C(1 + t_M + \log\frac{1}{k_0})\varepsilon \sum_{m=2}^{M} k_{m+1}\|\nabla\tilde{d_t^2}\overline{\overline{P}}^m\|^2$$
$$+ (1 + t_M + \log\frac{1}{k_0})^{-1}\varepsilon \sum_{m=2}^{M} \frac{1}{m+1}\|\nabla\tilde{d_t}\overline{\overline{Q}}^m\|^2 + C\varepsilon. \tag{10.102}$$

The last term here reflects the errors stemming from the initial time. It can be verified by an easy consideration that employs (10.95) and (10.96). — Now, we can apply Gronwall's lemma, leading to the desired result

$$\max_{0\leq m\leq M} \|\nabla\tilde{d_t}\overline{\overline{B}}_\varepsilon^m\| \leq C\{1 + \log\frac{1}{k_0}\}. \tag{10.103}$$

For our further investigations, this result is not sufficient, and we need the following auxiliary estimate: We test (10.101) with $\tilde{d_t}\overline{E}^{m+1/2}$ and afterwards sum over all iteration steps. This finally leads us to the following estimate, using integration by parts,

$$\frac{1}{4}\sum_{m=2}^{M}(m-1)k_{m+1}\|\nabla\tilde{d}_{t}\overline{\overline{E}}^{m}\|^{2} + \frac{1}{4}k_{0}^{2}\sum_{m=2}^{M}\|\nabla\tilde{d}_{t}\overline{E}^{m+1/2}\|^{2}$$

$$+\frac{1}{4}k_{0}\sqrt{\tau_{M+1}}\|\nabla\overline{E}^{M+1/2}\|^{2} + \varepsilon\|\nabla\tilde{d}_{t}\overline{\overline{\mathcal{Q}}}^{M}\|^{2}$$

$$\leq C\gamma\varepsilon\Big\{\|\nabla\tilde{d}_{t}\overline{\mathcal{P}}^{M+1/2}\|^{2} + \sum_{m=2}^{M}\frac{1}{m+1}\|\nabla\tilde{d}_{t}\overline{\mathcal{P}}^{m+1/2}\|^{2}\Big\}$$

$$+\frac{1}{\gamma}\varepsilon\sum_{m=2}^{M}k_{m+1}\|\nabla\tilde{d}_{t}^{2}\overline{\overline{\mathcal{Q}}}^{m}\|^{2}$$

$$\leq C\gamma\varepsilon\Big\{1+\log\frac{1}{k_{0}}\Big\} + \frac{1}{\gamma}\varepsilon\sum_{m=2}^{M}k_{m+1}\|\nabla\tilde{d}_{t}^{2}\overline{\overline{\mathcal{Q}}}^{m}\|^{2},$$

(10.104)

with $\gamma > 1$. This estimate will be used in the following consideration. For the following, it is important to have control over the second time-derivatives of the pressure function. In that respect, it is important to consider the following equations (in averaged form!),

$$-(m-2)k_{0}\Delta\tilde{d}_{t}^{2}\overline{\overline{E}}^{m} - \frac{1}{2}k_{0}\Delta\tilde{d}_{t}^{2}\overline{E}^{m+1/2}$$

$$-2\Delta\tilde{d}_{t}\overline{\overline{E}}^{m} + \nabla\tilde{d}_{t}^{2}\overline{\overline{\mathcal{Q}}}^{m} = 0.$$

(10.105)

Testing this equation with $(m+1)k_{0}\tilde{d}_{t}\overline{\overline{E}}^{m}$ before starting the summation over all iteration steps gives

$$\tau_{M+1}\|\nabla\tilde{d}_{t}\overline{\overline{E}}^{M}\|^{2} + k_{0}\sqrt{\tau_{M+1}}\|\nabla\tilde{d}_{t}\overline{E}^{M+1/2}\|^{2}$$

$$+\varepsilon\sum_{m=2}^{M}k_{m+1}\|\nabla\tilde{d}_{t}^{2}\overline{\overline{\mathcal{Q}}}^{m}\|^{2}$$

$$\leq C\varepsilon\{1+\log\frac{1}{k_{0}}\} + k_{0}\sum_{m=2}^{M}\|\nabla\tilde{d}_{t}\overline{E}^{m+1/2}\|^{2}.$$

(10.106)

Now, in order to control the last term, we can apply (10.104), choosing γ sufficiently large such that terms can be absorbed on the left hand side. — In the end, we obtain the second striking stability result with respect to the solution behavior of the pressure function,

$$\Big(\sum_{m=2}^{M}k_{m+1}\|\nabla\tilde{d}_{t}^{2}\overline{\overline{B}}_{\varepsilon}^{m}\|^{2}\Big)^{1/2} \leq C\{1+\log\frac{1}{k_{0}}\}.$$

(10.107)

Owing to the a-priori bounds (10.103) and (10.107), we are provided with sharp statements with respect to the dynamical behavior of the pressure function as the solution of (10.93), leading to sharp error bounds for the velocity field U_ε^{m+1}. For that purpose, we consider a dual incompressible Stokes problem of type: Given the triple of functions $\{W^m, R^m, g^{m+1/2}\}$, find the solution $\{W^{m+1}, R^{m+1}\}$ of the system of equations

$$- \Delta \overline{W}^{m+1/2} + \nabla \overline{R}^{m+1/2} = \overline{g}^{m+1/2},$$

$$\mathrm{div} W^{m+1} = 0, \hspace{5cm} (10.108)$$

$$W^0 = 0, \qquad R^0 = 0.$$

If we set $g^{m+1} = E^{m+1}$, apply the averaging operator to the first equation in (10.108), and finally test it with $\overline{\overline{E}}^m$, the scaling invariance of the incompressible problem leads us to

$$\|\overline{\overline{E}}^m\| \leq C\varepsilon \|\nabla \tilde{d}_t \overline{\overline{B}}_\varepsilon^m\| \leq C\varepsilon \{1 + \log \frac{1}{k_0}\}.$$

Correspondingly, owing to (10.107) we arrive at

$$\left(\sum_{m=2}^{M} k_{m+1} \|\tilde{d}_t \overline{\overline{E}}^m\|^2\right)^{1/2} \leq C\varepsilon \left(\sum_{m=2}^{M} k_{m+1} \|\nabla \tilde{d}_t^2 \overline{\overline{B}}_\varepsilon^m\|^2\right)^{1/2}$$

$$\leq C\varepsilon \{1 + \log \frac{1}{k_0}\}, \hspace{3cm} (10.109)$$

and the lemma is verified. $\hspace{8cm}$ □

Provided with the statement (10.109), we can proceed in our proof of Theorem 10.4 by analyzing the evolutionary error effects caused by the momentum equation. This will be done in the subsequent subsection.

10.4.3 The Propagation of Error Effects

This part of the proof is related to the following system of error equations, using the "common" notations $e_\varepsilon^{m+1} := u_\varepsilon^{m+1} - U_\varepsilon^{m+1}$, $\eta_\varepsilon^{m+1} := p_\varepsilon^{m+1} - P_\varepsilon^{m+1}$ and $Q_\varepsilon^{m+1} := (m+1) k_0 \eta_\varepsilon^{m+1}$,

$$\tilde{d}_t \overline{e}_\varepsilon^m - (m-1) k_0 \Delta \overline{e}_\varepsilon^{m-1/2} - k_0 \Delta \overline{e}_\varepsilon^m + \nabla \overline{\overline{Q}}_\varepsilon^{m-1/2} = \tilde{d}_t \overline{E}^m,$$

$$\mathrm{div} \overline{e}_\varepsilon^m - \varepsilon \Delta \tilde{d}_t \overline{\overline{Q}}_\varepsilon^m = 0,$$

$$\partial_n \tilde{d}_t \overline{\overline{Q}}_\varepsilon^m |_{\partial\Omega} = 0, \hspace{4cm} (10.110)$$

and initial data, with $\|\overline{\overline{e}}_\varepsilon^1\| \le C\sqrt{\varepsilon}k_0$ and $\|\overline{\overline{Q}}_\varepsilon^1\| \le Ck_0$. — We are provided with accurate initial data, so that we are allowed to test the first equation with $\overline{\overline{e}}_\varepsilon^{m-1/2}$. Summation over all time-steps then gives

$$\|\overline{\overline{e}}_\varepsilon^M\|^2 + \sum_{m=2}^m k_{m+1}\|\nabla\overline{\overline{e}}_\varepsilon^{m-1/2}\|^2 + k_0^2\|\nabla\overline{\overline{e}}_\varepsilon^M\|^2 + \varepsilon\|\nabla\overline{\overline{Q}}_\varepsilon^{M-1/2}\|^2$$

$$\le \varepsilon\|\nabla\overline{\overline{Q}}_\varepsilon^{2-1/2}\|^2 + k_0^2\|\nabla\overline{\overline{e}}_\varepsilon^2\|^2$$

$$+ C(1 + t_M + \log\frac{1}{k_0}) \sum_{m=2}^M k_{m+1}\|\tilde{d}_t\overline{\overline{E}}^m\|^2 \qquad (10.111)$$

$$+ \frac{1}{4}(1 + t_M + \log\frac{1}{k_0})^{-1} \sum_{m=2}^M \frac{1}{m+1}\|\overline{\overline{e}}_\varepsilon^m\|^2.$$

We omit the elementary considerations that verify the bound $C\varepsilon k_0^2$ for the initial errors on the right hand side of this inequality. Further, the normalized version of Gronwall's lemma furnishes bounding the right hand side by

$$\le C\varepsilon\Big\{k_0^2 + (1 + \log\frac{1}{k_0})\varepsilon\Big\}. \qquad (10.112)$$

Now, we are in a position to present optimal error estimates for the system (10.92) with solutions $\{u_\varepsilon^{m+1}, p_\varepsilon^{m+1}\}$, under application of the grid structure $\mathcal{G}_2(k_{m+1})$. As a recapitulation of the previous considerations, we have verified the following error statements that are valid for general flows and an arbitrary initial pressure function p_ε^0,

$$\max_{0\le m\le M}\Big\{\|\overline{\overline{u}}^m - \overline{\overline{u}}_\varepsilon^m\| + \min\{\sqrt{\varepsilon}\tau_{M+1}^{1/4}, k_0\}\|\overline{\overline{u}}^m - \overline{\overline{u}}_\varepsilon^m\|_1$$

$$+ \sqrt{\varepsilon}\tau_{M+1}^{1/2}\|\overline{\overline{p}}^{m-1/2} - \overline{\overline{P}}_\varepsilon^{m-1/2}\|\Big\} \qquad (10.113)$$

$$\le C\sqrt{\varepsilon}\Big\{k_0 + \sqrt{\varepsilon}(1 + \log\frac{1}{k_0})\Big\}.$$

In order to warrant the conservation of this error behavior if we pass from the implicit version of this nonstationary quasi-compressibility method (10.92) to a semi-explicit one (i.e., the revised Van Kan scheme), we have to assure that the solution of the implicit version exhibits a sufficient amount of stability. This means that we have to ascertain the following a-priori bound for the pressure function,

$$k_0\Big(\sum_{m=0}^M \|\nabla\tilde{d}_t^2\overline{\overline{P}}_\varepsilon^m\|^2\Big)^{1/2} \le C(1 + \log\frac{1}{k_0}). \qquad (10.114)$$

This can be seen by application of the following error equations, using the notations $e^{m+1} := u^{m+1} - u_\varepsilon^{m+1}$ and $\mathcal{K}^{m+1} := \mathcal{P}^{m+1} - \mathcal{P}_\varepsilon^{m+1}$,

$$\tilde{d}_t \bar{e}^{m+1/2} - mk_0 \Delta \bar{e}^m - \frac{1}{2}k_0 \Delta \bar{e}^{m+1/2} + \nabla \overline{\overline{\mathcal{K}}}^m = 0,$$

$$\text{div}\bar{e}^m - \varepsilon \Delta \tilde{d}_t \overline{\overline{\mathcal{K}}}^m = -\varepsilon \Delta \tilde{d}_t \overline{\overline{\mathcal{P}}}^m, \qquad\qquad (10.115)$$

$$\partial_n \tilde{d}_t \overline{\overline{\mathcal{K}}}^m |_{\partial\Omega} = \partial_n \tilde{d}_t \overline{\overline{\mathcal{P}}}^m |_{\partial\Omega}.$$

Additionally, we have $\|e^0\| = \mathcal{O}(\varepsilon)$. Then, the further proceeding to establish the inequality (10.114) corresponds to verifying part $ii)$ of Lemma 7.5 in Chapter 7.

Provided with the error statements (10.113) and the stability result (10.114) it is now possible to quantify the error that is committed by the transfer from the fully implicit version (10.92) of the nonstationary quasi-compressibility method to the semi-explicit version (10.85), with the underlying grid structure $\mathcal{G}_2(k_{m+1})$. This is subject of the following subsection.

10.4.4   The Semi-Explicit Quasi-Compressibility Method

The governing equations describing the transition from the fully implicit scheme (10.92) to the semi-explicit one (10.85) are the following (with setting $\varepsilon = \beta k_0^2$),

$$\tilde{d}_t e_{VK}^{m+1} - (m+1)k_0 \Delta \bar{e}_{VK}^{m+1/2} + \nabla \overline{\mathcal{Q}}_{VK}^{m+1/2} =$$

$$= \frac{1}{2}k_0^2 \nabla \tilde{d}_t^2 \mathcal{Q}_{VK}^{m+1} - \frac{1}{2}k_0^2 \nabla \tilde{d}_t^2 P_{\beta k_0^2}^{m+1},$$

$$\text{div}e_{VK}^{m+1} - \beta k_0^2 \Delta \tilde{d}_t \mathcal{Q}_{VK}^{m+1} = 0, \qquad \partial_n \tilde{d}_t \mathcal{Q}_{VK}^{m+1}|_{\partial\Omega} = 0, \qquad (10.116)$$

together with $e_{VK}^0 \equiv 0$ and $\eta_{VK}^0 \equiv 0$. — Again, we employed the notations $e_{VK}^{m+1} := u_{\beta k_0^2}^{m+1} - \tilde{u}_{VK}^{m+1}$, $\eta_{VK}^{m+1} := p_{\beta k_0^2}^{m+1} - p_{VK}^{m+1}$ and $\mathcal{Q}_{VK}^{m+1} := (m+1)k_0 \eta_{VK}^{m+1}$.

Owing to the preparatory work of the preceding subsections, we can now proceed in a fashion analogous to the one presented in Section 7.3 for the original Van Kan scheme. Of course, deviating from the investigations in the specified section, we presently need twice the application of the averaging operator onto the first equation in (10.116) (for the case $\beta \neq \frac{1}{2}$) or three times (for the case $\beta = \frac{1}{2}$), respectively, owing to (10.114). Owing to the fact that no new considerations are necessary any more to accomplish the proof of Theorem 10.4, we leave the elaboration of the given sketchy arguments as well as the discussion of the initial error behavior of the related functions to the interested reader.

Chapter 11

Summary and Outlook

The subject of this book is twofold: On one hand, we intend to present optimal error estimates for some well-known projection schemes (Chorin, the modification of Timmermans et al., Van Kan) and quasi-compressibility methods (penalty, artificial compressibility, pressure stabilization, pressure correction) that have not been available before. These numerical ansatzes are widely used in existing computer codes, because of their economical employment of computational resources and their easy implementation.

The second aim is the detection and study of the error mechanisms that are inherent to these schemes (boundary layers, the restriction to compatible flows) and the third one the construction of new optimized schemes that do not suffer from the drawbacks any more (e.g., methods using some damping time-weights in the quasi-compressibility constraint, multi-component schemes, projection schemes on certain time-grid structures).

Both groups of numerical methods, the quasi-compressibility methods and the projection schemes, have mainly been introduced for two reasons: to reduce the computational solution effort in a significant way and to circumvent the restrictive *LBB-condition* with respect to the choice of the stable tuple of discrete ansatz spaces for velocity field and pressure function. By means of re-interpreting the projection schemes as semi-explicit quasi-compressibility methods (in semi-discretized form of course), we have found a common basis for analyzing these different schemes. Out of this, we were able to detect distinct error effects that dominate the approximation properties of the iterates and to recognize the difficulties that pressure correction projection schemes like the one of Van Kan may have in case of incompatibly posed fluid flows.

For the many different schemes under consideration — quasi-compressibility schemes and projection schemes as well — we distinguished between the *stationary* (pressure stabilization, penalty; Chorin) and *nonstationary* (artificial compressibility, pressure correction; Van Kan, Chorin-Uzawa) quasi-compressibility methods, giving different accuracies of the iterates and stability features. One main subject was the focus on the stability problems arising for nonstationary quasi-compressibility methods or the Van Kan scheme — supported by numerical experiments.

As a second step, we constructed new projection schemes that do not suffer from a global reduction of the order of convergence in case of incompatibly posed fluid flow problems. A first idea to cope with the separated phases of different stability properties of the fluid flow was based on the application of different projection schemes for different time intervals. These so-called *multi-component schemes* possess better stability properties, but, in some aspect they are not quite satisfying for general flows at all. Therefore, we proposed another idea to enable the extension of pressure correction projection schemes to general fluid flows. It is presented in Chapter 10, using certain *time-grid structures* that improve the stability features of the resulting projection scheme. Again, the construction of the diverse time-grid structures presented there is oriented at the different stability properties of the solution in time of the continuous flow.

As we have already hinted at before, the extension of the former analyses of the diverse schemes to full space-time discretization is easily possible. In here, we only mention the decreased stability properties of a chosen tuple of ansatz spaces that does not satisfy the *LBB-condition* in case of nonstationary quasi-compressibility methods or pressure correction projection schemes like the schemes of Van Kan or the Chorin-Uzawa method. The idea of improving the stability by means of *mixed ansatzes* is elaborated on in Chapter 5.

The remainder of the chapter is devoted to the presentation of two outlooks that underline the deep connection between projection schemes and quasi-compressibility methods again, for the construction of new schemes and the analysis of existing ones as well. We start with the proposal of a composed projection scheme as a candidate to improve the accuracy of the pressure iterate to higher order. The presentation has to be sketchy, as the following points are not based on rigorous proofs.

A. A Higher Order Perturbation Ansatz:

Start with the revised Van Kan scheme to get iterates $\{\tilde{u}^{m_0}, u^{m_0}, p^{m_0}\}$ at a time $t_{m_0} = \mathcal{O}(1)$ that possess the following approximation properties,

$$\|u(t_{m_0}) - \overline{\tilde{u}}^{m_0}\| + k\|p(t_{m_0}) - \overline{\overline{p}}^{m_0-1/2}\| \leq Ck^2(1 + \log\frac{1}{k}), \tag{11.1}$$

with the identification $k = \mathcal{O}(k_0)$. — Now, we will give a family of projection schemes, each depending on the number of previous guesses. If we use the integer parameter $\ell \geq 1$ and define the algebraic sum

$$k^\ell \Phi(u^{m+1}, \tilde{u}^{m+1}; \ell) := u^{m+1} - \ell\tilde{u}^{m+1} + \sum_{i=2}^{\ell} (-1)^i \binom{\ell}{i} u^{m+2-i}, \tag{11.2}$$

the iterates $\{u^{m+1}, p^{m+1}\}\big|_{m \geq m_0}$ are determined as follows, using a uniform time-mesh that is parameterized by the number k,

1. Determine $\tilde{u}^{m+1} \in \mathbf{H}_0^1$ from

$$\frac{1}{k}\{\tilde{u}^{m+1} - \tilde{u}^m\} - k^{\ell-1}\Phi(u^m, \tilde{u}^m; \ell)$$
$$+ \frac{1}{2}(3 - k^{\ell-1})\nabla p^m - \frac{1}{2}(1 - k^{\ell-1})\nabla p^{m-1} = \overline{f}^{m+1/2}. \tag{11.3}$$

2. The subsequent projection step determines the tuple $\{u^m, p^m\}$ as the solution of

$$\Phi(u^{m+1}, \tilde{u}^{m+1}; \ell) + \frac{1}{2}\nabla\{p^{m+1} - p^m\} = 0,$$
$$\operatorname{div} u^{m+1} = 0, \qquad u^{m+1}\big|_{\partial\Omega} \cdot n = 0. \tag{11.4}$$

In order to start our analysis of the schemes that are parameterized by ℓ, we reinterpret them as a semi-explicit quasi-compressibility method, leading to the following constraint,

$$\operatorname{div}\tilde{u}^{m+1} - \frac{1}{2\ell}k^{\ell+1}\Delta d_t p^{m+1} = 0, \qquad \partial_n d_t p^{m+1}\big|_{\partial\Omega} = 0. \tag{11.5}$$

From this, essential parts for the analysis of scheme (11.3), (11.4) can be taken from the one given for the original Van Kan scheme in Chapter 7, and we expect the result

$$\max_{m_1 \leq m \leq M}\left\{\|u(t_m) - \overline{\tilde{u}}^m\| + k^{\psi(\ell)}\|p(t_{m+1/2}) - \overline{\overline{p}}^m\|\right\} \leq Ck^2\log\frac{1}{k}, \tag{11.6}$$

with $t_{m_1} \geq t_{m_0}$, where $|t_{m_1} - t_{m_0}| = \mathcal{O}(1)$, and the notation $\psi(\ell) = \max\{0, 2 - \frac{\ell+1}{2}\}$. We see that the choice $\ell = 3$ is even sufficient to obtain pressure iterates that converge of second order, pointwise in time.

Nevertheless, we emphasize that the error estimate (11.6) has to be considered as a conjecture. For proving it, we have to analyze certain damping properties of the scheme (11.3), (11.4) with respect to perturbed initial pressure data. Notice that the estimate (11.1) only assures first order convergence for the "iterate" $\overline{\overline{p}}^{m_0 - 1/2}$ as the initial pressure function. Therefore, the verification or rejection of this conjecture has to be left as an open question.

The previous considerations have led to the construction of a "formally" second order projection scheme with inherent boundary layers to the pressure approximation that vanish of the *same* order in the time-step k as the (consistent) order of discretization of the scheme. As a conclusion, the conservation as a second order scheme for both, velocity and pressure can be assured by introducing higher order perturbations to the incompressibility constraint. In the present case, we found a fourth order nonstationary perturbation to be sufficient.

In contrast to these considerations, giving rise to boundary layer structures that exhibit no striking influence on the global approximation properties of the pressure function, we can also combine the presented idea with a second order time-discretization approach to develop modifications of the Chorin-Uzawa scheme and its diverse modifications, getting second order of accuracy even for the pressure. In that aspect, we recall that the accuracy of the pressure function in the original Chorin-Uzawa scheme is now free from boundary layers, and its pointwise in time features depend merely on the nonstationary behavior of the continuous flow (up to the discretization error). For our purposes, let us come back to the modified Chorin-Uzawa scheme, that provides us with first order accuracy for the pressure-function. In the context of the above discussion, we can now construct a *second* order perturbation of the incompressibility constraint (together with a second order discretization of the momentum equation) giving second order accuracy for both velocity as well as pressure function. Note that this order of perturbing the incompressibility constraint is quite the same as the proclaimed order of convergence of the whole scheme, which is due to the regular character of perturbation. We leave the elaboration of these sketchy ideas to the interested reader.

B. The Fractional-θ-Step Scheme of Glowinski:

Glowinski's version of the "classical" Fractional-θ-Step method, applied to the incompressible Navier-Stokes equations is a special splitting scheme, separating the difficulties stemming from the incompressibility constraint from those arising from treating the nonlinearity in the solution process. This leads to an effective numerical scheme, which combines excellent accuracy properties with low computational costs. Despite the wide-spread application in computer codes, the underlying mathematical ingredients are not quite clear, and existing error analysis only confirms convergence of the scheme (see [22]) or (suboptimal) first order convergence (see [25]) — in contrast to computational observations. Here, the goal is to propose another analytical approach which is supposed to lead to a better understanding of the errors that dominate the scheme apart from the discretization errors that have already been investigated for the classical Fractional-θ-Step method.

In order to fix ideas, we start by recalling the scheme (cf. also [10]). The computation of the actual solution $\{u_{GL}^{m+1}, p_{GL}^{m+1}\}$ is decoupled into three sub-steps in each iteration step, with parameters $\theta = 1 - \sqrt{2}/2$, $\theta' = 1 - 2\theta$, and the free one $\alpha \in (1/2, 1]$, $\beta = 1 - \alpha$,

$$\frac{1}{\theta k}\{u_{GL}^{m+\theta} - u_{GL}^m\} - \alpha \Delta u_{GL}^{m+\theta} - \beta \Delta u_{GL}^m + \nabla p_{GL}^{m+\theta} = f^m,$$
$$\text{div} u_{GL}^{m+\theta} = 0, \tag{11.7}$$

$$\frac{1}{\theta' k}\{u_{GL}^{m+1-\theta} - u_{GL}^{m+\theta}\} - \beta \Delta u_{GL}^{m+1-\theta} - \alpha \Delta u_{GL}^{m+\theta} + \nabla p_{GL}^{m+\theta} = f^{m+1-\theta}, \tag{11.8}$$

$$\frac{1}{\theta k}\{u_{GL}^{m+1} - u_{GL}^{m+1-\theta}\} - \alpha \Delta u_{GL}^{m+1} - \beta \Delta u_{GL}^{m+1-\theta} + \nabla p_{GL}^{m+1} = f^{m+1-\theta},$$
$$\text{div} u_{GL}^{m+1} = 0. \tag{11.9}$$

For simplicity, we confine ourselves here to the presentation of the scheme for the "Stokes case" to give the main idea of the mathematical approach for getting optimal error statements for this scheme. For solving the Navier-Stokes equations, the equation of the intermediate sub-step (11.8) involves the nonlinear convection term.

Thus, this scheme employs two "solver"-components in the computational realization, one for the Stokes-problem and another for the intermediate sub-step (11.8). Note that the resulting velocity field in each iteration step is divergence-free (in the strong sense) — which is in contrast to the L^2-projection schemes of Chorin, Van Kan, and their modifications that have been investigated in the main part of this book.

The key idea for an analytical approach to Glowinski's scheme in the form (11.7) through (11.9) is the re-interpretation of the intermediate step (11.8) as an "inversely shifted" semi-implicit pressure-stabilization method with "problem-consistent" boundary conditions for the pressure function, which is as follows: Subtraction of equation (11.8) from the first one in (11.9) leads to

$$\frac{1}{\theta k}\left\{u_{GL}^{m+1} - u_{GL}^{m+1-\theta}\right\} - \frac{1}{\theta' k}\left\{u_{GL}^{m+1-\theta} - u_{GL}^{m+\theta}\right\}$$
$$- \alpha\Delta\left\{u_{GL}^{m+1} - u_{GL}^{m+\theta}\right\} + \nabla\left\{p_{GL}^{m+1} - p_{GL}^{m+\theta}\right\} = 0. \tag{11.10}$$

Now, the application of the divergence operator leads to

$$\mathrm{div}u_{GL}^{m+1-\theta} - \left\{\frac{1}{\theta} + \frac{1}{\theta'}\right\}^{-1}k\Delta\left\{p_{GL}^{m+1} - p_{GL}^{m+\theta}\right\} = 0,$$
$$\partial_n p_{GL}^{m+1}|_{\partial\Omega} = \partial_n p_{GL}^{m+\theta}|_{\partial\Omega} + \alpha\Delta\left\{u_{GL}^{m+1} - u_{GL}^{m+\theta}\right\}|_{\partial\Omega} \cdot n. \tag{11.11}$$

This is an "implicitly shifted" pressure-correction formulation, with a non-stationary perturbation of the incompressibility constraint being of second order. We emphasize, that in contrast to the projection schemes of Chorin and Van Kan no unnatural boundary conditions are imposed on the pressure function. This leads to the conjecture that this scheme will provide better rates of convergence, even in stronger norms.

This reformulation establishes the connection of Glowinski's scheme with the pressure-correction methods that have been thoroughly investigated in this book. However, in contrast to the singularly perturbed stabilization methods that are caused by the prescription of homogeneous boundary data for the actual pressure function, the present scheme is consistent in the way that the Neumann boundary data for the pressure function correspond to the actual flow. This is the reason for the observation that no boundary layers are present in Glowinski's scheme.

Bibliography

[1] Blum, H., *Asymptotic error expansion and defect correction in the finite element method*, Habilitationsschrift, University of Heidelberg, June 1990

[2] Brezzi, F., Fortin, M., *Mixed and Hybrid Finite Element Methods*, Springer Series in Computational Mathematics, 1991

[3] Chorin, A. J., *Numerical solution of the Navier-Stokes Equations*, Math. Comp. 22 (1968), pp. 745-762

[4] Chorin, A. J., *On the convergence of discrete approximations of the Navier-Stokes Equations*, Math. Comp. 23 (1969), pp. 341-353

[5] Constantin, P., Foias, C., *Navier-Stokes Equations*, Chicago Lectures in Math. Series, University of Chicago Press, 1988

[6] Blum, H., Harig, J., Mueller, S., Turek, S., *FEAT2D, Finite Element Analysis Tools, User Manual, Release 1.3*, Preprint 92 - 18, University of Heidelberg, 1992

[7] E, W., Liu, J. G., *Projection method I: Convergence and numerical boundary layers*, SIAM J. Num. Anal. 32 (1995), pp. 1017-1057

[8] E, W., Liu, J. G., *Projection method II: Godunov-Ryabenki Analysis*, to appear in SIAM J. Num. Anal.

[9] Girault, V., Raviart, P. A., *Finite Element Methods for Navier-Stokes Equations*, Springer, Berlin-Heidelberg, 1986

[10] Glowinski, R., *Le θ-scheme*, in: *Numerical methods for the Navier-Stokes equations*, M.O. Bristeau, R. Glowinski and J. Perieux, eds., Comp. Phys. report, 6 (1987), pp. 73-87

[11] Gresho, P. M., *Incompressible fluid dynamics: some fundamental formulation issues*, Ann. Rev. Fluid Mech. 23 (1991), pp. 413-453

[12] Gresho, P. M., *On the theory of semi-implicit projection methods for viscous incompressible flow and its implementation via a finite element method that also introduces a nearly consistent mass matrix. Part 1: Theory*, Int. J. Numer. Meth. Fluids 11 (1990), pp. 621-649

[13] Harig, J., *Eine robuste und effiziente Finite-Elemente Methode zur Loesung der inkompressiblen 3D-Navier-Stokes Gleichungen auf Vektorrechnern*, Ph.D. Thesis, University of Heidelberg, 1991

[14] Haraoutunian, V., Engelman, M. S., Hasbani, I., *Segregated Finite Element Algorithms for the Numerical Solution of Large Scale incompressible Flow Problems*, Int. J. Num. Meth. Fluids, 17, Nr. 4 (1993), pp. 324-348

[15] Hebeker, F. K., *The penalty method applied to the instationary Stokes equations*, Appl. Anal., 14 (1982), pp. 137-154

[16] Heywood, J. G., Rannacher, R., *Finite element approximation of the nonstationary Navier-Stokes Problem. I. Regularity of solutions and second order error estimates for spatial discretization*, SIAM J. Numer. Anal. 19 (1982), pp. 275-311

[17] Heywood, J. G., Rannacher, R., *Finite element approximation of the nonstationary Navier-Stokes Problem. II. Stability of solutions and error estimates uniform in time*, SIAM J. Numer. Anal. 23 (1986), pp. 750-777

[18] Heywood, J. G., Rannacher, R., *Finite element approximation of the nonstationary Navier-Stokes Problem. IV. Error analysis for second-order time discretization*, SIAM J. Numer. Anal. 27 (1990), pp. 353-384

[19] Heywood, J., Rannacher, R., Turek, S., *Artificial boundaries and flux and pressure conditions for the incompressible Navier-Stokes Equations*, to appear in: Int. J. Numer. Math. Fluids

[20] Hughes, T. J. R., Franca, L. P., Balestra, M., *A new finite element formulation for computational fluid mechanics: V. Circumventing the Babuska-Brezzi condition: A stable Petrov-Galerkin formulation of the*

Stokes problem accommodating equal order interpolation, Comp. Meth. Appl. Mech. Eng. 59 (1986), pp. 85-99

[21] Kim, J., Moin, P., *Application of a fractional-step method to incompressible Navier-Stokes*, J. Comp. Phys. 59 (1985), pp. 308-323

[22] Kloucek, P., Rys, F., *Stability of the Fractional-θ-Scheme for the Nonstationary Navier-Stokes Equations*, SIAM J. Numer. Anal. 31 (1994), pp. 1312-1336

[23] Luskin, M., Rannacher, R., *On the smoothing property of the Galerkin method for parabolic equations*, SIAM J. Numer. Anal. 19 (1981), pp. 93-113

[24] Luskin, M., Rannacher, R., *On the smoothing property of the Crank-Nicolson scheme*, Applicable Anal., 14 (1982), pp. 117-135

[25] Mueller-Urbaniak, S., *Eine Analyse des Zwischenschritt-θ-Verfahrens zur Loesung der instationaeren Navier-Stokes-Gleichungen*, IWR-Heidelberg, Preprint 94-01 (1994)

[26] Prohl, A., *Ueber die Chorin'sche Methode zur Loesung der inkompressiblen Navier-Stokes Gleichungen*, Diploma thesis, 1992

[27] Prohl, A., *Projektions- und Quasi-Kompressibilitaetsmethoden zur Loesung der inkompressiblen Navier-Stokes Gleichungen*, Ph.D. thesis, 1995

[28] Quartapelle, L., *Numerical Solution of the incompressible Navier-Stokes Equations*, ISNM, vol. 113, Birkhaeuser, Basel-Boston-Berlin 1993

[29] Quarteroni, A., Valli, A., *Numerical Approximation of Partial Differential Equations*, Springer Series in Computational Mathematics, vol. 23, Springer, Berlin-Heidelberg-New York, 1994

[30] Rannacher, R., *On Chorin's projection method for the incompressible Navier-Stokes Equations*, in: Proceedings of the Oberwolfach Conference: *Navier-Stokes Equations: Theory and Numerical Methods*, September 1991

[31] Rannacher, R., *On the stabilization of the Crank-Nicolson Scheme for long time calculations*, Preprint, Universitaet des Saarlandes, 1986

[32] Schwab, C., *Remarks on Pressure Approximation in Projection Methods for Viscous Incompressible Flow*, in: *Proc. 9th Int. Conf. on Finite Elements in Fluids*, Oct. 1995

[33] Shen, J., *On error estimates of projection methods for the Navier-Stokes Equations: First order schemes*, SIAM J. Numer. Anal. 29 (1992), pp. 57-77

[34] Shen, J., *On error estimates of higher order projection and penalty-projection-schemes for the Navier-Stokes equations*, Num. Math. 62 (1992), pp. 49-73

[35] Shen, J., *Remarks on the pressure error estimates for the projection methods*, Num. Math. 67 (1994), pp. 513-520

[36] Shen, J., *On error estimates of the projection methods for the Navier-Stokes Equations: 2nd order schemes*, Preprint, Penn State University, 1993

[37] Shen, J. *On error estimates of the projection methods for the Navier-Stokes equations: Second-order schemes*, Math. Comp., vol 65, no. 215, July 1996, pp. 1039-1065

[38] Shen, J., *A new pseudo-compressibility method for the Navier-Stokes equations*, Preprint, Penn State University, 1995

[39] Shen, J., *On error estimates of the Penalty Method for unsteady Navier-Stokes equations*, SIAM J. Num. Anal., vol. 32, no. 2, April 1995, pp. 386-403

[40] Temam, R., *Sur l'approximation de la solution des equations de Navier-Stokes par la methode des pas fractionnaires II*, Arch. Rat. Mech. Anal. 33 (1969), pp. 377-385

[41] Temam, R., *Navier-Stokes Equations. Theory and numerical analysis*, North Holland, Amsterdam, 2nd edition 1977

[42] Temam, R., *Navier-Stokes Equations and Nonlinear Functional Analysis*, in: NSF/CBMS Regional Conference series in Applied Mathematics. SIAM, Philadelphia, 1983

[43] Thomee, V., *Galerkin Finite Element Methods for Parabolic Problems*, Lecture Notes in Math. 1054, Springer, Berlin, New York, 1984

[44] Timmermans, L. J. P., Minev P. D., Van De Vosse, F. N., *A Spectral element Projection Scheme for Incompressible Flow*, submitted to Int. J. Num. Meth. Fluids

[45] Van Kan, *A second-order accurate pressure-correction scheme for viscous incompressible flow*, SIAM J. Sci. Stat. Comp. 7 (1986), pp. 870-891

[46] Zheng, S., *Nonlinear Parabolic Equations and Hyperbolic-Parabolic Coupled Systems*, Pitman Monographs and Surveys in Pure and Applied Mathematics 76, Longman, 1995

Index